Impacts of Diesel-Powered Light-Duty Vehicles

Diesel Technology

Report of the
Technology Panel
   of the
Diesel Impacts Study Committee
National Research Council

NATIONAL ACADEMY PRESS
Washington, D.C.   1982

This study by the Diesel Impacts Study Committee was conducted under Contract No. 68-01-5972 with the U.S. Environmental Protection Agency, the U.S. Department of Energy, and the U.S. Department of Transportation.

Library of Congress Catalog Card Number 82-81125

International Standard Book Number 0-309-03243-1

Available from

NATIONAL ACADEMY PRESS
2101 Constitution Avenue, N.W.
Washington, D.C. 20418

Printed in the United States of America

TECHNOLOGY PANEL
DIESEL IMPACTS STUDY COMMITTEE

ROBERT F. SAWYER, Professor of Mechanical Engineering, University of
California at Berkeley, Chairperson
FREDERICK L. DRYER, Associate Professor of Mechanical and Aerospace
Engineering, Princeton University
JAMES R. JOHNSON, Executive Scientist, 3M Company, Retired
JAMES R. KLIEGEL, President, KVB, Incorporated
PAUL KOTIN, Senior Vice-President for Health, Safety, and Environment,
Johns-Manville Corporation
WILLIAM J. LUX, Director of Product Engineering Center, John Deere
Industrial Equipment Division, Deere & Company
PHILLIP S. MYERS, Professor of Mechanical Engineering, University of
Wisconsin
JAMES WEI, Professor of Chemical Engineering, Massachusetts Institute
of Technology

CONSULTANTS TO THE TECHNOLOGY PANEL
DIESEL IMPACTS STUDY COMMITTEE

ALFRED G. CATTANEO, Consulting Engineer
NICHOLAS P. CERNANSKY, Associate Professor of Mechanical Engineering,
Department of Mechanical Engineering and Mechanics, Drexel
University
DAVID CHRISTISON, Manager of Planning, Technology and Financial
Analysis for Manufacturing, Mobil Corporation, retired
JOHN H. JOHNSON, Professor of Mechanical Engineering, Michigan
Technological University
LLOYD JOHNSON, President of Leejon Engineering Company, East Peoria,
Illinois
RONALD D. MATTHEWS, Assistant Professor of Mechanical Engineering,
University of Texas
OWEN I. SMITH, Assistant Professor of Engineering, School of
Engineering & Applied Science, University of California at Los
Angeles

STAFF
TECHNOLOGY PANEL AND
DIESEL IMPACTS STUDY COMMITTEE

L.F. BARRY BARRINGTON, Executive Director
KENNETH R. SMIKLE, Staff Officer
CLAUDE E. OWRE, Editor
SHARON J. RASMUSSEN, Consulting Editor
S. ANN ANSARY, Program Coordinator
JULIA TORRENCE, Secretary

DIESEL IMPACTS STUDY COMMITTEE

HENRY S. ROWEN, Professor of Public Management, Stanford University,
Chairman
WILLIAM M. CAPRON, Professor of Economics, Boston University
KENNETH S. CRUMP, President, Science Research Systems, Incorporated
ALAN Q. ESCHENROEDER, Senior Staff Scientist, Arthur D. Little,
Incorporated
SHELDON K. FRIEDLANDER, Department of Chemical, Nuclear, and Thermal
Engineering, University of California at Los Angeles, Vice-Chairman
BERNARD M. GOLDSCHMIDT, Associate Professor of Environmental Medicine,
New York University Medical Center
HERSCHEL E. GRIFFIN, Dean of the Graduate School of Public Health,
San Diego State University
JACK D. HACKNEY, Professor of Medicine, Rancho Los Amigos Hospital
(University of Southern California)
IAN T. HIGGINS, Professor of Epidemiology, University of Michigan
FRED S. HOFFMAN, Senior Economist, Rand Corporation
WILLIAM W. HOGAN, Professor of Political Economy, Harvard University
JAMES R. JOHNSON, Executive Scientist (Retired), 3M Company
DAVID B. KITTELSON, Professor of Mechanical Engineering, University of
Minnesota
PAUL KOTIN, Senior Vice-President for Health, Safety and Environment,
Johns-Manville Corporation
WILLIAM J. LUX, Director of the Product Engineering Center, John Deere
Industrial Equipment Division, Deere & Company
ROBERT F. SAWYER, Professor of Mechanical Engineering, University of
California at Berkeley
RICHARD O. SIMPSON, President, Litek International, Incorporated
BRUCE O. STUART, Manager of the Inhalation Toxicology Section, Stauffer
Chemical Company, and Clinical Associate, University of Connecticut
Medical School
JAMES P. WALLACE, Vice-President and Division Manager, The Chase
Manhattan Bank
JAMES WEI, Professor of Chemical Engineering, Massachusetts Institute
of Technology

CONTENTS

In 1979 the Environmental Protection Agency, the Department of Energy, and the Department of Transportation requested that the National Research Council undertake an evaluation of the research and public policy issues associated with the prospective widespread use of diesel-powered light-duty vehicles in the United States. Because the Environmental Protection Agency is required under the Clean Air Act to regulate the amounts of chemical compounds and particulate concentrations emitted by all sources, including diesel vehicles, it is concerned about the possible adverse human health effects arising from an anticipated increase in the number of diesel-powered vehicles. Both the Department of Energy and the Department of Transportation have responsibilities for encouraging improved fuel economy of passenger cars, including examining the potential benefits of diesel engines in such vehicles. According to forecasts of the nation's largest automobile manufacturer, General Motors, diesel-powered light-duty vehicles are likely to constitute 18 percent of its light-duty vehicle sales by 1985 and as much as 25 percent by the end of the century.

The National Research Council's study began in May 1979 with the formation of the Diesel Impacts Study Committee, whose members were selected in accordance with its policy of providing competent experts with balanced viewpoints. Because the scope of the study involved a complex range of concerns and interactions, the committee established four panels to examine the technological, environmental, human health effect, and public policy issues. Each of the four panels consisted of specialists drawn from the relevant area of concern and some members of the committee.

The Technology Panel was formed to provide the committee with a comprehensive discussion of the boundaries of technology and engineering design that might accommodate both fuel economy goals and requirements for reduction of environment and health hazards identified with the operation of diesel-powered cars. Four of the original nine members, including the chairman, are associated with universities; four with private industry; and the ninth member, Dr. Jane V. Hall, was--before her June 1980 resignation from the panel--a staff member of the California Air Resources Board. The expertise of the panel members covers combustion, engine design, health science, material sciences, and emission control. Panelists' disciplines were supplemented by staff and seven consultants, who brought a range of needed specialized knowledge to the study.

The Technology Panel was charged by the committee with assembling information to characterize the major emissions of diesel-powered vehicles, qualitatively comparing them with emissions from spark ignition engines, and examining the cost and effectiveness of various emission control technologies and devices proposed for use in diesel-powered vehicles. The panel also reviewed plans for future diesel engine designs and future operating adjustments for either emission reduction or fuel economy. In agreement with the committee's emphasis on diesel fuel economy, the panel also reviewed information on fuels appropriate for diesel engines, taking into account both costs and future producibility.

These tasks were accomplished at four workshops (two at facilities of General Motors and two at the Environmental Protection Agency), at three panel meetings, and during visits to 54 organizations concerned with the future expansion and improvement in diesel engine use. Contacts were made with automakers, fuel refiners, control device suppliers, research institutes, universities, and agency facilities. In addition, 18 site visits were made by staff, consultants, and panelists to manufacturers in Japan and Western Europe. (Appendix H provides a listing of Technology Panel site visits and consultations.) Beyond the first-hand collection of information, the staff abstracted the pertinent technical literature and placed it in the working library of the study. The research activities provided an understanding of the current and the future state of diesel engineering technology and an appreciation of the research still needed to reach the objectives of emission control and fuel economy.

This report reflects the panel's understanding of diesel engine technology and offers precepts on current technical knowledge about diesel vehicles. Much of the text of this report was prepared by the consultants to the panel, though the panel members and consultants established a close interaction. It should be noted that the principal authors of some sections have paraphrased and borrowed from their own earlier, written work. In acknowledging this form of composition, the panelists made the observation that the authors are, in those cases, world-class experts who have been the principal creators of those segments of diesel engine literature that deal with noise, odor, particulate, and gaseous emission control. These are primary issues of concern to the committee.

The Technology Panel's report does not dwell at length on evaluation or cost-benefit analysis. Such assessments were undertaken by the committee's Analytic Panel, which has provided the committee an integrated perspective on the benefits and risks of greater availability of diesel-powered light-duty vehicles. Thus, the findings and conclusions of the Technology Panel have been taken into account and meshed with those of the Environmental Impacts Panel and the Health Effects Panel. All four panels contributed to the committee's final report, which contains the ultimate conclusions and recommendations of the overall study.

SUMMARY AND CONCLUSIONS

The number of diesel engines being used to power light-duty cars, vans, and pickup trucks, and heavy-duty trucks and buses is increasing in the United States and worldwide. This trend is the result of increased energy costs, a favorable pricing of diesel fuel relative to gasoline, and the fuel economy advantage that the diesel engine has over the spark ignition engine. Within the next 10 years as many as 25 percent of all new cars sold may be powered by diesel engines. If this switch from gasoline-powered cars does occur, the additional diesel fuel demand can be met by existing petroleum capacity with a 10 percent reduction in refinery energy consumption, assuming the same crude quality mix of the present. Using the highest projections, the fuel demand that would result from increased dieselization would not represent more than 20 percent of the total distillate fuel supply in 1990. Increased light-duty dieselization appears to be a favorable national strategy for conserving crude oil, although a rigorous net energy study has not yet been done to quantify this assertion.

Fuel conservation would be an obvious advantage of increased dieselization; an increase in the amount of some pollutants in the ambient air would be a disadvantage. The diesel engine emits higher levels of particulate and oxides of nitrogen (NO_x), but lower levels of carbon monoxide and hydrocarbons than the conventional catalyst-equipped gasoline engine.

The report of the Technology Panel is principally concerned with the technology of the light-duty diesel. However, for the remainder of this century, heavy-duty diesel vehicles will continue to dominate diesel fuel consumption. Although particulate emission controls for light-duty diesels are scheduled to begin in 1982, particulate emission standards for heavy-duty vehicles are not scheduled until 1986.

From a national viewpoint, relaxing NO_x control for heavy-duty diesels in return for obtaining increased particulate control would be a better strategy than imposing stringent particulate standards for light-duty diesels. NO_x control might be better obtained from gasoline-powered vehicles which already have very low particulate emissions. In addition, heavy-duty vehicles might be better able to bear the cost of particulate control than light-duty vehicles. Such a strategy would be effective for getting both fuel economy and emission reductions in the overall transportation fleet.

FUEL ECONOMY

Primarily because of its higher density, a gallon of diesel fuel has approximately 13 percent greater energy (calorific) content than a gallon of gasoline. This energy content difference is one of the factors in the diesel fuel economy advantage. Light-duty indirect injection diesels have a clear fuel economy advantage over comparable gasoline-powered vehicles. On average, the fuel mile-per-gallon economy advantage of the diesel is about 35 percent; however, this figure varies depending on driving conditions, vehicle engine design, and vehicle use. Under light-load city driving, diesels can have a 100 percent mile-per-gallon fuel economy advantage over their gasoline-powered counterparts. However, under high-speed, high-load highway driving, the advantage may be only 15 percent and primarily due to the greater energy context of the diesel fuel. Comparison of a variety of vehicles and engine types is provided in Table 1. As can be seen, the fuel economy advantages of the diesel as presently designed are consistent and substantial.

Test Methods

Vehicle efficiency is generally expressed in terms of a ratio of distance traveled to fuel consumed (miles per gallon (mpg), kilometers per liter, etc.). Most fuel economy data for the various engine/vehicle configurations come from vehicle certification tests conducted by the Environmental Protection Agency (EPA). These data are the result of chassis dynamometer testing (vehicle testing in laboratories) during specified driving cycles. The Federal Test Procedure (FTP), which is considered to be typical of urban driving, is used in the vehicle certification tests. The adequacy of chassis dynamometer testing for fuel economy has been questioned. Critics claim that track, road, and in-use tests provide better data. In addition to criticism of the test methods, industry has stated that the FTP cycle underestimates in-use fuel economy. In response the EPA added a higher-speed cycle, the Fuel Economy Test (FET). A weighted combination of the FTP (urban or city) and FET (highway) fuel economy figures is used in determining compliance with the Corporate Average Fuel Economy (CAFE) requirements. In most cases, these composite figures are used for comparisons in this report. Several comparisons between the EPA test procedure and road or in-use (usually fleet) testing have been conducted. These comparisons have shown that the FTP fuel economy number is probably closer to actual in-use fuel economy but that some biases are present. More importantly, they showed that the fuel economy advantage of diesel vehicles over gasoline vehicles is systematically underestimated by the FTP by about 5 percent.

An additional advantage of the diesel is an improved fuel economy on start-up and during short trips (trips of less than 10 miles). Because of the high percentage of trips that are under 10 miles, this factor may represent a significant energy savings for a large portion of vehicle operators in the United States. The fuel saving effect

TABLE 1 Fuel Economy Data for Light-Duty Vehicles, 1981 Model Year, EPA FTP (Urban) Cycle

Vehicle Class	Vehicle Make and Model	Estimated mpg	Engine CID	Trans- mission	Fuel System No. bbl or Fuel Injection	Fuel Econ. Advantage, mpg
Subcompact	VW Rabbit	28	105	M4	FI	
cars	VW Rabbit Diesel	42	97	M4	FI	50%
Compact	Audi 5000	19	131	M5	FI	
cars	Audi 5000 Diesel	27	121	M5	FI	42%
	Peugeot 505[a]	19	120	A3	FI	
	Peugeot 505[a]	16	120	M5	FI	
	Peugeot 505 SD Diesel[a]	29	141	M&A4	FI	53-81%
	Mercedes 280E[a]	16	168	A4	--	
	Mercedes 240D Diesel[a]	24	183	A4	FI	50%
Mid-size	Olds Cutlass	21	231	A3	2	
cars	Olds Cutlass	19	260	A3	2	
	Olds Cutlass Diesel	23	350	A3	FI	10-21%
Large Cars	Buick LeSabre	19	231	A3	2	
	Buick LeSabre	18	252	A3	4	
	Buick LeSabre	16	307	A3	4	
	Buick LeSabre Diesel	22	350	A3	FI	16-38%
	Cadillac deVille/ Brougham	18	252	A4	FI	
	Cadillac deVille/ Brougham Diesel	21	350	A3	FI	17%
Mid-size	Olds Cutlass	21	231	A3	2	
station	Olds Cutlass	17	260	A3	2	
wagons	Olds Cutlass	16	307	A3	4	
	Olds Cutlass Diesel	23	350	A3	FI	10-44%

TABLE 1 (continued)

Vehicle Class	Vehicle Make and Model	Estimated mpg	Engine CID	Trans- mission	Fuel System No. bbl or Fuel Injection	Fuel Econ. Advantage, mpg
Large station wagons	Pontiac Catalina/ Bonneville Safari	16	307	A4	4	
	Pontiac Catalina/ Bonneville Safari Diesel	21	350	A3	FI	31%
Standard pickup trucks	Chevrolet C10	17	250	A3	2	
	Chevrolet C10	17	305	A3	2	
	Chevrolet C10 Diesel	20	350	A3	FI	18%

[a]1981 certification data are not available. Data are for the 1980 model year.

SOURCE: EPA 1979a,b; 1980a,b.

stems from the fact that the diesel reaches normal operating conditions quickly and does not use overall fuel-rich mixtures during initial operation.

Fuel Economy Data

The city, highway, and combined ("35/45 mpg") fuel economy of EPA certification vehicles, sales-weighted for the passenger car fleet, is shown in Figure 1. The in-use fuel economy ("road mpg") and the average inertia weight of vehicles, which indicates the downsizing that started in 1976, are also shown. Figure 2 shows the fleet fuel economy normalized to the 1978 inertia weight mix. The data indicate that control technology does not have to degrade fuel economy. The data show that since 1977, downsizing has been the main factor in fuel economy improvements.

Projections

Table 2 shows, in summary form, a comparison of estimated fuel economies among current gasoline light-duty vehicles, indirect injection diesel engines, direct injection stratified charge engines, and direct injection diesel engines, including improved gasoline and indirect injection diesel engines.

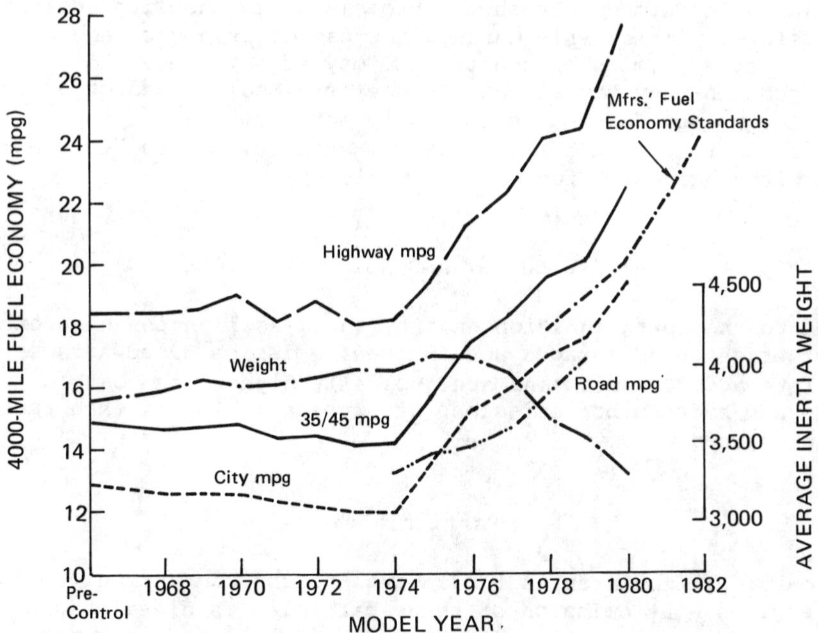

FIGURE 1 Sales-weighted fleet fuel economy trend.

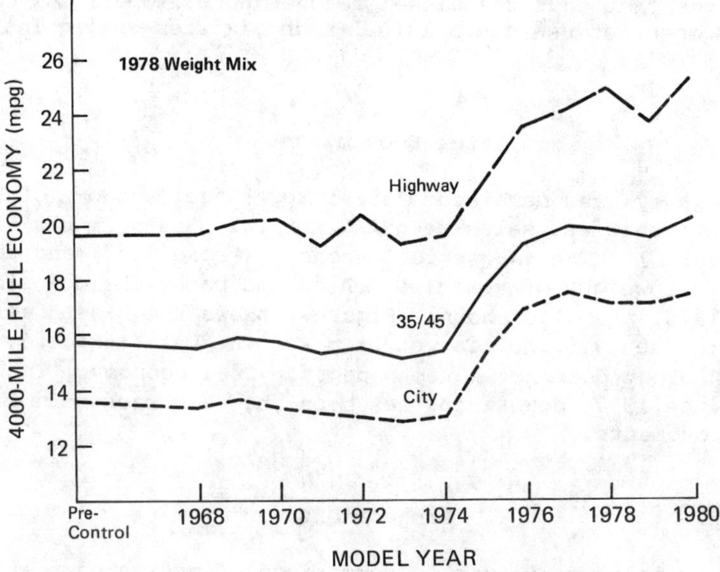

FIGURE 2 Weight-normalized fleet fuel economy trend (1978 TW Mix).

Because of its fuel economy advantage, the introduction of the light-duty direct injection diesel seems likely within 5 to 10 years, particularly in Europe. Further improvements in gasoline engines are also possible, particularly through the use of lean-burn and stratified charge concepts. Emission constraints may affect the use of these engines just as they may affect the use of diesels. Reduction catalyst control technology for NO_x is currently not adaptable to the stratified charge, lean-burn, or diesel engines, because they all operate with an overall fuel-lean (oxidizing) mixture.

AIR POLLUTANT AND NOISE EMISSIONS

Compared to the spark ignition engine, the diesel engine has both inherent emission advantages and inherent emission disadvantages. Particulate and NO_x emissions are typically higher, and carbon monoxide and hydrocarbon emissions are typically lower, than those from gasoline engines.

Particulates

Formation of particulates is fundamental to the combustion process in diesel engines, but emission of these particles in diesel exhaust is not, because the particulate finally emitted in diesel exhaust is the result of both formation and subsequent oxidation processes.

TABLE 2 Fuel Economy Comparison of Light-Duty Vehicles: Comparison at Constant Performance in Percent With Respect to Current Gasoline Engine Vehicles[a]

Period	Engine Type	Volume Basis, mpg	Energy Basis
Current (1980)	Gasoline	0	0
	Indirect injection diesel	+35	+19
Intermediate (1980s)	Improved gasoline	+20	+20
	Direct injection stratified charge gasoline	+25	+25
	Improved indirect injection diesel	+45	+28
	Direct injection diesel	+55	+37
Long range (after 1990)	Advanced stratified charge gasoline	+42	+42
	Advanced ignition-assisted diesel	+60	+42
	Advanced fuel-tolerant engine	+40 to +65	+42

[a]Estimated uncertainties to any absolute value are +10 percent of value. Difference between volume basis and energy basis is based on +13 percent greater volumetric energy content of diesel fuel.

Experimental evidence from a variety of combustion systems (premixed and diffusion flames, perfectly stirred reactors, etc.) and fuels indicates that chemical kinetics is the dominant factor governing the formation and oxidation of particulate. (See Appendix B.) Well-mixed systems emit particulate when the carbon/oxygen molar ratio in the fuel-oxidizer mixture exceeds 0.5. The diesel is neither a premixed nor a homogeneous system.

A combination of high mixing rate and long ignition delay to approach premixed, homogeneous conditions has been shown to reduce particulate emissions. However, ignition delay times characteristic of current diesel fuels are too short to accomplish this using known mixing technology, and if longer mixing and ignition times were permitted, the increased peak cylinder pressures that resulted would cause serious structural problems. Thus, current approaches optimize particulate emissions through programmed fuel injection and system components that control the rate and intensity of combustion. The kinetics of NO_x formation and the methods used to minimize the amount of NO_x emitted are in direct opposition to the most effective means for reducing production of particulates in the combustion chamber.

After 1984, achieving particulate emission levels below 0.5 g/mi
(for 2,000- to 4,000-pound-inertia-weight cars and the over 2,500-pound-
inertia-weight trucks) and simultaneously controlling NO_x to 1.0 g/mi
may result in a further cost disadvantage for diesel engines relative
to spark ignition engines. This problem will become even more acute
with the probable equalization of diesel and gasoline fuel prices.
Additional cost for particulate control and the associated materials
and manufacturing energy use will also decrease the energy advantage of
dieselization.

Test Methods

During the past 5 years, improved methods for characterizing diesel
engine emissions have been developed. Figure 3 shows the laboratory
system (dilution tunnel) developed to simulate the exhaust dilution
processes that occur in the atmosphere. With this system a particulate
probe samples a diluted mixture of exhaust and air. During the engine
operating cycle, particulates are collected on a filter substrate. The
resulting sample is used for physical, chemical, and biological char-

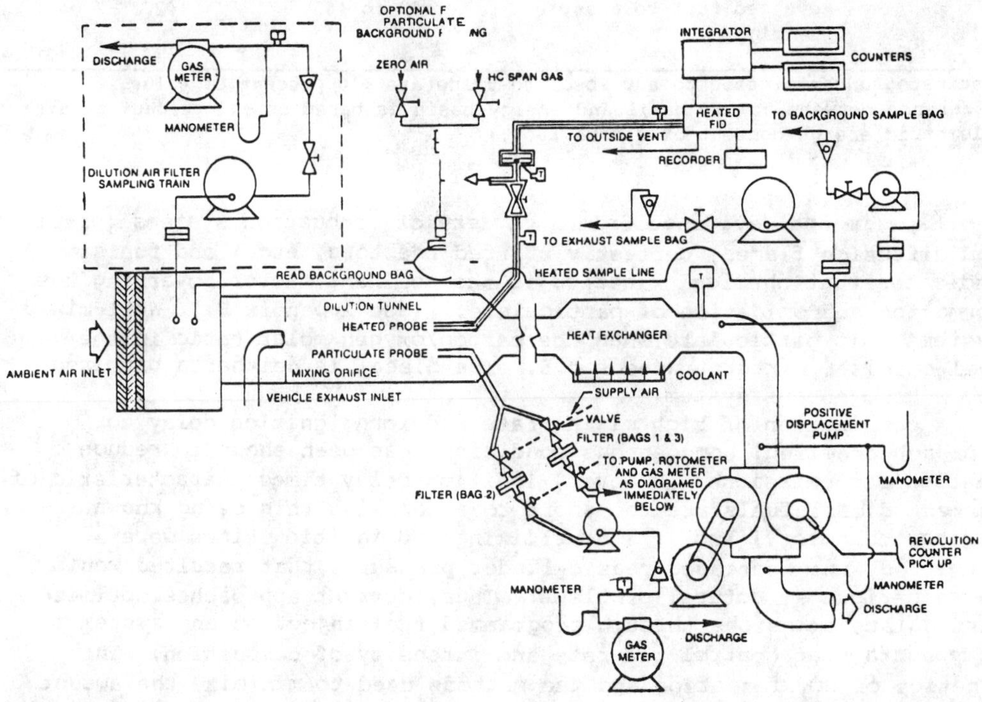

FIGURE 3 Gaseous and particulate emission sampling system (PDP-CVS)
(diesel vehicles only).

acterization. Current particulate standards relate only to the total mass of the particulates collected. However, this laboratory system makes it possible to characterize the various chemical fractions contained in the total mass sample. The material that can be extracted from the particles with a conventional solvent is called the soluble organic fraction. This fraction is important because it contains biologically active components that can be further characterized chemically.

In addition to particulate determination, a separate probe heated to 190°C (375°F) is used to extract gaseous emissions. Hydrocarbons are continuously monitored, but NO_x and carbon monoxide are determined later from a bag sample collected during the test cycle.

Because the heated gaseous sample is taken upstream of the particulate sampling filter, a portion of the hydrocarbon emissions are accounted for once as condensate on the particulate filter and again as gaseous hydrocarbon emissions. This procedure was originally justified by EPA on the basis that the particulate measurement was related to potential respiratory health effects and the hydrocarbon measurement was primarily related to smog effects. Thus, double-counting permitted assessment of the maximum potential effect for each. Hydrocarbons adsorbed on the particulate will participate predominantly in one effect or the other. They should, therefore, be counted only once.

Current EPA certification procedures specify that a maximum temperature of 52°C (125°F) should not be exceeded in the dilution tunnel sample zone. With different sized cars operating at various emission levels, a simple maximum temperature limitation does not provide mixture temperature and dilution ratio control throughout the complete FTP driving cycle. Integrating the mixture temperature throughout the complete cycle, holding to a set limit (such as 90° ± 10°F) for the average mixture temperature would provide more accurate results. In addition, although there would be excursions above and below a mean temperature and dilution ratio, different-sized cars would follow a similar mixture temperature profile throughout the FTP driving cycle. This should provide more comparable data for the different vehicles.

Because the dilution ratio is important for determining hydrocarbon adsorption and nitrogen dioxide levels (and in turn the chemical and biological character of the particulate sample), some temperature specification, such as the one suggested above, would be better for generating comparable data. Cycle temperature variations below 140 F do not appear to be significant relative to particulate mass measurements.

Hydrocarbons

Existing atmospheric modeling studies show that methane does not participate in smog formation chemistry. Because methane is a nonreactive hydrocarbon, any methane emissions should be excluded from all hydrocarbon emission measurements for both diesel and gasoline engines. This approach would provide a larger adjustment for gasoline vehicle emissions. Hydrocarbon emissions from diesel vehicles are inherently low, and with the proper measurement technique (using the particulate filter to divide the particulate and gaseous hydrocarbon phases), diesels can easily meet the 1980 federal 0.41-g/mi standard (including methane) or California's 0.39-g/mi (nonmethane) standard. Diesel fuel is less volatile than gasoline, and the evaporative emissions are extremely low--even without the addition of a control system. Diesel vehicles should be given credit for this fact in the federal hydrocarbon emission standard as they are in the California standard.

Carbon Monoxide

Carbon monoxide emissions from diesel engines are inherently low. Hence, even without emission control technology, diesels should always be able to meet the federal 3.4-g/mi carbon monoxide standard. In addition, carbon monoxide emissions from diesels do not increase significantly with the age of engines as is the case for catalyst-equipped gasoline engines.

Other Gaseous Species

Diesel engines emit more aldehydes, nitrogen dioxide, sulfur dioxide (due to the higher percentage of sulfur in diesel fuel relative to gasoline), and odor or irritants than gasoline engines do. Of these four types of pollutants, nitrogen dioxide and the aldehydes are direct irritants. Nitrogen dioxide and other nitrogen compounds can react with the hydrocarbons to form biologically active nitroorganic compounds. Better measurement methods to characterize the tail pipe nitrogen dioxide levels are needed along with a better data base. The odorous compounds consist of a variety of oxygenates and unburned fuel species. Both odor and irritants are increased by cold-start effects. Aldehydes, which are unregulated, have high photochemical reactivities and are largely unaccounted for by current hydrocarbon emission assessments.

Diesels are typically noisier than gasoline engines, particularly at the idling point and low load. Furthermore, the frequency distribution of the noise emitted is more irritating to most persons than that from gasoline engines. Available technology can be applied effectively to the engine compartment to make the pass-by, driver compartment, and curbside idling noise levels nearly as low as those of the gasoline engine. Efforts in noise reduction suggest that diesel noise is not a major long-term problem, but that reducing noise levels may result in higher costs or performance penalties.

Fuel Effects

Data indicate that there is a relationship between the composition of diesel fuel and emissions of particulates and hydrocarbons. For instance, increasing aromatic content and the 90 percent distillation temperature will increase particulate and odor emissions. Using a minimum cetane number and a seasonably adjusted cloud point will reduce cold start hydrocarbon emissions, noise, odor, irritant, and fuel system wax separation problems.

The feasibility of producing an improved diesel fuel should be studied to determine whether the trade-off between cost and avail- ability of such a fuel and the engine/emission advantages that would result are justified. Data indicate that the following general property limits for an improved automotive (cars, trucks, and buses) fuel may be desirable:

Cetane number	>48
Aromatics	<20 percent
90 percent distillation temperature	<316°C (600°F)
Sulfur	<0.25 percent by mass
Cloud point temperature	seasonably adjusted

Unfortunately, although improved specifications for an automotive diesel fuel would be beneficial for improving particulate, sulfur dioxide, sulfate, and hydrocarbon emission levels, and cold-start characteristics, the more restrictive specifications might limit the available diesel fuel stock sources and increase refining costs and energy requirements. Furthermore, there might be an indirect limitation imposed on other distillate fuel production.

IN-USE CHARACTERISTICS OF DIESEL ENGINES

The durability of both diesel and gasoline engines is primarily a function of design and development and is not an inherent character- istic or difference. Some early models of light-duty diesels had fuel injection system problems. Water separators were inadequately designed and lacked dashboard warning lights to alert drivers of water contamination. In addition, excessive water was found in some local distributors' fuel storage tanks. Some structural problems were also experienced--problems typical of new product introduction. They can all be solved by improved designs, materials, and operational practices. There are no fundamental reasons why reliable, trouble-free light-duty diesel engines cannot be produced in large numbers.

Lubrication

In addition to water contamination, diesel engines also have problems with lubricating oil contamination. The lubricating oil becomes contaminated when the products of combustion and the oil come into

contact on the cylinder walls. Particulates appear to adsorb the
additives that are used in lubricating oils to protect surfaces that
are subject to wear. The additives are substantially depleted when the
particulate carbon content in the oil is at 2 to 3 percent by mass
level. This problem is currently being handled through shortened oil
change intervals and increased oil sump capacity. Until recently
diesel engine manufacturers have suggested oil changes at intervals of
every 3,500 miles; however, Volkswagen now recommends a 7,000 mile
interval, and General Motors is recommending 5,000 miles for some of
its engines. Other manufacturers are attempting to increase the oil
change interval to at least 5,000 miles, their goal being a 7,500-mile
interval. Similar programs are underway to develop oils for gasoline
engines that have 15,000-mile, or once-a-year change, intervals. The
additional maintenance cost for oil changes will be a disadvantage for
the diesel, unless a new oil additive that is not adsorbed by particulates is found or particulate control methods that reduce the oil
contamination rate are developed.

Safety

In contrast to gasoline, the vapor pressure of diesel fuel at
temperatures below 125°F (52°C) is insufficient to produce flammable
vapor-air mixtures. Thus, diesel fuel is safer than gasoline during
normal fuel handling operations or when it leaks from the fuel tank or
system.

CONTROL TECHNOLOGY, AND RESEARCH AND DEVELOPMENT OPPORTUNITIES

With few exceptions, diesel automobiles had little difficulty in
meeting federal exhaust emission standards through the 1979 model year.
The standards for the 1980 model year and beyond, however, require
substantial reductions in hydrocarbons, NO_x, and particulates--
especially for larger vehicles and for vehicles sold in California.
Compliance with these standards will necessitate the addition of engine
emission control systems. Basic emission control approaches fall into
three major classes: engine modifications, exhaust aftertreatment, and
fuel modifications.

Engine modifications include modifications of combustion chamber
design, fuel injection timing and characteristics, introduction of
blowing or turbocharging, and use of exhaust gas recirculation.
Exhaust aftertreatment methods include traps, trap-oxidizers, and
catalysts. Fuel modifications include control of fuel properties, fuel
additives, and nonconventional fuels and fueling configurations--such
as emulsions, alcohol blends, fumigation, and synthetic fuels.

Before introduction into the marketplace, diesel emission control
technologies must be evaluated for:

* effects on NO_x, hydrocarbons, carbon monoxide, and
particulates;

* effects on unregulated emissions;
* effects on fuel economy;
* effects on vehicle acceleration and drivability;
* effects on engine durability;
* need for active control of concept, and degree of sensitivity of control;
* complexity;
* degree of maintenance required;
* relative cost;
* ease of integration within the engine.

Engine Modifications

Combustion System

The combustion chamber configuration and the fuel injection system determine the combustion system of a diesel engine. Both must be simultaneously optimized for an engine system to achieve high performance. Table 3 shows combustion chamber design and fuel injection factors that influence emissions. (In the table, particulates are broken down into hydrocarbons and smoke because hydrocarbons contribute

TABLE 3 The Effects of Combustion Chamber Design and Fuel Injection Variables on Emissions from an Indirect Injection Diesel Engine

Factors[a]	Gaseous Emissions			Particulates	
	HC	NO_x	CO	Unburned Hydrocarbons	Smoke
Injection timing	+++	+++	+	+++	++
Injection rate	++	++	+	++	++
Spray cone angle	+	+	+	+	+
Secondary injection	+++	+	+	+++	+
Combustion chamber geometry	+++	+++	+	+	+++
Injector location	++	++	+	+	+

[a]All factors hae been optimized as much as possible.

+++ Strongly dependent.
++ Moderately dependent.
+ Slightly dependent.

SOURCE: Toyota, 1980.

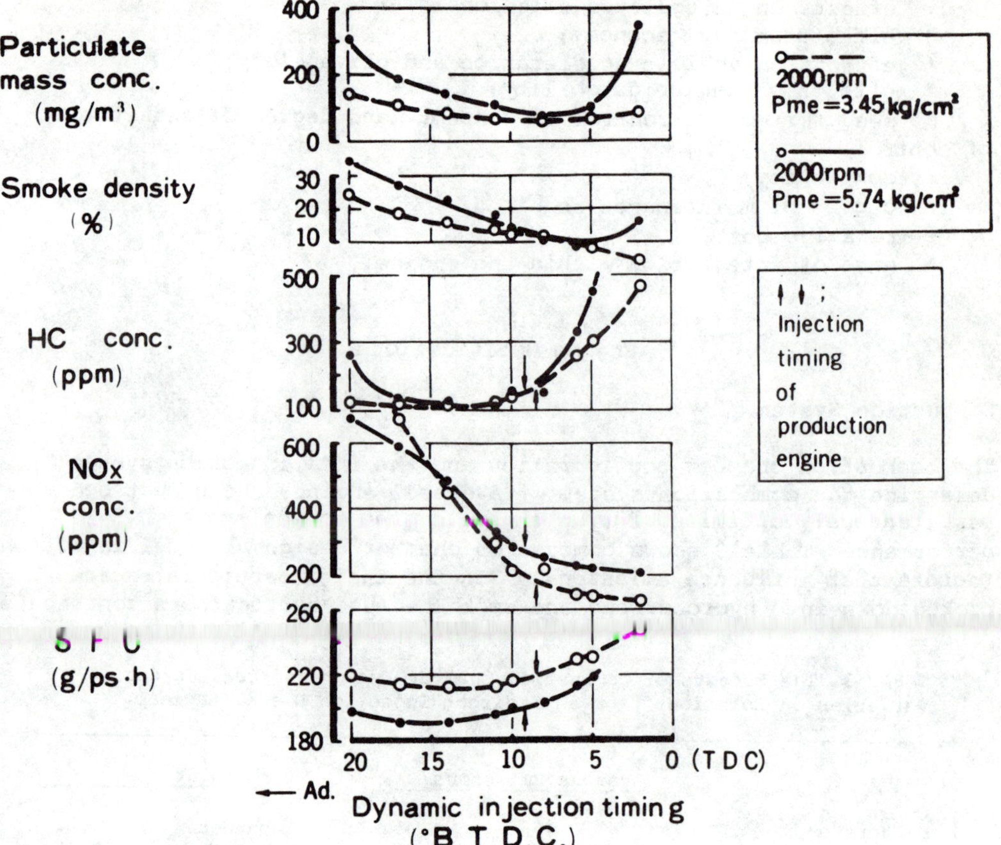

FIGURE 4 Effect of injection timing on emissions from an indirect injection diesel engine.

to the soluble organic fraction and smoke is an indicator of the solid particulate fraction.)

Figure 4 shows the effect of injection timing on fuel consumption and emissions for an indirect injection engine speed of 2,000 rpm and loads of 345 and 574 kPa. Retarded timings resulted in reduced NO_x emissions, but higher hydrocarbon emissions. These data demonstrate the importance of injection timing in minimizing fuel consumption, hydrocarbons, NO_x, and particulates. Engine emissions and fuel consumption cannot be simultaneously minimized at a single timing point for this engine.

The pump-line-nozzle injection system, modified for improved timing and injection control as a function of load and speed, will continue to be used in light-duty vehicles through 1984. Other control functions on current light-duty-vehicle engines are: an automatic cold-start timing advance, a temperature-controlled idle stop (for fuel quantity change), and a temperature controlled starting fuel quantity. Turbo-

charged engines also use aneroid devices, which sense the reduced
engine boost pressure during acceleration and lower the full-load fuel
delivery rate until the compressor boost pressure reaches a preset
level.

Electronically controlled pumps, which improve hydrocarbon, NO_x,
and particulate control, are presently undergoing research and development and should become available in 1985. With recent developments in
electronic microprocessors, electronic control of fuel injection equipment is an attractive proposition and is compatible with the 1984 all-
altitude emission requirement. In addition to the technical advantages,
the electronically controlled fuel pump requires a smaller capital
investment than the advanced technology fully controlled mechanical
systems. It, therefore, has an economic advantage as well.

With this system, signals proportional to fuel rate and piston
advance position are measured by sensors and are electronically
processed by the electronic control system to determine the optimum
fuel rate and timing. A block diagram of the control system is shown
in Figure 5. The maximum fuel delivery curve as a function of speed,
and the timing map as a function of speed and load (or derivatives),
are stored in the programmed memory of the microprocessor. This makes
possible the close matching of the fuel injection with engine demands
or past operating conditions.

The electronic technology for such a system is available. A number
of prototype cars are currently running with electronic fuel injection
control systems. However, further development needs to be carried out
in the sensor and actuator area in order for this technology to be
available for production by 1985. Thus, the all-altitude system needed

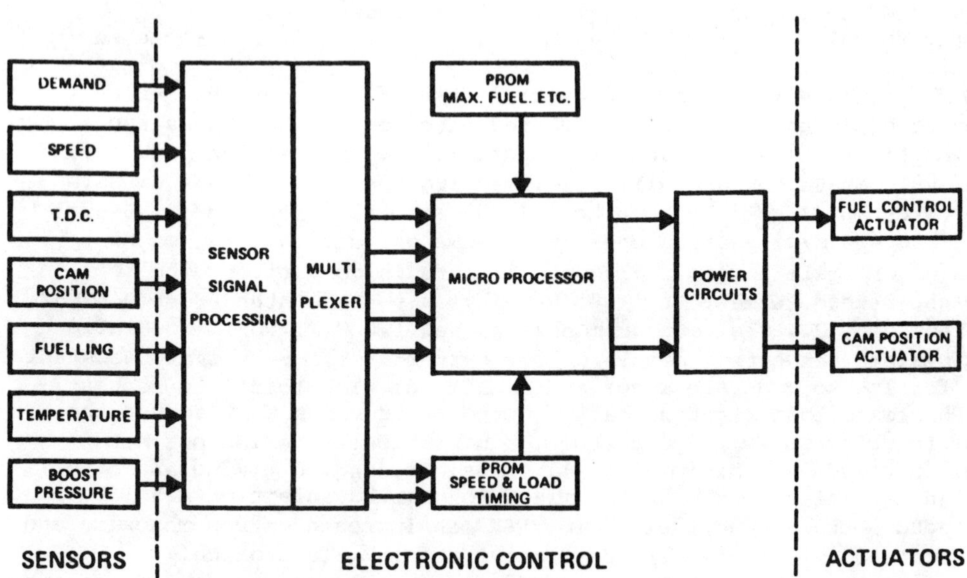

FIGURE 5 Block diagram of electronic control. SOURCE: Gliken _et al._,
1979.

to meet the 1984 federal standard would be a mechanical interim system that would be used one year only. It would, therefore, appear reasonable to delay implementation of the altitude requirement until 1985.

Turbocharging

Blowing or turbocharging diesel engines is an effective way to increase the power output from a given displacement engine. This technology is attractive because it allows manufacturers to add additional power levels to a basic displacement engine. If turbochargers are properly designed and developed, in conjunction with vehicle design, i.e., engine installation and rear axle ratio, improvements in emissions can also be achieved.

Turbocharger technology will become more widely used for automotive diesel engines in the early 1980s. One of the limiting factors in current turbochargers is their poor low-speed efficiencies. Improving low-speed efficiency, which can improve turbocharged indirect injection engine acceleration performance, is an important research and development area. Turbocharging is also important for direct injection engines, because the smoke-limited load/speed range of naturally aspirated direct injection systems is usually less than that of indirect injection engines.

Exhaust Gas Recirculation

Exhaust gas recirculation (EGR) is the principal engine technology being used to reduce NO_x emissions. EGR reduces combustion temperatures and thereby NO_x emissions; however, EGR can increase particulates.

EGR systems fall into two general categories, constant and modulated. In the long run, modulated electronic EGR systems appear to be the preferable, but simpler mechanical systems will be used during the early 1980s. These will allow meeting the 1982 to 1984 emission standards (0.41 g/mi hydrocarbon, 3.4 g/mi carbon monoxide, 1.0 g/mi NO_x, and 0.6 g/mi particulate), with some waivers to 1.5 g/mi NO_x, with no or small incremental control costs in comparison with the expense needed to meet the 1981 model year vehicle standards (0.41 g/mi hydrocarbon, 3.4 g/mi carbon monoxide, and 1.0 g/mi NO_x). Modulated EGR systems range from simple on/off systems (with relatively constant EGR for low to moderate loads and no EGR for high loads) to sophisticated closed-loop electronically controlled devices that adjust the EGR rate to optimize NO_x, hydrocarbon, and particulate emissions, and fuel economy. For minimizing all emissions, engine timing and other design variables must be reoptimized to take advantage of EGR. Because it reduces the oxygen/fuel ratio, EGR can increase carbon monoxide and particulate emissions. Figure 6 shows a fully electronically controlled system for a turbocharged engine. This system uses sensors for airflow, engine speed, and fuel pump load to determine accurately the air/fuel ratio under all operating conditions. An accurate carbon

dioxide sensor would be another approach to measuring a single variable
for controlling the air/fuel ratio and EGR, although no feasible sensor
is now available.

Exhaust Aftertreatment

The development of aftertreatment technology for particulate control in
diesel engines must take into account the low exhaust temperatures in
light-duty vehicles. Typically, exhaust temperatures range from 150°
to 300°C during the FTP cycle. Exhaust temperature is approximately
linear with load, increases with engine speed, and is highly transient
during FTP cycle operation. This is because the diesel is a variable
air/fuel ratio engine and the FTP cycle requires substantial low-speed,
part-load operation. In contrast, the gasoline engine, which is
basically a constant air/fuel ratio engine, produces exhaust temperatures that range from from 600° to 700°C varying little during the FTP
cycle.

Three basic approaches to exhaust aftertreatment are under
investigation: reactors/thermal in-stream oxidation; catalysts; and
traps, catalyzed traps, and trap-oxidizers. Of these approaches,
catalyzed or uncatalyzed trap systems appear to be the most promising
technology, although they will need to be regenerated every 50 to 100
miles. A trap is generally only thought of as being used to remove
particles from the flowstream; it also can be used to concentrate
particulate matter. The three principal types of materials under
consideration are: metal meshes, ceramic monoliths and foams, and
fibers. Paper and other fiber materials are the most recent types to
be investigated. Figure 7 shows two prevalent design approaches to
aftertreatment filters.

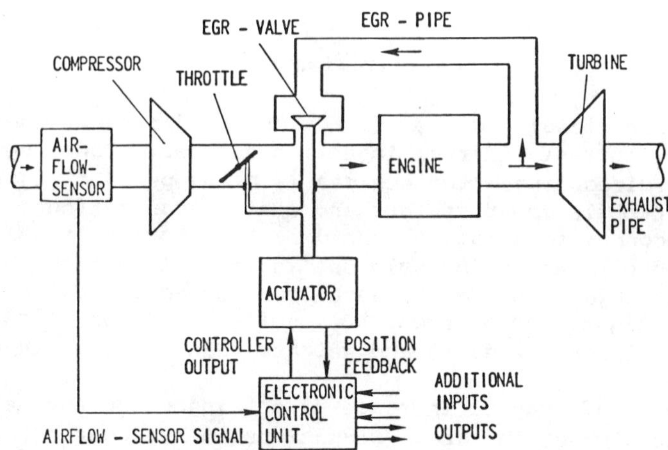

FIGURE 6 Electronically controlled EGR system
for a turbocharged engine. SOURCE: Robert
Bosch, 1980.

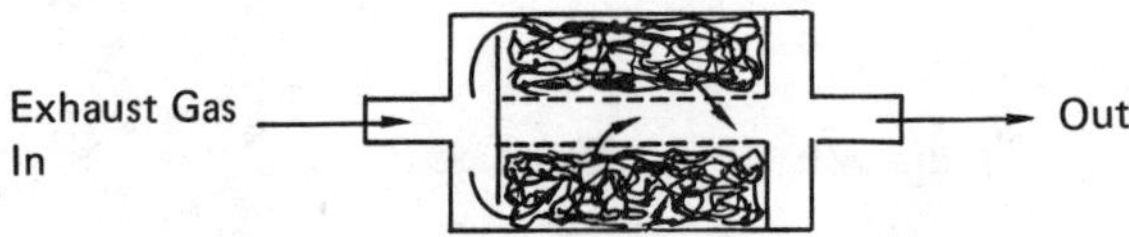

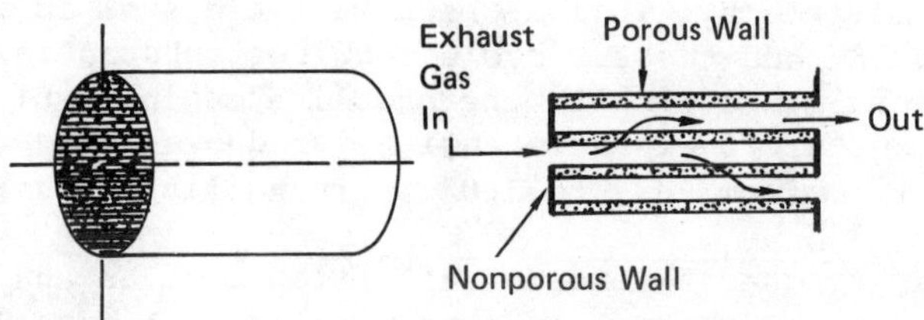

FIGURE 7   Schematic: steel mesh and ceramic exhaust aftertreatment filters.

An overall engine trap system that shows promise for post-1985 application is illustrated in Figure 8. This system uses an intake throttle regenerative system with electronic fuel injection and two traps in series. The present stage of development does not allow definition of an optimum system. Because 5 years lead time is required for production from tested technology, it is unlikely that exhaust aftertreatment devices will be in place nationwide before 1986 or 1987.

Direct Injection Engines

Direct injection diesel engines offer about a 15 percent fuel economy advantage over current indirect injection engines. Thus, development of light-duty direct injection engines is being pursued vigorously by a number of automobile manufacturers and other organizations. In spite of the fuel economy advantage of the direct injection diesel, all diesel-powered cars now being sold use indirect injection engines. Current direct injection engines have lower speed and power output capabilities; higher hydrocarbon, NO_x, noise, and odor emissions; and higher peak cylinder pressures and weights than indirect injection engines.

Efforts are underway to increase direct injection engine output by increasing the combustion rate. Reductions in hydrocarbon, odor, and noise emissions are being sought through increasing the turbulence level, retarding injection timing, minimizing the injection nozzle sac volume, and increasing the compression ratio. It appears that turbocharging may be needed to increase the smoke-limited load/speed range, which is generally higher in indirect injection engines.

In spite of current limitations, the development of direct injection diesel engines should be accelerated to obtain the advantages projected in Table 4.

FUEL PROPERTIES AND DEMAND

Diesel Fuels

Commercial diesel fuels are mixtures of hydrocarbons derived from crude oil. The properties of diesel fuel depend on the types of crude oils used as raw material, the refinery process involved, and the properties and mix of the refinery stocks from which the fuel is blended. Variations in these factors affect ignition quality, volatility, heating value, hydrocarbon composition (e.g., paraffins, naphthenes, olefins, or aromatics), sulfur content, and other properties.

Two grades of automotive diesel fuel (No. 1 and No. 2, established under ASTM 975 specification) are currently produced by the petroleum industry. No. 1 diesel fuel is primarily produced for circumstances in which No. 2 causes cold-weather handling and engine starting problems. No. 1 diesel fuel is also used in some city bus fleets to reduce smoke emissions. The price of No. 1 fuel is presently higher than that of

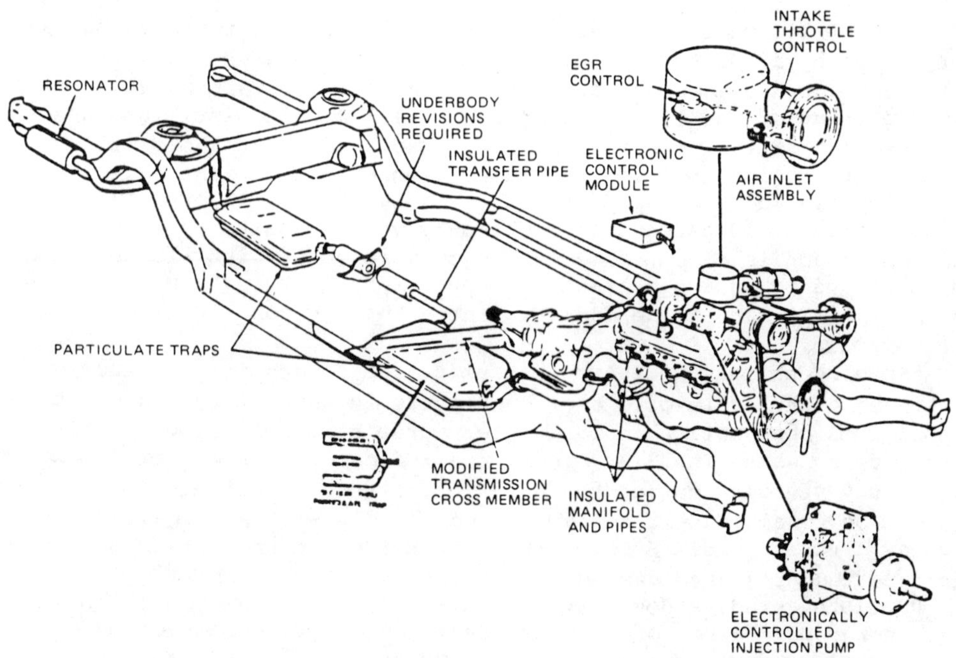

FIGURE 8 Diesel emission control system. SOURCE: GM, 1980.

TABLE 4 Summary Evaluation of Direct Injection in Relation to Indirect Injection

	Better	Equal	Inferior
Fuel economy	15-20%		
Heat reduction to coolant	-30%		
Starting	___		
Suitable for turbocharging	___		
Speed range		___	
BMEP potential		___	
Smoke, full load			___
Smoke part load	___		
Low NO$_x$/HC potential (USA only)			___
Particulates	___		
Noise (combustion)			
Full load			___
Idle	___		
Durability		___	
Production cost		___	

SOURCE: Cartellieri, 1980.

No. 2. The availability of No. 1 fuel is expected to decrease in the future because of competing demands for use as blending stock and jet fuel. Thus, No. 2 diesel fuel is the most common truck and passenger car fuel and in the future may be the only diesel fuel available in large quantities.

Demand and Supply

If sales of light-duty diesels grow to 25 percent of total annual light-duty sales by 1990, the diesel fuel demand for this use in that year will be about 440 thousand barrels per day (TBD). This quantity of fuel is 13 percent of the 1979 consumption of all middle distillate fuels.

A survey of the petroleum industry indicates that there will be no major problems in supplying increased demands for current ASTM 975 specification No. 2 diesel fuel, even if the diesel light-duty-vehicle sales represent 50 percent of light-duty vehicle sales by 1990. This will be accomplished by reducing the conversion of distillates to gasoline and by adding a limited amount of refinery equipment. Even further supplies could be made available through conservation of heating oils.

Assuming sharply increased dieselization of medium-duty trucks, and continued high levels of diesel engine use in heavy-duty trucks, highway diesel fuel consumption in 1990 is expected to double from the 850 TBD that were consumed in 1979. Thus, the indicated diesel fuel demand for light-duty vehicles during the next 10 years is relatively small in the context of total distillate demand and the diesel fuel requirements for highway trucks. Table 5 shows the 1978 petroleum product demand compared to the projected demand in 1990.

Published studies demonstrate that a modern refinery will consume less energy per barrel of capacity as the proportion and volume of diesel fuel production are increased. Most of this reduction in refinery energy consumption will occur because of reduced catalytic

TABLE 5 Domestic Petroleum Product Demand (in TBD)

	Actual 1978	Projected 1990
Motor gasoline	7,412	6,124
Jet fuel		
Naphtha type	199	124
Kerosene type	858	1,203
Distillates		
Automotive diesel[a]	988	1,988
Other[b]	2,619	2,287
Residual fuel oil	3,023	2,344
Other[c]	3,748	4,827
Total Demand	18,847	18,897

[a]Includes off-highway use.
[b]Includes heating oils for households, industry, electric utilities, railroads, vessels, military, and miscellaneous.
[c]Includes aviation gasoline, naphthas, liquefied gases, petrochemical feed stocks, lubricants, waxes, coke, asphalts, road oils, still gases for fuel, and miscellaneous.

SOURCE: National Petroleum Council, preliminary unpublished data.

cracking operations. As a typical modern fuel refinery shifts from a maximum gasoline production mode to a maximum distillate mode, the energy output in product (fuel), as a percentage of energy input as raw material, will increase by 1 percent, from about 92 to 93 percent of input energy.

Fuel Price

The cost of crude oil is the major factor determining the cost of producing diesel fuel. Typically the crude oil raw material cost may represent 85 to 90 percent of the cost of producing diesel fuel. Although a minor part of the total, the other significant element is the refining cost, which is inherently lower for diesel fuels than for gasoline.

Because of the uncertainties of crude oil availability and producing country pricing policies, no attempt to forecast the effect of future raw material costs on diesel fuel prices was made.

The relative prices of diesel fuel and gasoline are an important consideration to the Diesel Impacts Study. Although price differential has narrowed in the past 3 years, diesel fuel in the previous 10 years (1967 to 1976) sold at a price that was about 15 percent less than the price of regular gasoline.

In the face of declining gasoline demand and increasing distillate demand, the petroleum companies predict a narrowing of the differential between gasoline and distillate prices. Assuming free market conditions, refiners predict that sometime in the 1980s the price of diesel fuel will equal the price of unleaded gasoline. During the same period, kerosene jet fuel prices are expected to rise above the price of gasoline. To maintain refinery profit margins, the refiner will shift a larger proportion of total refining costs onto the increasing middle distillate production. This shift will lessen the current operating cost advantage of diesel-powered light-duty and heavy-duty vehicles relative to gasoline-powered vehicles.

MAJOR CONCLUSIONS

Fuel Economy

• A gallon of diesel fuel has 13 percent greater energy content than a gallon of gasoline. The volume approach, rather than the energy approach, is the more logical fuel economy comparison basis from the consumer's viewpoint so long as fuel is purchased by the gallon. Comparisons on a mile-per-gallon (mpg) basis must take into consideration the present and future prices of the fuel, including any price differential between gasoline and diesel fuel to the consumer, because the economics of dieselization are sensitive both to these prices and to the total annual mileage driven.

• On the basis of comparable performance, diesel engines have an approximate 35 percent fuel economy (mpg) advantage over gasoline engines in typical light-duty applications. On an energy basis the advantage is approximately 20 percent. The fuel economy advantage can be considerably greater for urban stop-and-go driving and cold-climate operation. The initial start-up and running characteristics of diesel engines (as opposed to choke operation of spark ignition engines) contribute to this fuel economy advantage.

Emissions and Control Technology

• With conventional diesel fuels, the formation of particulates is fundamental to the diesel engine combustion processes; however, the actual exhaust emission of these particulates is not. The development of low-particulate-emission diesel engines will be constrained by fuel quality, engine design and control system requirements, NO_x standards, materials, and conflicting operational considerations.

• Important nonregulated air pollutants from diesel engines are: aldehydes, nitrogen dioxide, sulfur dioxide, the particulate soluble organic fraction, and odor emissions. Aldehydes and nitrogen dioxide, which are significantly higher in diesels, are photochemically active and act as direct irritants. The particulate soluble organic fraction contains biologically active agents. Widespread dieselization of light-duty vehicles may result in a significant localized odor problem. Better measurements of tailpipe nitrogen dioxide compounds are needed.

* Diesel engine emissions of sulfur dioxide and sulfates are proportional to fuel sulfur levels. Some proposed controls for other exhaust components decrease sulfur dioxide emissions and increase sulfate emissions. The technology for further desulfurization of diesel fuel exists, but fuel desulfurization is a complex national issue.

* Cold-start conditions and some control technologies will increase diesel irritants (aldehydes and nitrogen dioxide) and odorants.

* Visibility of emissions is not measured in the Federal Test Procedure. Additional characterization and control of visible smoke emissions may be required.

* Control technologies using catalysts may prove effective for soluble organic fraction of particulate, hydrocarbon, aldehyde, and sulfur dioxide reduction, but increased levels of sulfates and nitrogen dioxide are likely to result.

* Deterioration factors for regulated emissions from uncontrolled diesel engines are close to unity and are consistently better than those for gasoline engines. Hydrocarbon, NO_x, and especially particulate deterioration factors could worsen markedly with controls.

* The diesel engine has inherently low carbon monoxide emissions and should easily meet the 3.4-g/mi carbon monoxide emission standard in all light-duty applications.

* Further reduction of particulates will await development of technology for traps, engine modifications, and fuels. The vehicle weight will in large measure determine the minimum level achievable, which, for the smallest vehicles, could approach 0.1 g/mi.

Fuel Characteristics and Availability

* The specification of a minimum quality automotive diesel fuel for cars, trucks, and buses would improve the ability of engine designers to optimize diesel engines and would insure that the performance of existing engines would not degrade. Important specifications to be considered are: a minimum cetane index; a maximum sulfur level; a maximum aromatic content; a maximum 90 percent boiling point; and a seasonably adjusted cloud point. Competition for a growing light-duty-fuel market may result in the availability of a more uniformly high-quality diesel fuel.

* Current total refinery capacity for producing diesel fuels that meet ASTM specifications is sufficient to accommodate projected dieselization of the light-duty vehicle fleet at least through 1990.

* Marketing economics suggest that increased demand for diesel fuel will likely result in equal pricing for diesel and gasoline fuels in the near future.

- Lower cetane numbers, higher aromatic and sulfur contents, and higher end points would normally be expected as the future quality of crude feedstocks falls. The establishment of tighter specifications for automotive diesel fuel will result in restricting refining flexibility to manufacture a range of products from crudes of uncertain quality. To optimize the motor fuel energy system, efforts should be made to improve the fuel tolerance of the diesel engine.

- There is wide variation in the properties of diesel fuels marketed in the United States. Improved analysis of the extent and importance of these variations is needed.

- Research is required to determine the adequacy of the cetane number as a measure of diesel fuel ignition quality in modern, light-duty engines.

- The very low volatility of diesel fuel in comparison with gasoline represents a fire safety advantage.

Engine Technology

- Some automotive diesel engines have experienced a higher frequency of required lubricating oil changes, fuel injection system damage by water contaminated fuel, and valve-train wear. These are all problems that have engineering (design) solutions.

- Current diesel engines are typically noisier than gasoline engines, particularly at idling and light load. The frequency distribution of diesel noise is more irritating to most individuals. Diesel-powered vehicles can be made with noise levels comparable to those of gasoline-powered vehicles, but the design changes needed for noise reduction may entail economic or performance costs.

- The light-duty indirect injection diesel engine is currently less developed than the gasoline engine and probably has a greater potential for further improvements. The development of the direct injection light-duty diesel engine is anticipated to provide additional fuel economy improvements.

- High-payoff diesel technologies that should be developed during the 1980 to 1990 period include: improved fuel injection systems, electronic control systems and sensors, exhaust gas recirculation systems, blowers and turbochargers, transmissions and transmission control, lubricant contamination and loss control, oil additives that are not influenced by particulates, and regenerative trap-oxidizers.

- Programmed fuel injection technology is critical to the performance of the diesel engine and to the limitation of emissions. However, the level of development effort on this technology appears inadequate.

- The development of lean-burn and direct injection stratified charge spark ignition engines should continue as a means to achieve a fuel efficient national automotive fleet designed to use gasoline and diesel fuel in refinery energy-efficient proportions. These engines may provide the advantages of diesel engines with minimum particulate problems.

REGULATORY ISSUES

- The EPA particulate test procedure is adequate for total
mass   and soluble organic fraction determinations. The adequacy of
this test method for determination of individual hydrocarbon species
and the biological activity of the particulate soluble organic fraction
remains to be established.
- Current emission test procedures result in "double counting"
of the heavy portion of hydrocarbon emissions from diesel engines.
This should be rectified by changing the test procedure to measure the
heavy hydrocarbons as particulates only.
- Significant differences exist in the characteristics of hydrocarbon emissions from diesel and gasoline engines. These differences
should be reflected in establishing appropriate hydrocarbon emission
regulations. A nonmethane hydrocarbon emission standard for both
engine types and credit for lower evaporative hydrocarbon emissions
from diesel vehicles should be considered. Current test methods do not
correctly account for aldehydes as hydrocarbon emissions.
- Meeting the 1982 to 1984 federal light-duty-vehicle emission
standards is technologically feasible on the assumption that some NO_x
waivers are granted for larger vehicles in 1983 and 1984.
- Meeting the 1985 light-duty-vehicle 1.0-g/mi NO_x and 0.2-g/mi
particulate emission standards in light-duty diesels is technologically
feasible for only the smallest (approximately 2,000-pound) vehicles.
- Control of particulate at the 0.2-g/mi level is likely to
require the use of exhaust aftertreatment devices. The most promising
is the regenerative trap-oxidizer, although this has not yet been
successfully demonstrated in field durability tests of 50,000 miles for
each device. Historical precedent would suggest about a 5-year lag
between such test demonstration and road operations.
- Diesel engines will probably not be able to meet a 0.4 g/mi
NO_x standard except in the smallest (approximately 2,000-pound)
vehicles.
- Opportunities for emphasizing particulate control for
heavy-duty diesel vehicles and NO_x control for gasoline-powered
vehicles should be considered in developing optimum emission control
strategies for light-duty diesel vehicles.
- Delaying the 1984 altitude emission requirement for
diesel-powered light-duty vehicles until 1985 would eliminate the need
for developing an interim fuel injection system that would be good for
only one model year.

1

INTRODUCTION AND BACKGROUND

INTRODUCTION TO THE ISSUES

Diesel engines are widely used in transportation and stationary applications where fuel economy has been of primary importance. Historically, they have powered marine, railroad, and heavy-duty off-road and highway vehicles. In the United States, however, the substantial use of diesel engines as power plants for light-duty vehicles, both passenger cars and trucks, only began in 1978. The principal reason for increasing dieselization of light-duty vehicles is the better fuel economy that the diesel provides.

Growth in Diesel Use

The use of diesel engines in light- and heavy-duty vehicles has increased over the past 20 years. In 1979, sales of diesels increased to 2.6 percent of passenger cars. Table 1.1 shows the 1984 to 1988 projected levels of dieselization for light-duty vehicles as projected by a number of automobile manufacturers (EPA, 1980c). A recent study investigated the dieselization of trucks by using a computer model (Jambekar and Johnson, 1978). Figure 1.1 shows three rates of dieselization that were used in the model. Each vehicle class had separate and different rates corresponding with the general market trends. Figure 1.2 shows the total truck population by years for these three levels of dieselization (Jambekar and Johnson, 1980). There is clearly a delayed effect on the diesel population growth because of the long life of the truck.

Figure 1.3 shows cumulative fuel usage by years for various vehicle classes for the moderate level of dieselization. These data show that dieselization of the light- and medium-duty fleet tends to increase the fraction of the total truck fuel used by heavy-duty trucks.

Critical Issues

A number of technological issues appear to be critical to the future of the diesel engine in light-duty applications. The following are of primary importance:

TABLE 1.1 Projected Percentage of Diesel-Powered Light-Duty
Vehicles Sold by Manufacturers (All Values in Percent)

| | Year | | | | |
Manufacturer	1984	1985	1986	1987	1988
General Motors	12	13.8	17.5	22	23
Ford	8	10	11	12	13
Chrysler	8	10	12	14	16
AMC	8	10	12	14	16
VW[a]	42	42	46	50	55
Mercedes-Benz	66	70	74	78	82
Peugeot	66	70	74	78	82
Volvo, Fiat, IHC, BMW	8	10	12	14	16

[a]Includes Audi.

SOURCE: EPA, 1980c.

• <u>Fuel economy</u>: What are the real fuel economy advantages of
the diesel engine? What will be the effect of new technologies,
particularly turbocharging, electronic control systems, and the small
direct injection diesel engine, on fuel economy?

• <u>In-use characteristics</u>: How do diesel engines behave in
actual use compared with gasoline engines? Are there safety advantages
or disadvantages? What are the true lifetime cost comparisons?

• <u>Diesel particulate fundamentals</u>: Are particulate emissions
fundamental to diesel engine design? Do possibilities exist through
combustion, engine, or fuel modification to obtain substantial
reductions in diesel particulate emissions?

• <u>Diesel particulate emissions</u>: Are the test methods for diesel
particulates adequate in terms of total mass, soluble organic fraction,
and specific toxic substances or groups of substances? What are the
chemical and physical characteristics of particulates produced by
diesel engines? What is the effect of the fuel properties on diesel
particulates?

• <u>Gaseous pollutants</u>: What are the emission characteristics for
regulated and unregulated gaseous pollutants? How do these compare
with those of gasoline engines?

• <u>Noise, odor, irritants</u>: Are the test methods of these effects
adequate? How can the diesel be improved with respect to these
irritants?

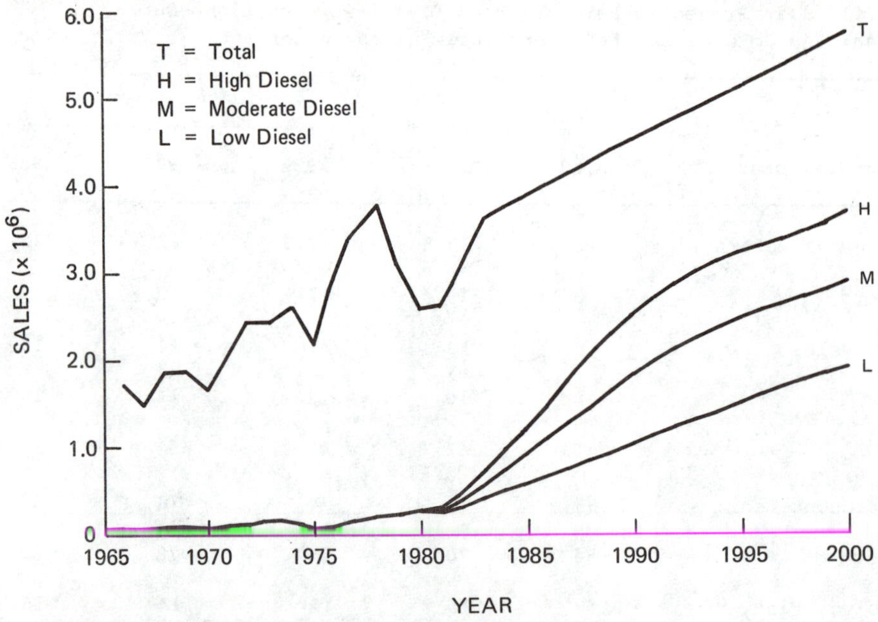

FIGURE 1.1   Annual sales--total truck.
SOURCE: Jambekar and Johnson, 1980.

• Control technologies: What are the control technologies that
will be applied to diesel engines to meet environmental regulations?
What is the effect of engine design? What are the characteristics of
exhaust gas recirculation (EGR) systems for diesels? What gains are to
be realized from electronically controlled fuel injection systems or
electronic control of the engine? What is the effect of turbocharging?
What are the characteristics of direct injection diesel engines? What
are the most promising control technologies? When will they be
available? What are the costs?

• Fuel availability: What is the likely fuel availability for
light-duty diesels? What are the likely fuel properties? How do
alternative fuels affect the diesel? Is a special light-duty automotive diesel fuel required? What are the refining and distribution
considerations of making such a fuel?

• Research and development opportunities: What is the future of
diesel engine improvements? Which advanced concepts have the highest
potential payoff? How will diesel improvements compare with gasoline
engine improvements? Are there alternatives to the diesel engine that
may overtake it?

This report treats the evolving technology of the diesel engine
through focusing on these topics. There is no attempt to provide a
complete review of diesel technology; several studies have already
admirably taken on that task. Rather, the report is intended to
provide technological information and analysis that will assist in the

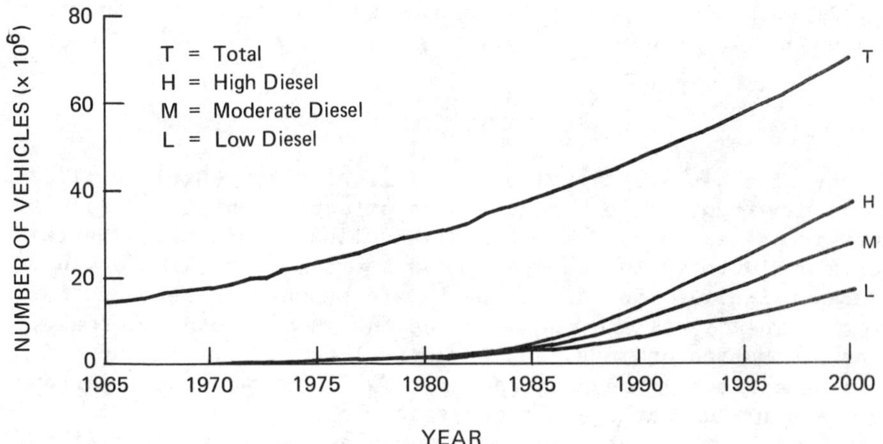

FIGURE 1.2 Population--total truck.
SOURCE: Jambekar and Johnson, 1980.

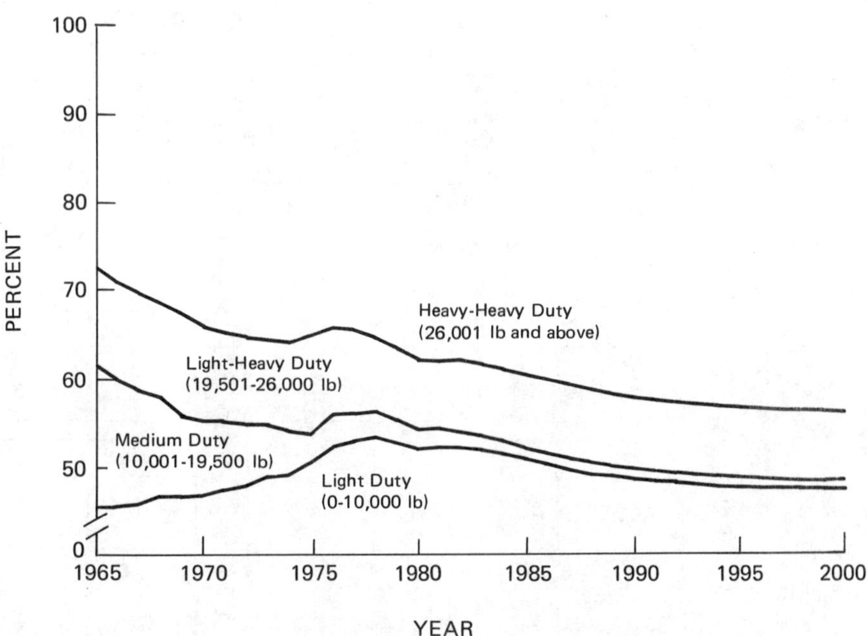

FIGURE 1.3 Fuel usage (gasoline and diesel). Percent
cumulative distribution moderate sale, moderate diesel.
SOURCE: Jambekar and Johnson, 1980.

Diesel Impacts Study Committee's broad examination of the future of the diesel engine light-duty vehicle.

BACKGROUND

A variety of engines are available for light-duty vehicle application. Although many types have been tried on an experimental basis, most engines in use are one of two types: gasoline or diesel. (Spark ignition and Otto cycle are synonymous terms for gasoline engines. Compression ignition and diesel cycle are synonymous terms for diesel engines.) The diesel and the gasoline engine are both internal, intermittent combustion engines, and their fuel economy is influenced by the same fundamental thermodynamic parameters. The two most important of these are air/fuel ratio and compression ratio.

The variety of other light-duty vehicle power plants, available or being designed, is shown in Figure 1.4. Beyond gasoline and diesel engines, the only automotive power plant that seems likely for widespread applications in light-duty vehicles during this century is the battery-powered electric motor. The potential efficiency advantage of Stirling and Brayton (gas turbine) cycle engines ensures that they will

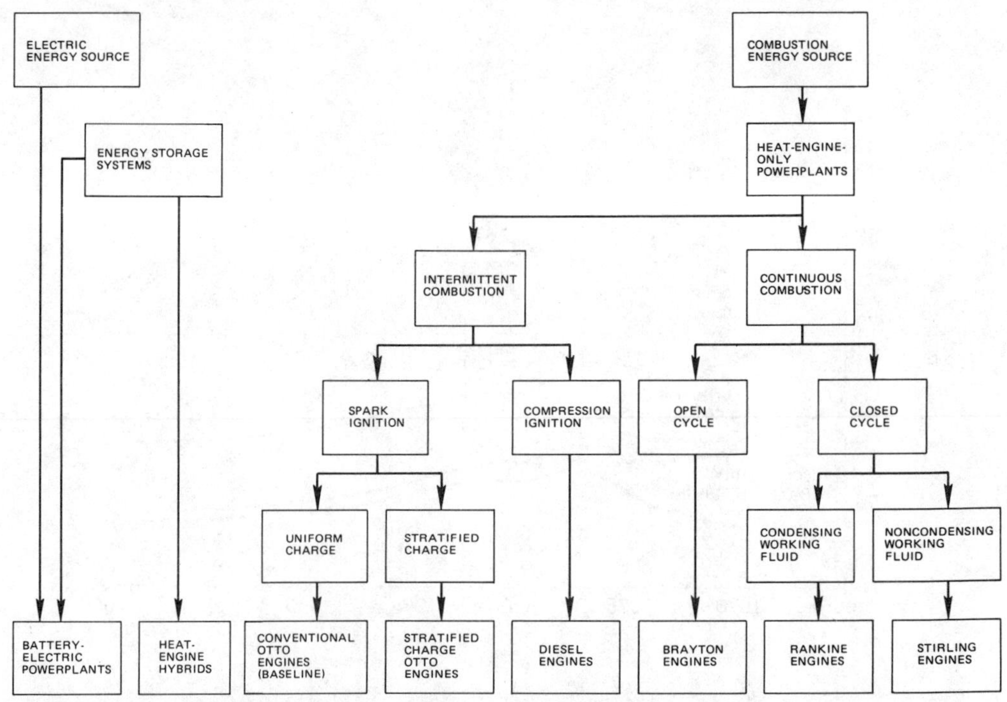

FIGURE 1.4 Internal, intermittent combustion engines.
SOURCE: Adapted from Jet Propulsion Laboratory, California Institute of Technology, 1975.

be of continuing interest and will undergo continued development during
this period, but it is unlikely that either of these engines will have
an important impact in light-duty applications during the next 20 years.

As shown in Figure 1.5, both gasoline and diesel engines have been
designed with a number of variations and combinations. The most common
engines used in light-duty applications are the single-chamber gasoline
engine and the prechamber diesel engine. Variations include: two-
chamber stratified charge engines, direct injection stratified charge
engines, lean-burn homogeneous charge engines, spark-assisted diesels,
and direct injection diesels. Some of these engines exist only as
prototypes; others are commercially available.

The intermittent, internal combustion engines, dominated by the
gasoline and diesel versions and variations thereon, are compact,
reliable, inexpensive power plants that are now well integrated in the
fuel supply system for light-duty vehicles and transportation.
Potential improvements in these engines are sufficient to insure that
their performance will remain competitive with other engine types
during the next 20 years.

Diesel Types and Characteristics

There are two categories of diesel engines: the open chamber, or direct
injection engine; and the divided chamber, or indirect injection
engine. Each of these engine configurations is shown in Figure 1.6.
Traditionally, open-chamber engines have been preferred for heavy-duty

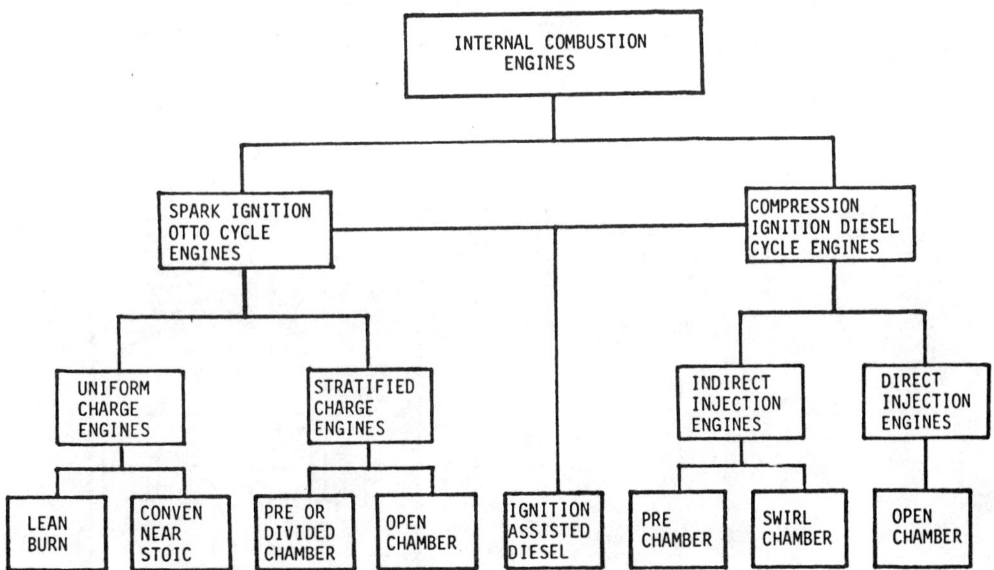

FIGURE 1.5 Types of spark ignition and diesel engines.
SOURCE: Adapted from Jet Propulsion Laboratory, California Institute
of Technology, 1975.

a) Indirect Injection,
    Pre-combustion Chamber

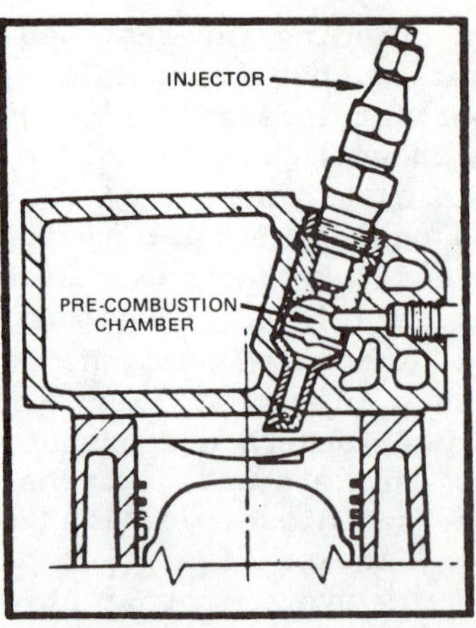

b) Indirect Injection,
    Swirl Combustion
    Chamber

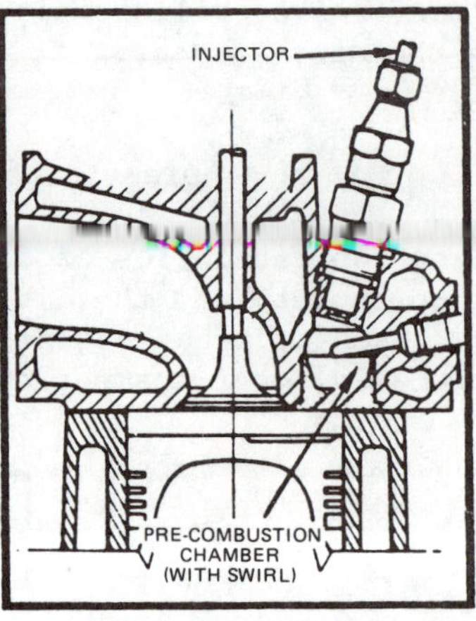

c) Direct Injection
    Combustion Chamber

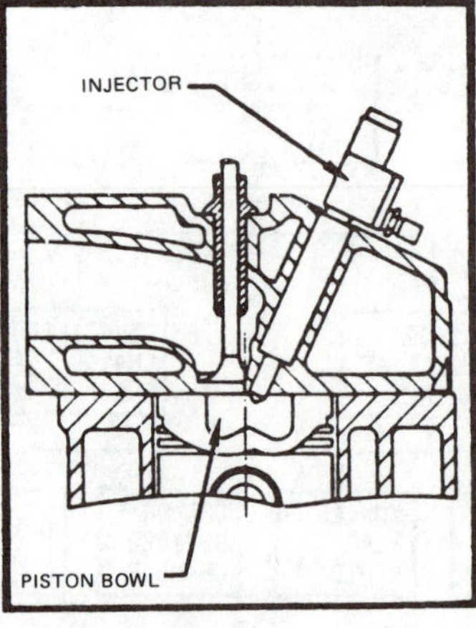

FIGURE 1.6 Diesel engine combustion chamber geometries.
SOURCE: Wade, 1980.

applications because they offer the best fuel economy. Divided chamber engines have been preferred for light-duty applications because they are less fuel sensitive, have a wider speed range (and therefore greater power/weight ratio), are smaller, run more quietly, have lower pressure fuel injection systems, are less odorous, and emit fewer pollutants.

In addition to the two basic engine categories, diesels can be further subdivided according to the type of combustion chamber, air induction system, fuel injection system, number of strokes per cycle, and mechanical configuration. The diesel engine currently used in light-duty applications is a compression ignition, four-stroke, reciprocating engine with a divided chamber; however, the potential fuel economy advantages of the direct injection engine will ensure that its development is pursued. Both naturally aspirated and turbocharged air induction systems are currently used in diesel engines.

The following characteristics distinguish diesels from gasoline engines:

• In the diesel engine, the fuel is self-ignited as it is injected into air that has been heated through the compression process. In the gasoline engine the fuel is ignited by spark plugs.

• In the diesel engine, load is controlled by varying the flow rate of fuel to the engine. The mixture in the combustion chamber of a diesel, therefore, has a varying air/fuel ratio. This feature allows unthrottled operation of the engine, which provides a reduction of pumping losses, but more importantly, it allows lean operation with accompanying better fuel economy and an improvement of performance, particularly during low-load operation. In the gasoline engine, the air/fuel ratio remains the same, but the rates of both are varied.

• In the light-duty diesel engine, a compression ratio exceeding 20 to 1 is advantageous from a thermodynamic standpoint; it is needed to facilitate starting at low ambient temperatures; and it is required to achieve a self-ignition temperature that will enhance combustion and thus lower the level of hydrocarbon emissions. The resultant high chamber pressure necessitates a stronger, heavier mechanical design than is needed in gasoline engines. Current gasoline engines have compression ratios of about 8.5 to 1.

• In the diesel engine, the fuel injection process is critical to ignition, efficiency, and emissions. Good mixing between the fuel and air is essential and must occur over a period of a few milliseconds. The air induction and fuel injection systems, therefore, must be designed to achieve optimum interaction. Injection timing, pressure, spray geometry, and delivery rate are critical to good performance and are difficult to optimize over all engine operating conditions. In the gasoline engine, air and fuel are mixed in the carburetor before they enter the combustion chamber.

The Combustion Process

The combustion process begins when liquid fuel jets are injected at high pressure into the combustion chamber. Within the chamber the jets disintegrate into a core of fuel surrounded by an envelope of air and fuel droplets. By this time the compression process has caused the air temperature within the chamber to increase to the level required to ignite the fuel. The combustion process is usually divided into four stages:

 • an ignition delay period, during which time the fuel is atomized, vaporized, partially mixed with air, raised in temperature, and ignited;
 • a period of rapid pressure rise, which is largely the result of the combustion of the premixed air and fuel vapor formed during ignition delay period;
 • a period of controlled pressure rise, which is associated with fuel vaporization and the diffusion controlled burning of the fuel; and
 • a late burning period, which occurs during the expansion stroke and is critical in determining the final pollutant emission levels. The control and tailoring of the combustion process can affect efficiency, noise, and pollutant formation. The overall effect of the combustion process may be described in terms of an energy release diagram, such as Figure 1.7, which also shows the four stages of combustion.

Pollutant emissions are directly related to the combustion process. Diesel engines operate with an excess of air, and as a result, carbon monoxide emission levels are generally very low. In a system such as this, when fuel is mixed in regions where there is too much air and the temperature is too low to promote complete oxidation, hydrocarbon and partially oxygenated hydrocarbon emissions sometimes result. A late delivery of fuel, particularly fuel coming from the injector sac volume (the volume between the needle seat and the nozzle holes), also contributes to unburned hydrocarbons in the exhaust. In this case, the last

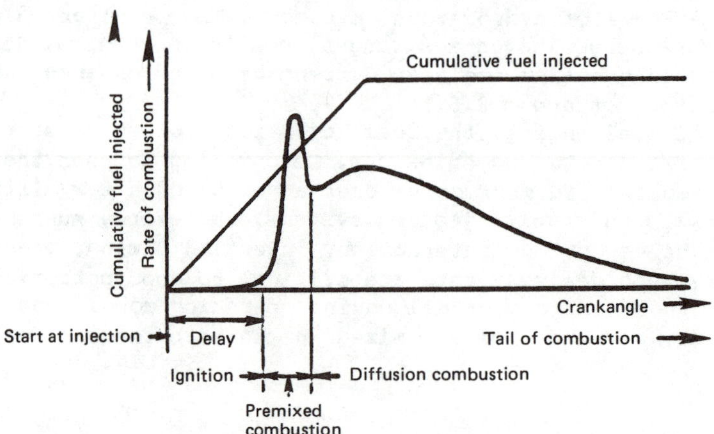

FIGURE 1.7 Typical rate of energy release for diesel engine combustion. SOURCE: Whitehouse et al., 1980.

portion of the fuel being injected is more poorly atomized because the injection pressure is lower and the combustion chamber pressure is higher; thus locally there is not enough air to completely oxidize the hydrocarbons. Wall impingement by the fuel, which is a special problem in small, high-speed direct injection engines, also contributes to unburned hydrocarbons and particulate emissions. Nitric oxide is formed in the fuel-lean, high-temperature regions of combustion through the thermal fixation of the molecular nitrogen in the air. Current fuels are low in nitrogen; hence this source (fuel nitrogen) is not an important contributor to emissions of oxides of nitrogen. Excess air during combustion and in the exhaust system is conducive to the formation of nitrogen dioxide, and, as a result, this species represents a higher fraction of the oxides of nitrogen emitted from diesel engines than from gasoline engines.

The highly stratified mixture of fuel and air in the diesel engine results in fuel-rich zones within the combustion chamber. The combination of these fuel-rich zones and high temperatures leads to the formation of carbonaceous particulate, often referred to as soot. The details of the formation and subsequent oxidation of this carbonaceous particulate are not well established. Particulate emissions of this type are not a problem in conventional gasoline engines, where fuel and air are premixed before they enter the combustion chamber. In addition, it is known that the introduction of small amounts of engine lubricants into the combustion chamber can also contribute to this type of diesel engine particulate emissions.

Fuel Injection Systems

The fuel injection system is the key mechanical component of the diesel engine. Extremely high injection pressures (130 to approximately 1,300 atmospheres) are needed to assure that injection, atomization, and vaporization occur within the short time available. The injection system must:

- meter the quantity of fuel needed for different engine speeds and loads;
- distribute the fuel equally among the cylinders;
- inject the fuel at the proper time in each cylinder cycle;
- inject the fuel at the proper rate;
- produce the spray pattern and atomization required by the combustion chamber design; and
- complete the injection cleanly without dribbling or after-injections.

Current injection systems use several mechanical elements to perform these functions. For instance, two basic methods of metering can be employed: the "common rail" system, which uses a single, steady high-pressure fuel supply to feed all of the injector nozzles, and the "jerk pump" system, which delivers a high-pressure pulse to each individual injector nozzle. The jerk pump system may consist of an individual pump nozzle assembly for each cylinder (unit injector) or a pump-line-nozzle system, which consists of a single pump with a distributor valve or a multiplunger pump that delivers the fuel to individual cylinders. Light-duty vehicles typically use pump-line-nozzle systems. Work is progressing on unit injectors and the introduction of electrically controlled and/or actuated components.

Comparison with Spark Ignition Engines

Although the main advantage of diesel vehicles is their better fuel economy, they also may be more reliable, serviceable, durable, and drivable than gasoline vehicles. Diesels disadvantages include: higher NO_x and particulate exhaust emissions, cold-start problems, exhaust odor and irritancy, noise, vibration, engine weight, and cost. Each of these factors is examined in detail in this report.

Procedures Used for Comparing Vehicles

Comparisons of light-duty vehicles are generally made by using data on measurements of engine efficiency, vehicle efficiency, and emissions. Each of these parameters requires different test procedures and comparative scales. Engine efficiency is measured in terms of fuel per work output, vehicle efficiency is measured in terms of fuel consumed per distance traveled, and emissions are measured in terms of mass of pollutant per distance traveled.

Engine Measurement Methods   Engine efficiency is measured on laboratory dynamometers for steady state operating conditions. It is expressed as a quantity of fuel consumed per unit of work output by the engine (lb/Bhp-hr, g/W-hr, etc.). Engine performance is usually mapped in terms of fuel consumption contours as a function of engine load and speed. Figure 1.8 shows typical fuel consumption maps for comparable diesel and gasoline engines. Engine emissions can be measured and mapped in a similar fashion. Emissions are usually expressed as a mass rate per work output rate (lb/Bhp-hr, g/kW-hr, etc.) or as mass per volume of fuel (kg/liter, etc.). These data may be used to predict the performance of vehicles over arbitrary driving conditions. The accuracy of this approach is limited by the use of steady state operating data to predict transient behavior.

Vehicle Measurement Methods   Vehicle efficiency is generally expressed as a ratio of fuel consumed per distance traveled (mpg, km/liter, etc.). Although miles per gallon (mpg) is the conventional measure used in the United States (and will be used in this report), the quantity of fuel consumed per distance traveled is a more direct efficiency parameter. Vehicle fuel economy can be measured in several ways. Most of the available data come from the data bank generated by the Environmental Protection Agency (EPA) from its vehicle emission certification procedure. These data are the results of chassis dynamometer testing (vehicle testing in laboratories) during specified driving cycles. Two cycles are used in these tests: the Federal Test Procedure (FTP) and the Fuel Economy Test (FET).

The FTP is also referred to as both the LA-4 and the Urban Cycle. It is used to obtain the fuel economy figures reported in annual EPA mileage guides and to measure exhaust emissions. The FTP cycle was designed to approximate urban driving conditions; hence the vehicle is driven at an average of 19.7 mph and makes 2.4 stops per mile. In response to industry criticism that the FTP underestimated in-use fuel economy, the FET (a higher speed cycle) was added to the EPA test procedures. This cycle is used to determine fuel economy during highway driving; it is not used for emission certification data. Other test cycles, such as the European and New York City (NYCC), are used to obtain local emission and fuel economy data; however, only the FTP and FET are used by the federal government.

Figure 1.9 shows the vehicle cycles used in chassis dynamometer tests. The debate over the adequacy of chassis dynamometer testing for fuel economy continues. Many people believe that track, road, and in-use tests are more accurate. For comparative purposes, however, chassis dynamometer testing is considered adequate. Therefore, with some qualification, FTP data are used extensively in this report. A

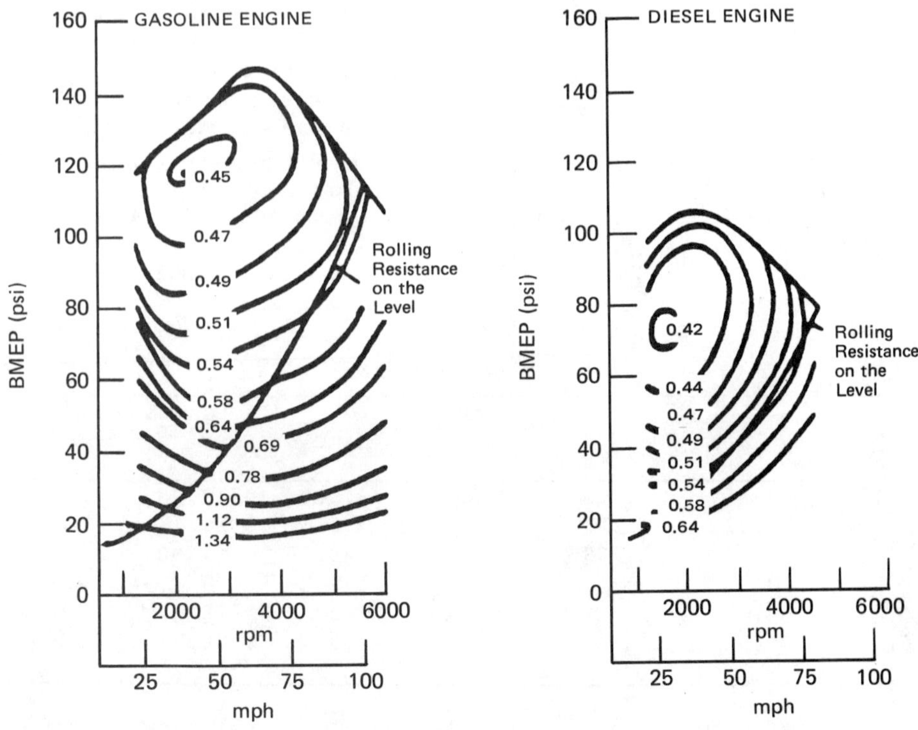

FIGURE 1.8 Fuel consumption maps for comparable diesel and gasoline engines.

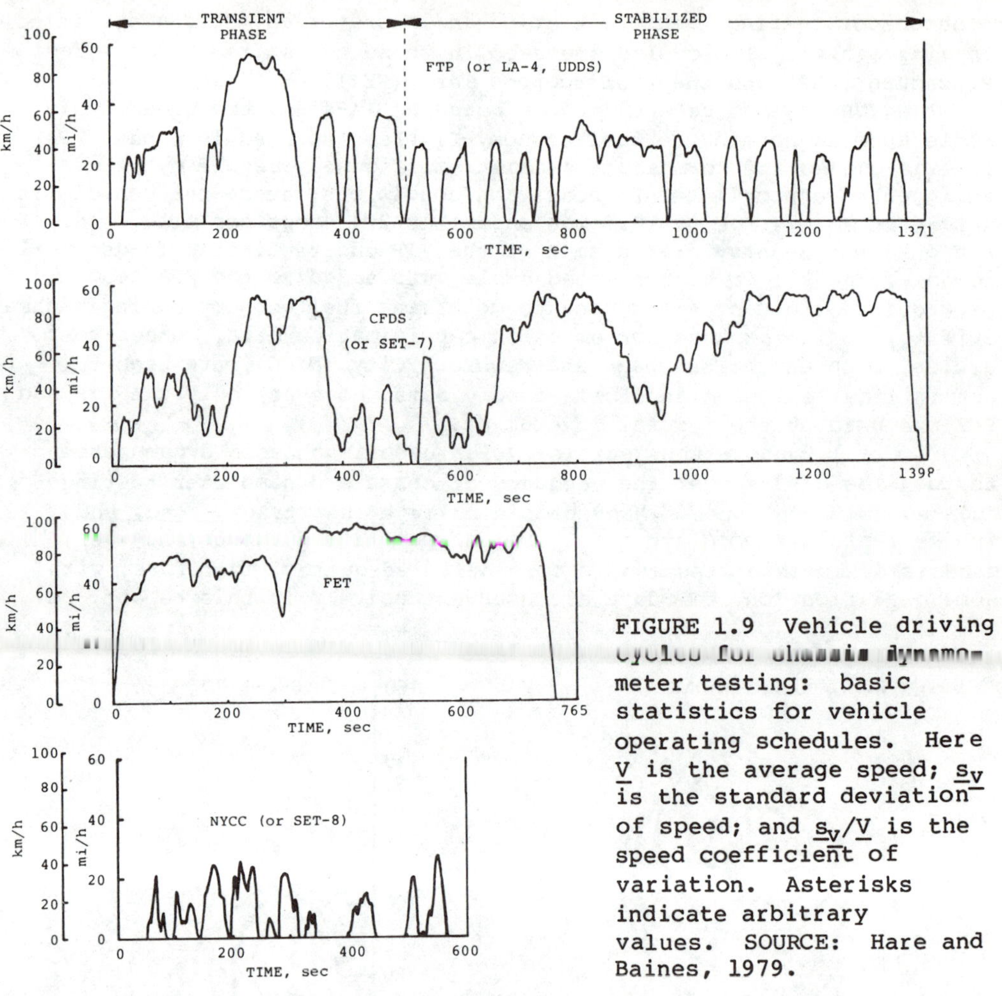

FIGURE 1.9 Vehicle driving cycles for chassis dynamometer testing: basic statistics for vehicle operating schedules. Here V is the average speed; s_V is the standard deviation of speed; and s_V/V is the speed coefficient of variation. Asterisks indicate arbitrary values. SOURCE: Hare and Baines, 1979.

VALUE BY STATISTICS

Schedule	V km/h[a]	s_V/V[a]	Stops, km	Idle Time, %	Length, km	Time, s
FTP	31.46	0.752	1.42	19.0	11.98	1371
FET	77.52	0.213	0.06	0.8	16.47	765
NYCC	11.37	1.129	5.79	40.2	1.90	600
IDLE	0	–	–	100.0	0	1200[b]
50 km/h	50.0	0	0	0	16.67[b]	1200[b]
85 km/h	85.0	0	0	0	28.33[b]	1200[b]

[a] V = Average speed
s_V = Standard deviation of speed
s_V/V = Speed coefficient of variation

[b] = Arbitrary

weighted combination of the FTP and FET fuel economy is used in determining compliance with the Corporate Average Fuel Economy (CAFE) requirements specified in Table 1.2.

Several comparisons between the EPA test procedure and road or in-use (usually fleet) testing have been conducted. The comparisons showed that the FTP fuel economy number is probably close to actual in-use fuel economy but that some biases are present. EPA annual mileage guides currently list the FTP fuel economy number and use it in annual operating cost estimates published therein. The fuel economy of diesel vehicles appears to be systematically underestimated by about 5 percent in comparison with that of gasoline vehicles in this procedure. This is an important fact to remember in comparing EPA gasoline and diesel light-duty-vehicle mileage and fuel economies.

<u>Emissions Measurement Methods</u> Emissions are measured during the FTP. In accord with the requirements of the 1982 federal particulate standard, new test methods have been adopted for the measurement of particulates. Difficulties appear to exist in the measurement of hydrocarbons for emission certification. Some of the hydrocarbons are counted twice (once in the gaseous hydrocarbon measurement and again in the particulate measurement) because the samples are taken simultaneously rather than sequentially. Because diesel engines have inherently lower evaporative emission levels than gasoline engines, it may be reasonable to provide, as California has done, some credit in terms of a higher exhaust hydrocarbon emission standard.

TABLE 1.2 Corporate Average Fuel Economy (CAFE) Standard for Passenger Cars

Year	Average mpg
1978	18.0
1979	19.0
1980	20.0
1981	22.0[a]
1982	24.0[a]
1983	26.0[a]
1984	27.0[a]
1985	27.5[a]

[a] Set by the Secretary of DOT.

SOURCE: Energy Policy and Conservation Act of 1975.

Fuel Economy Comparison

On a volume basis, indirect injection light-duty diesel-powered vehicles
have a fuel economy advantage of about 25 to 45 percent over comparable
gasoline-powered vehicles. The amount of this advantage strongly
depends on driving conditions and vehicle use. For instance, under
light-load, short-duration city driving conditions the fuel economy
advantage of the diesel engine may be 100 percent or more, but under
high-speed, high-load, long-duration highway driving conditions the
advantage is considerably less, perhaps 15 percent.

Fuel Economy Characteristics

The combined urban and highway fuel economy of 1978 and 1979 EPA
certification vehicles is shown in Figure 1.10. The clear and
consistent advantage of the diesel vehicles is evident for all vehicle
inertia weights. The most efficient diesel vehicles are from 33 to 65
percent more efficient than the most efficient gasoline vehicles of
comparable inertia weight. The least efficient diesels are from 76 to
109 percent more efficient than the least efficient gasoline vehicles
of comparable inertia weight. It should be noted, however, that the
diesel vehicles are not truly comparable to the gasoline vehicles,

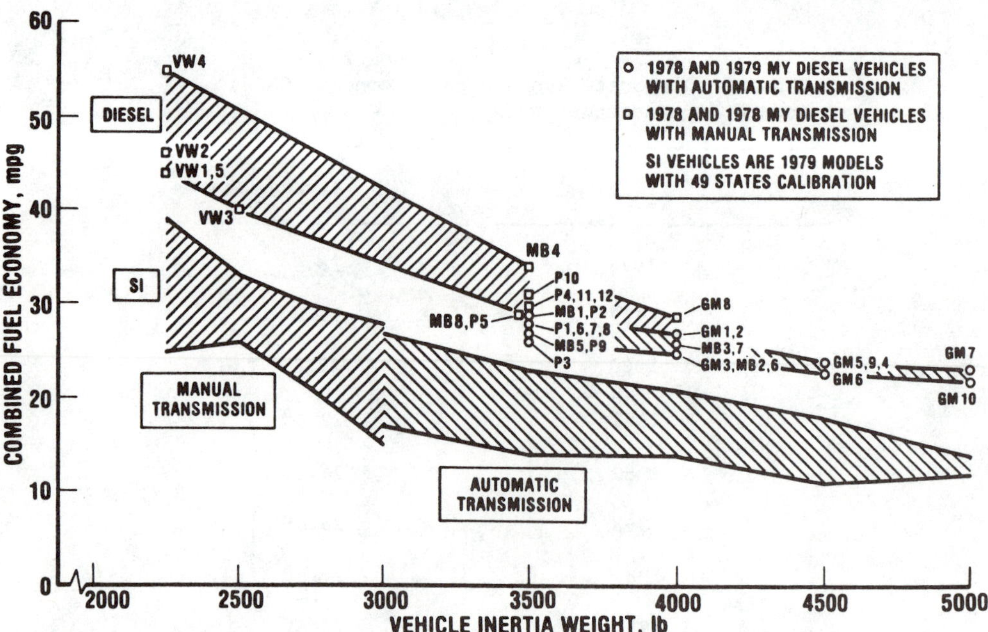

FIGURE 1.10 Combined urban and highway fuel economy for 1978 and 1979
passenger cars. For a description of the data used in this figure see
Appendix A. SOURCE: Roessler et al., 1980.

because the rated power to inertia weight ratios of the most fuel efficient vehicles are in the range of 0.018 to 0.030 hp/lb for the diesels and 0.028 to 0.035 hp/lb for the gasoline vehicles (Roessler, 1979).

The advantage directly attributable to the diesel engine is shown in the engine maps of Figure 1.8. Fuel consumption is lower for the diesel at all operating points. The greatest difference occurs at the low-load operating conditions typical of urban driving.

Additional comparisons for a variety of vehicles and diesel engine types are provided in Table 1.3. As can be seen, the fuel economy advantages of the diesel over the gasoline engine are consistent and substantial, although it should be clear that a portion of the advantage is the result of lower vehicle performance.

Within the next 5 to 10 years it is likely that direct injection diesel engines will be introduced for light-duty vehicle applications. When this occurs, the new diesels will offer substantial fuel economy advantages over today's indirect injection diesels and gasoline engines. It should be recognized, however, that further improvements in indirect injection diesels and gasoline engines are also possible, particularly with the use of lean-burn and stratified charge technologies. In the latter case, emission constraints may affect the use of these engines just as they may affect the use of diesels.

Catalyst control technology for oxides of nitrogen is currently not adaptable to stratified charge, lean-burn, or diesel engines used for transportation purposes. Therefore, these engines may not be available for large vehicles should a stringent oxides of nitrogen (NO_x) standard be adopted. The projected fuel economy advantage of diesel and other advanced power plant vehicles is given in Table 1.4.

Emission Comparison

The pollutant species emitted in diesel exhaust are summarized in Figure 1.11. The uncontrolled diesel engine emits smaller amounts of carbon monoxide and hydrocarbons, comparable or slightly higher amounts of oxides of nitrogen, and considerably greater amounts of particulate than the gasoline engine. It is anticipated that both engines will be required to meet the same emission standards, although some differences now exist in terms of waiver options and test procedures. Federal and California automotive and light-duty-truck emission standards are summarized in Table 1.5. A comparison of emission data for 1979 model year light-duty vehicles is provided in Table 1.6.

All diesel engines appear capable of meeting the federal emission standard of 3.4 g/mi for carbon monoxide and of 0.41 g/mi for hydrocarbons. The major difficulty is in meeting the standards for oxides of nitrogen and particulates. Current techniques to control the oxides of nitrogen emissions from diesels often increase particulate levels and may increase hydrocarbon emissions to a point that compliance with the hydrocarbon standard is difficult. There is a definite strong trade-off between NO_x and particulate control, and a much weaker trade-off between NO_x and hydrocarbon control, in diesel engines.

TABLE 1.3 Fuel Economy Data for Light-Duty Vehicles, 1981 Model Year, EPA FTP Cycle

Vehicle Class	Vehicle Make and Model	Estimated mpg	Engine CID	Trans-mission	Fuel System No. bbl or Fuel Injection	Fuel Econ. Advantage, mpg
Subcompact cars	VW Rabbit	28	105	M4	FI	
	VW Rabbit Diesel	42	97	M4	FI	50%
	VW Rabbit	25	105	M5	FI	
	VW Rabbit Diesel	38	97	M5	FI	52%
Compact cars	Audi 5000	19	131	M5	FI	
	Audi 5000	19	131	A3	FI	
	Audi 5000 Diesel	27	121	M5	FI	42%
	Peugeot 505[a]	19	120	A3	FI	
	Peugeot 505[a]	16	120	M5	FI	
	Peugeot 505 SD Diesel[a]	29	141	M&A4	FI	53-81%
	Mercedes 280E Diesel[a]	16	168	A4	--	
	Mercedes 240D Diesel[a]	24	183	A4	FI	50%
Mid-size cars	Olds Cutlass	21	231	A3	2	
	Olds Cutlass	19	260	A3	2	
	Olds Cutlass Diesel	23	350	A3	FI	10-21%
Large cars	Buick Electra	18	252	A4	4	
	Buick Electra	16	307	A4	4	
	Buick Electra Diesel	21	350	A3	FI	17-31%
	Buick LeSabre	19	231	A3	2	
	Buick LeSabre	18	252	A3	4	
	Buick LeSabre	16	307	A3	4	
	Buick LeSabre	22	350	A3	FI	16-38%

	Diesel					
	Cadillac deVille/ Brougham	18	252	A4	FI	
	Cadillac deVille/ Brougham Diesel	21	350	A3	FI	17%
	Olds Delta 88	19	231	A3	2	
	Olds Delta 88	17	260	A3	2	
	Olds Delta 88	16	307	A3	4	
	Olds Delta 88 Diesel	23	350	A3	FI	21-44%
Mid-size station wagons	Olds Cutlass	21	231	A3	2	
	Olds Cutlass	17	260	A3	2	
	Olds Cutlass	16	307	A3	4	
	Olds Cutlass Diesel	23	350	A3	FI	10-44%
	Peugeot 504[a,b] Diesel	29	141	M&A4	FI	
	Peugeot 504[a,b] Diesel	28	141	A3	FI	
Large station wagons	Olds Cutlass Cruiser	16	307	A4	4	
	Olds Cutlass Cruiser Diesel	21	350	A3	FI	31%
	Pontiac Catalina/ Bonneville Safari	16	307	A4	4	
	Pontiac Catalina/ Bonneville Safari Diesel	21	350	A3	FI	31%
Standard pickup trucks	Chevrolet C10	17	250	A3	2	
	Chevrolet C10	17	305	A3	2	
	Chevrolet C10 Diesel	20	350	A3	FI	18%
	GMC C-15	17	250	A3	2	
	GMC C-15	17	305	A3	2	
	GMC C-15 Diesel	20	350	A3	FI	18%

[a]Certification data for 1981 are are not available. Data are for the 1980 model year.
[b]Vehicles have the same engines as Peugeot 505 SD.

SOURCE: EPA 1979a,b, 1980a,b.

TABLE 1.4 Fuel Economy Comparison of Light-Duty Vehicles: Comparison at Constant Performance in Percent with Respect to Current Gasoline Engine Vehicles[a]

Period	Engine Type	Volume Basis, mpg	Energy Basis
Current (1980)	Gasoline	0	0
	Indirect injection diesel	+35	+19
Intermediate (1980s)	Improved gasoline	+20	+20
	Direct injection stratified charge gasoline	+25	+25
	Improved indirect injection diesel	+45	+28
	Direct injection diesel	+55	+37
Long range (after 1990)	Advanced stratified charge gasoline	+42	+10
	Advanced ignition-assisted diesel	+60	+42
	Advanced fuel-tolerant engine	+40 to +65	+42

[a]Estimated uncertainties to any absolute value are +10 percent of value. Difference between volume basis and energy basis is based on +13 percent greater volumetric energy content of diesel fuel.

Of the unregulated gaseous pollutants, only emissions of sulfur dioxide, nitrogen dioxide, and aldehydes appear to be higher for diesels than for gasoline engines. Sulfur dioxide emissions are directly related to the fuel sulfur content, which is currently 10 times higher in diesel fuel than in gasoline.

Comparison of In-Use Characteristics

In general, diesels have a reputation for sturdiness and long life. The life of diesel engines used to power heavy-duty trucks reaches well into the hundreds of thousands of miles. This type of service is characterized by few cold starts and by long steady operation periods at constant engine speed and power. These conditions approach those in stationary engines and are most conducive to a minimum of particulate formation (that portion due to transients), engine deposits, corrosion, and wear.

Passenger cars, in contrast to heavy-duty vehicles, are often subject to many cold starts each day with trip lengths well below 10 miles. This type of operation, particularly at engine temperatures not at steady state, results in a high volume of wet blow-by gases that carry products of incomplete combustion past the pistons into the crankcase. Within the crankcase these blow-by gases can contaminate the oil, causing corrosive wear of bearing surfaces and the formation of heavy deposits, which in turn leads to piston ring and valve sticking.

Design and metallurgical improvements and the incorporation of specific detergent, antioxidant, anticorrosion, and antiwear additives in automotive engine oils have largely eliminated these problems in gasoline engines. The SAE/API Engine Oil Performance and Engine Service Classification system, which is based on a series of laboratory engine tests, is used to assess performance of oils with respect to these engine problems. Current premium automotive engine oils provide all these additives. The result is that the average car life in the United States currently exceeds 10 years, or 100,000 miles.

The blow-by gases in diesel engines contain larger amounts of particulate than those in gasoline engines. The particulate enters the lubricating oil and greatly contributes to wear and engine deposits.

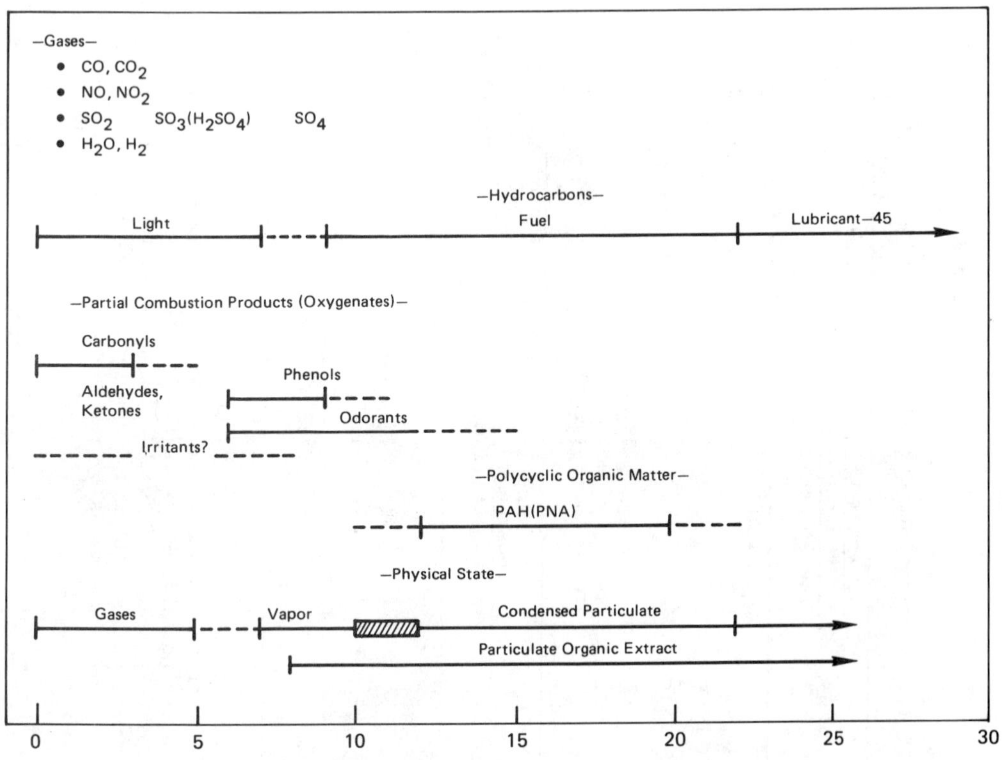

FIGURE 1.11 Species of pollutants observed in relation to carbon and physical state. SOURCE: Levins, 1979.

TABLE 1.5 Emission Standards for Light-Duty Vehicles (in g/mi)

| Year | Passenger Cars | | | | | | | |
| | Hydrocarbons | | Carbon Monoxide | | Nitrogen Oxides | | Particulates | |
	Federal	California	Federal	California	Federal	California	Federal	California
1960	10.6[a]		84[a]		4[a]		--	--
1975-1977	1.5	0.9	15.0	9.0	3.1	2.0	--	--
1977-1979	1.5	0.41	15.0	9.0	2.0	1.5	--	--
1980	0.41	0.39[b]	7.0	9.0	2.0	1.0	--	--
1981	0.41	0.39[b]	3.4[c]	7.0	1.0[d]	0.7[e]	--	--
1982	0.41	0.39[b]	3.4[c]	7.0	1.0[d]	0.7[e]	0.6	0.6
1983-1984	0.41	0.39[b]	3.4	7.0	1.0[d]	0.4	0.6	0.6
1985	0.41	0.39[b]	3.4	7.0	1.0	0.4	0.2	0.2

| Year | Light-Duty Trucks (Less Than 6,000 Pounds GVW Prior to 1979 and Less Than 8,500 Pounds After) | | | | | | | |
| | Hydrocarbons | | Carbon Monoxide | | Nitrogen Oxides | | Particulates | |
	Federal	California[f]	Federal	California[f]	Federal	California[f]	Federal	California
1975	2.0	2.0	20	20	3.1	3.1	--	--
1976-1978	2.0	0.9	20	17	3.1	2.0	--	--
1979	1.7	0.41/0.50[g]	18	9/9	2.3	1.5/2.0	--	--
1980-1981	1.7	0.39[g]/0.50[g]	18	9/9	2.3	1/5/2.0	--	--
1982-1984	1.7	0.39[g]/0.50[g]	18	9/9	2.3	1.5/2.0	0.6	0.6
1985	1.7	0.39[g]/0.50[g]	18	9/9	2.3	1.5/2.0	0.26	0.26

[a] Average values with no controls.
[b] Nonmethane hydrocarbon standard.
[c] Waiver possible--between 3.4 and 7.0 g/mi.
[d] Waiver possible--between 1.0 and 1.5 g/mi.
[e] 1 g/mi option with 100,000 miles durability also possible.
[f] California standards by inertia weight <4,000 lb/>4,000 lb.
[g] Nonmethane standard.

TABLE 1.6 Comparison of Diesel and Gasoline Vehicle Emissions[a]

Make of Car[b]	Inertia Weight, lbs	Power, hp	HC, g/mi	CO, g/mi	NO_x, g/mi	Particulates, g/mi
VW Dasher	2,000	48	0.52	1.19	0.98	0.32
VW Rabbit	2,250	48	0.51	1.01	0.87	0.23
Avg. S.I.[c]	2,250		0.80	8.00	1.40	0.008[d]
Mercedes 240D	3,500	62	0.18	0.98	1.60	0.53
Peugeot 504D	3,500	71	0.87	1.69	1.16	0.29
Avg. S.I.	3,500		0.60	9.00	1.50	0.008[d]
Mercedes 300 SD	4,000	110	0.14	0.81	1.71	0.45
Mercedes 300D	4,000	77	0.28	1.42	2.00	0.83
Avg. S.I.	4,000		0.55	8.00	1.60	0.008[d]
Oldsmobile 350	4,500	120	0.59	1.51	1.49	0.84
GMC Truck	4,500	120	0.69	1.91	1.62	0.59–0.61
Avg. S.I.	4,500		0.50	8.00	1.50	0.008[d]

[a]Unless otherwise indicated, all vehicles are diesel-powered.
[b]Diesel emission figures are for model year 1979 certification vehicles (see Particulate Measurement--Light-Duty Diesel Particulate Baseline Test Results, EPA, January 1979d). Inertia weights and gasoline emission figures are for model year 1978 certification vehicles (see Diesel Engine Research and Development Status and Needs prepared by Roessler et al., the Aerospace Corporation, for U.S. Department of Energy, September 1978, pp. 2-5 to 2-11).
[c]S.I. denotes "spark ignition" (gasoline) vehicle of the listed inertia weight with needed emission control devices.
[d]Average for all catalyst-equipped gasoline vehicles was 0.008 g/mi, according to the Regulatory Analysis.

To date, engine manufacturers have attempted to solve this problem by increasing the crankcase oil capacity and by prescribing, as new-car warranty conditions, shorter oil and filter change intervals than are required for gasoline engines. Lubricating oil contamination could be further alleviated by modifying the combustion process to reduce particulate formation and by redesigning the pistons, the combustion chamber, and the piston rings to reduce the quantity of fuel combustion products that reach the crankcase. In addition, ring sticking may respond to improved distribution of oil flow through the piston ring region; improved piston cooling might prevent the formation of hard deposits; and metallurgical improvements, which are already in use in heavy-duty engines, could alleviate abrasive wear. (Engine wear and deposit problems would also benefit from improved lubricating oil formulations.) A special oil for diesel engines could have a substantially higher concentration of phosphorus and so provide much improved and prolonged wear protection. The concentration of some additives, particularly phosphorus (which, in the form of zinc dithio-phosphate (ZDP), is the major antiwear agent), is limited by the adverse effect they have on the gasoline engine catalytic exhaust converter system.

Over the past 20 years, a high level of maintainability and durability has been achieved in the automotive gasoline engine. If the extra maintenance currently required (mainly more frequent oil changes) is faithfully performed, today's diesel engines probably match or exceed the useful life of most gasoline engines.

The individual aspects of maintainability of the two types of engines and their auxiliaries are compared and summarized in Table 1.7. The basic areas of comparison are the systems for the following: ignition; starting; fuel handling and metering; lubrication; engine cooling; and exhaust emission control.

Ignition System

In gasoline engines the ignition system requires considerable main-tenance, although this is much less so with electronic ignition and unleaded fuels. If the spark timing is not correct or if the spark advance mechanism, the distributor, the sparkplugs, and the wiring are not functioning properly, the performance of the car, its fuel economy, and its exhaust emissions will be adversely affected. No equivalent system is used in diesel engines.

Starting System

Although the starting systems for gasoline and diesel engines are quite similar, diesel engines are more difficult to start in colder tempera-tures. Ignition in the diesel relies on elevated temperature and on the fuel's forming a proper injection spray pattern. The engine, therefore, is sensitive to a minimum cranking speed. As the ambient temperature decreases, starting problems become more pronounced. To

TABLE 1.7   Maintenance Requirements: Effect of Replacing Gasoline
Engine by Diesel Engine

System Component or Servicing Problem	Affected Engine Type		Maintenance Requirement Cost With Diesel Engine
	Gasoline	Diesel	
Ignition system	x		---
Starting system			
Carburetor and choke	x		-
Glow plug system		x	+
Fuel dilution at			
low temperature		x	+
Battery	x	x	++
Fuel handling and metering			
Carburetor adjustments	x		---
Manifold injection system	x		-
Cylinder injection system		x	+
Fuel contamination		x	+
Lubrication system			
Oil and filter changes	x	x	+++
Cooling system	x	x	=
Emission control system			
PCV valve	x	x	+
EGR	x	x	+
Oxygen sensor system	x		-
Catalytic converter	x		-
Particulate trap		x	++

+++ or --- indicates high maintenance differential.
++ or -- indicates medium maintenance differential.
+ or - indicates low maintenance differential.
= indicates no maintenance differential

alleviate these problems, glow plugs are added to each combustion
chamber as standard equipment in automotive diesels. The glow plugs
are turned on to heat the chamber before a cold start is attempted.
The extra power needed to achieve the minimum cranking speed at higher
compression ratios and to power the glow plugs requires both a larger
battery and a larger starting motor.

In addition to the glow plugs, in cold climates, diesel fuel may
have to be diluted with kerosene or gasoline to maintain the fluid
state needed for proper pumping, injection, and ignition. In arctic
climates, special starting fluids (ethers) are available. Still, diesel
starting becomes quite difficult at temperatures below 0°F.

The gasoline engine overcomes cold-start problems through the use of a temperature-controlled air inlet passage choking system, which provides the cold engine with an ignitable air-fuel vapor mixture. This system requires adjustments and maintenance.

Under mild temperature conditions a comparison would seem to favor the diesel engine with regard to performance and ease of maintenance of the starting system. The diesel is, nevertheless, more sensitive to lower ambient temperatures than the gasoline engine. Where winters are long and severe, the light-duty diesel may be less suitable for general use, although additional developments of cold-start technology can probably improve it further.

Fuel Handling and Metering

In a gasoline engine, the fuel handling and metering equipment consists primarily of a carburetor, throttle body injection system, or manifold injection system, which adjusts the air/fuel ratio in accordance with operating requirements, i.e., rich for idle and acceleration, and lean for cruising. Carburetor maladjustments lead to poor performance and fuel economy, and excessive emissions; they do not harm the engine.

The effect of carburetor maladjustments on carbon monoxide and hydrocarbon emissions has been studied by the California Air Resources Board's (CARB) Haagen-Smit Laboratory (CARB, 1979b) and others. The CARB study showed that an overly fuel-rich idle adjustment is the most frequent reason why cars on the road exceed emission limits.

In new gasoline-powered cars, oxygen sensors placed in the exhaust are commonly being used to control the mixture ratio. The precise adjustment provided by the sensors is often used in conjunction with an injection system because of its better controllability. Such systems are initially more expensive than carburetors but have become necessary as regulatory emission limits have been tightened. They will largely eliminate the need for periodic tune-ups. The use of unleaded fuel is important not only for the catalysts but also for the proper operation of the oxygen sensors.

The only control over the amount of fuel in the air-fuel mixture in diesels is that exerted by the driver through the accelerator pedal. Injection timing is, however, automatically adjusted as a function of engine speed and load. This timing system requires occasional servicing. Injectors also must be inspected periodically to ensure that they operate at the proper injector opening pressure. In practice, injectors seem to hold their original setting well.

A potential problem with fuel injection equipment, particularly the high-pressure type used in diesel engines, is its sensitivity to fuel contaminants (mainly water) that can damage some of the expensive precision parts used. The viscosity of diesel fuel, relative to that of gasoline, greatly increases the time that water droplets require to settle out. Through condensation excessive water can reach the fuel tank and enter the fuel line to the engine. Difficulties occur if any organic diaphragm or sealing material in the fuel system is not water resistant, if the fuel filter becomes clogged, or if the water reaches

the fuel pump and injectors. If water reaches the fuel pump and
injectors, the water may settle out when the engine stops and corrode
finely finished steel surfaces. When this occurs, an expensive
overhaul of the fuel injection system and replacement of some parts are
necessary. In comparison, if water enters the fuel system of a
gasoline engine, it usually passes through and causes no damage.

Manufacturers are attempting to prevent water contamination
problems by installing water separators in the fuel tank or in the fuel
line, between the tank and the injection pump. Operators must take the
responsibility for ensuring that the separator is serviced at appropriate intervals. Water level sensors and dashboard warning lights,
which signal the need for servicing, are also becoming standard equipment in some newly manufactured diesel automobiles. Overall, the fuel
system of a diesel engine normally requires much less attention than
that of a gasoline engine and should last the life of the engine. The
problem of water in diesel fuel can be solved; it should not be an
important problem in the future.

Lubrication System

The lubrication systems in gasoline and diesel engines are basically
the same, except for internal adaptations to individual cooling and
lubrication requirements. In both types of engines, contamination of
the engine oil by water and the products of incomplete combustion
carried by the blow-by gases is the result of typical urban stop-and-go
driving. Current engine oils, formulated to meet SAE/API classifica-
tions, can handle the contaminants adequately in gasoline engines and
permit oil change intervals of 7,500 miles and more. There are no
engine oils available today that provide the same protection to diesel
engines; hence, the oil must be changed about twice as often in diesel
engines as in gasoline engines.

In diesel engines, combustion products contain larger quantities of
particulates. These particles enter the oil and are kept suspended by
the detergent additives present. The finer particles are generally too
small to be removed by a clean oil filter, and if they are removed (by
a mat of larger particles already on the filter surfaces), they will
clog the filter, and the oil will flow through the bypass. The
suspended particulate will eventually inhibit the operation of engine
oil antiwear additives (principally ZDP) and lead to accelerated wear
on surfaces subject to friction (Sapienza et al., 1980). This type of
accelerated wear is primarily observed at sites in the valve train,
particularly between cams and followers. Particulate-contaminated oil
also contributes to most of the deposits in piston ring grooves that
lead to early ring sluggishness and sticking, and in turn to increased
volumes of blow-by gases.

Abrasive wear can be kept to a minimum by increasing the oil
capacity of the crankcase and by increasing the frequency of oil
changes. Because the engine design requires obvious limits on oil
capacity, manufacturers specify oil change intervals in their warranty
service requirements that will assure that the wear rate will not be so

high as to shorten the life of the most affected parts below that of the engine as a whole.

Most manufacturers prescribe oil change intervals at every 3,000 miles for diesel-powered cars subject to average service conditions and shorter intervals for severe conditions with multiple cold starts and short trip lengths. Diesel owners, therefore, must pay the extra maintenance cost and accept the inconvenience of frequent oil changes. This situation is changing--the recommended oil change interval for a 1980 diesel engine model produced by Volkswagen is 7,000 miles, which is close to the interval recommended for gasoline engines (VW, 1980). The longer oil change interval is partly due to the larger oil capacity to engine displacement ratio in the small engine.

Historically, owners of gasoline-powered cars have adhered only very loosely to oil change interval schedules. In the gasoline engine the oil added between changes replenishes the additives that may have been used up, and little or no damage results even if the change is put off indefinitely. In the diesel engine any substantial postponement of a scheduled oil change will cause the level of wear and deposits to increase beyond acceptable limits and adversely affect the life of the engine. Complete neglect can result in the need for a major overhaul or replacement of the engine.

Cooling System

There are no substantial differences between the engine cooling systems used in gasoline and diesel engines. Consequently, the maintenance requirements and durability of the two systems are virtually the same. The diesel engine, because of its greater expansion ratio and lower exhaust gas temperatures, generally requires less cooling. It is unlikely, therefore, that a diesel engine will overheat, even when idling on a hot day on a clogged freeway.

Exhaust Emission Control Equipment

Diesel engines operate on the fuel-lean side of the stoichiometric ratio. The engine should, in principle, produce no carbon monoxide or hydrocarbons as exhaust pollutants. In practice, small amounts of carbon monoxide, partially burned hydrocarbons, and particulates are produced. Diesels tend to produce more NO_x and particulate emissions than gasoline engines do. In general, emissions from diesels do not significantly increase with the age of engines. A study by Hergenrother (1979) compared NO_x emission levels in a diesel-powered taxicab fleet of 66 cars run 120,000 miles each, with those from a similar gasoline-powered fleet. The fact that NO_x emissions from diesels do not increase with the age of engines, while NO_x emissions from gasoline-powered vehicles do, means that the lifetime NO_x emissions of diesels may be comparable to, or even lower than, the lifetime NO_x emissions of gasoline engines, even though the initial NO_x emissions of new diesels are higher.

All manufacturers of diesel automobiles anticipate using exhaust
gas recirculation (EGR) to meet new federal NO_x limits. With
gasoline engines, deposits tend to accumulate on the EGR control valves
and cause sticking problems. As a result of this problem the valve
must be replaced. Because of the high particulate content in diesel
exhaust, control valves and lines for light-duty diesels will have to
be carefully designed to prevent deposit and sticking problems. In
addition, EGR increases the quantity of particulate reaching the
crankcase oil, as shown by the results of tests conducted by Toyota.
(See Figure 1.12).

Both diesel and gasoline engines are equipped with a positive
crankcase ventilation device: the systems are identical. It is
expected that the high particulate content of the blow-by gases in
diesel engines will make it difficult for all emission control
equipment to last over the desired 50,000-mile lifetime of the vehicle
without intermediate maintenance.

Overall Maintenance

For the present, it appears that diesel engines require less maintenance
than gasoline engines but that the maintenance cost is equal to, or
higher than, that required for gasoline engines. In addition, diesels

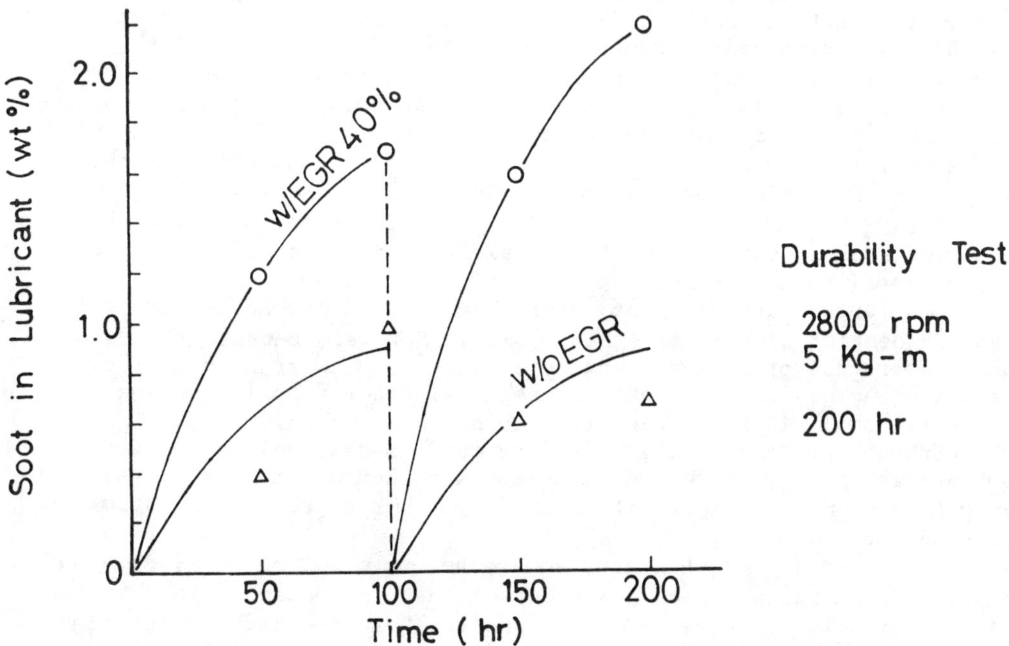

FIGURE 1.12 Effect of exhaust gas recirculation on the amount of soot
in the lubricating oil. SOURCE: Toyota, 1980.

require stricter adherence to the prescribed schedule for oil and filter changes. The diesel engine is generally free of the minor failures and maladjustments of the carburetion and ignition systems that make adjustments and tune-ups of gasoline engines necessary.

Quantitative data on the maintenance experience with comparable gasoline- and diesel-powered cars do not exist. In the few reports available, the cost of maintenance is often lumped with the cost of the fuel and the tires, and with other operating expenses. Several state transportation departments have recently converted parts of their service fleets to diesel power, and good maintenance data should become available within a few years. However, these cars and light-duty trucks are likely to be used in conditions quite different from those of most general use cars.

Safety Comparison

Diesel fuel has 8 to 20 carbon atoms per molecule compared with the 4 to 10 carbon atoms per molecule in gasoline. The resulting vapor pressure of diesel fuels is much lower than that of gasoline at all temperatures. Diesel fuel is therefore less flammable than gasoline.

At room temperature, the vapor pressure of diesel fuel is so low that insufficient vapor is present on its surface to form an ignitable mixture with air. The vapor pressure increases when the fuel is heated, and at certain temperatures the vapor-air mixture above the liquid surface will burn if a spark or small flame is applied, but combustion at this temperature is not self-sustaining. The temperature at which the vapor-air mixture burns is called the flash point. If the fuel is heated further, the fire point is reached, and combustion will continue after ignition. The flash and fire points of gasoline are well above normal ambient temperatures.

By far the largest number of service station and vehicle fires, and even many home fires, are due to leakage or accidental spillage of gasoline. Gasoline vapor will ignite if it comes in contact with any ignition source, such as an electrical spark or open flame. In most cases diesel fuel will not.

In an automotive accident, diesel fuel could become heated if it came in contact with a hot exhaust pipe. However, because the exhaust gas temperature of a diesel engine is very low compared to that of a gasoline engine, the chance of diesel fuel vapor reaching an ignition source is small in any accident, unless the other car burns.

Although it is possible to ignite cold diesel fuel if the fuel surface is greatly increased either by wick action or by dispersion in very fine droplets (vaporization), this is not a problem in automotive applications.

If, however, a crash starts some other material on fire, such as gasoline from the other car, any diesel fuel in contact with this fire will eventually vaporize and burn, and the heat fed back to the liquid diesel fuel pool would eventually cause it to vaporize and ignite. A diesel fuel fire requires a much longer time to reach the same volume and intensity that a gasoline fire reaches in seconds. Thus, because of its ignition characteristics, diesel fuel provides for better fire safety than gasoline does. Diesel fuel, although flammable, is less of a fire risk than gasoline during fueling, normal operation, collisions, and other accidents.

Two other safety aspects of diesel engines should not be overlooked. First, as a consequence of the extremely low carbon monoxide content in diesel exhaust, leakage into the passenger compartment does not lead to anoxemia and drowsiness, and their potentially fatal consequences. Both of these effects can be caused by gasoline engine exhaust, particularly when the engine is idling. Second, diesel engines are more reliable under adverse weather or flood conditions. Where gasoline engines would stall if water short-circuited some part of the electrical ignition system, a diesel will operate as long as air can enter the inlet, even if it carries quite large amounts of water with it.

Diesel Fuel

Diesel fuel is a light petroleum distillate composed of a mixture of aliphatic and aromatic hydrocarbons. The properties of diesel fuel can vary depending on the type of crude oil used as raw material, the refinery process used, and the properties and mix of refinery stocks from which the fuel is blended. Variation in these factors can affect ignition, volatility, heating value, hydrocarbon composition (e.g., paraffins, naphthenes, olefins, or aromatics), sulfur content, and other properties. Additives are used to reduce engine deposits and corrosion and to improve flow characteristics and ignition quality. In general, diesel fuels produced primarily from straight-run stocks (distilled directly from crude oil) have better burning qualities than fuels containing cracked stocks (produced by conversion processes), and distillates from paraffinic crudes produce better-quality fuels than distillates from naphthenic or mixed-base crudes.

Fuel Properties

Two grades of automotive diesel fuel are currently produced by the petroleum industry--No. 1 and No. 2. The diesel fuel in general use for trucks and passenger cars is No. 2. No. 1 diesel fuel is primarily produced for use in cold climates where No. 2 causes cold weather handling and engine starting problems. It is also used in some city bus fleets to reduce smoke emissions. No. 1 diesel fuel represents about 3 percent of the total diesel fuel demand for automotive use. The four properties of diesel fuel that are of concern are cetane number, cloud point, heating value, and aromatic content.

<u>Cetane Number</u> The cetane number, sometimes estimated by the cetane index, describes the ignitability of diesel fuel. For instance, fuels with high cetane numbers have low self-ignition temperatures and short ignition delay periods. Fuels with low cetane numbers cause engine roughness, but fuel cetane numbers higher than the minimum requirement

for a particular engine will not measurably improve engine performance. Cetane numbers range from zero to 100; typical cetane numbers of diesel fuels used in the United States are from 45 to 55. The cetane number is influenced by the source of the crude, by the refining process, and occasionally by the use of cetane-improving additives.

Although the terms "cetane number" and "cetane index" both describe ignitability, they are sometimes erroneously used interchangeably. The cetane number is determined by comparing the ignition delay of a blend of two reference fuels to that of the test fuel in a special single-cylinder engine. The cetane index is used in engineering practice to approximate the cetane number. The basic variable used in calculating the index is the aniline point temperature, i.e., the temperature at which the fuel and aniline (an aromatic amine) are completely miscible. The cetane index is equal to the aniline point temperature multiplied by the API gravity divided by 100. The cetane index is usually higher than the cetane number by 2 to 4 numbers.

Cloud Point The cloud point, which is seasonably specified according to geographic area, is the temperature at which wax crystals appear in the fuel. Wax can block fuel flow and plug filters. To prevent these problems, the cloud point must be lower than the temperature of the vehicle's fuel system. This is accomplished by blending compounds through the refining processes or by local blending of No. 1 diesel fuel or kerosene fuels.

Heating Value and Aromatic Content Two nonspecification properties of diesel fuel are the heating value and the aromatic content. On a volume basis, typical No. 2 diesel fuel has heating value that is approximately 13 percent greater (Btus per gallon) than that of typical low-octane gasoline. No. 1 diesel fuel, being less dense than No. 2, has heating value that is approximately 10 percent greater than that of gasoline. Typical commercial No. 2 diesel fuel has an aromatic content in the range of 25 to 40 percent by mass.

Diesel Fuel Specifications

Voluntary specifications for No. 1 and No. 2 diesel fuel have been set by the American Society for Testing and Materials (ASTM). For the most part, meeting ASTM specifications is not required by law, although some states have adopted ASTM specifications as law. The individual states have power to govern the properties of diesel fuels sold within their boundaries. The principal components of ASTM D975, the specification for automotive diesel fuels, are provided in Table 1.8. Because the ASTM specifications for No. 1 and No. 2 heating oils are very similar to those for No. 1 and No. 2 diesel fuels, most petroleum refiners produce and market diesel fuels and heating oil as a common product. This permits the use of common storage and distribution facilities.

ASTM boiling range and sulfur content specifications are the same for No. 2 fuel oil and No. 2 diesel fuel. The minimum flash point for No. 2 fuel oil is lower than that for diesel fuel (100° versus 125°F).

TABLE 1.8 ASTM D975 Specifications for Automotive Diesel Fuels

Fuel Characteristic	No. 1	No. 2
Flash point, °F, minimum	100	125
Cloud point	a	a
Carbon residue 10 percent Btm, percent maximum	0.15	0.35
Distillation 90 percent point, °F, maximum	550	640
Viscosity at 100°F, CST	1.3–2.4	1.9–4.1
Sulfur, wt %, maximum	0.50	0.50
Cetane number, minimum	40	40

a To local requirements.

Although the minimum viscosity is the same (2.0), the maximum viscosity for No. 2 fuel is lower (3.6 versus 4.3 centistokes (CST) at 100°F, kinematic). No. 2 fuel oil does not have a cetane number specification. The fact that cetane number is an important property for diesel fuel and not for heating oil may lead to the future marketing of these fuels as separate products rather than as a single product.

A few states have sulfur limitations between 0.25 and 1.0 percent by mass. Several states have minimum flash point and limitations on boiling range, and 16 states require a minimum cetane number of 40. Two states, Iowa and Nevada, require that all ASTM fuel specifications be met.

Although the ASTM standard for No. 2 fuel oil sets a 0.5 percent by mass limit on sulfur content, many states have more restrictive standards. Because of the general practice of supplying a common product for No. 2 fuel oil and diesel, the diesel fuel sulfur content largely meets the fuel oil sulfur level requirements in any given area.

Properties of Diesel Fuels Produced The Department of Energy (DOE) publishes an annual survey of the properties of diesel fuel produced by 29 petroleum refining companies. For automotive No. 2 diesel fuel produced in 1979, the survey showed the following average properties (DOE, 1979): a sulfur content of 0.23 percent mass, a cetane number of 46, and a 90 percent distillation point of 578°F. Over the past 10 years the average sulfur content and 90 percent distillation point have remained essentially unchanged, but the cetane number has dropped by three numbers.

Diesel fuel properties vary considerably among refining companies and regions of the United States. In 1979, sulfur content varied from zero to 0.89 percent by mass, cetane number ranged from 39 to 54, and the 90 percent distillation point ranged from 448° to 635°F across the country. The DOE survey did not include aromatic content. Although the average diesel fuel falls well within ASTM specifications, there are instances where the sulfur content and cetane numbers do not.

   <u>Future Trends</u> At the present time, attention of the petroleum refining industry is focused on the problems of maintaining existing product specifications and projected demand requirements in the face of declining crude oil quality and availability. Declining oil quality will result in a declining diesel fuel cetane number. Petroleum industry spokesmen have made diesel engine manufacturers aware of this trend and have indicated that efforts will be made to maintain a minimum cetane number of 40.
   The trend toward higher sulfur crude oil will tend to result in higher-sulfur diesel fuels. The need to meet regulated sulfur levels in distillate heating oils will require more desulfurization capacity for distillates. As a consequence, diesel fuel sulfur content is not expected to increase significantly in those areas where heating oil sulfur levels are restricted.

Refinery Processing

In a typical U.S. refinery, a substantial part of the No. 2 diesel fuel is obtained by distillation of crude oil to separate the portion with the proper boiling range. The amount of diesel fuel obtained varies according to type of crude oil used as a raw material. Historically, crude sources, such as the Mid-Continent and Arabian Light crudes, contained about 35 percent by volume of petroleum within the diesel fuel boiling range. The newer crude oil sources, such as the Arabian Heavy and Alaska North Slope crudes, contain about 25 percent by volume of material in the diesel fuel boiling range.
   The hydrocarbon composition of diesel fuel distilled directly from crude oil depends entirely on the crude oil composition. The aromatic content ranges between about 12 and 30 percent by volume. A highly paraffinic crude such as Arabian Light will produce a low-aromatic, high-paraffin-content distillate with a high cetane number; North Slope or typical California crudes will produce distillates having higher aromatic and naphthene contents.
   In a modern refinery, additional diesel fuel is obtained by catalytic cracking of higher-boiling crude oil fractions to produce lower-boiling materials in the gasoline and diesel fuel range. The diesel fuel obtained from this type of processing is relatively high in aromatic content. Combining diesel fuels from both processes--catalytic cracking and direct distillation--produces a blended fuel that is about 33 percent aromatics by volume.
   Hydrocracking is another process used to produce gasoline and diesel fuel from higher-boiling-range hydrocarbons. Although the process produces a diesel fuel component with a higher cetane number and lower aromatic content than are contained in catalytically cracked

material, the availability of hydrocracking capacity is presently
limited.

Fuel Demand Projections

If 25 percent of the passenger cars sold in 1990 are diesel-powered,
the diesel fuel demand for that year will be about 440 thousand barrels
per day (TBD). This quantity of fuel represents only 13 percent of the
1979 consumption of all middle distillate fuels. Even if 50 percent of
the cars sold in 1990 were diesel-powered, the fuel requirements would
represent only 20 percent of the total distillate market. Assuming
sharply increased dieselization of medium-duty trucks and continued
high levels of diesel engine use in heavy-duty trucks, highway diesel
fuel use is expected to double to about 1,700 TBD by 1990. Thus, the
indicated demand for passenger car diesel fuel during the next 10 years
is relatively small when compared to the total distillate demand and
diesel fuel requirements for highway trucks.
   A survey of the petroleum industry indicates that the projected
demand for current specification diesel fuel can be met, even if 50
percent of the cars sold in 1990 are diesel-powered. Basically, this
will be accomplished by reducing the conversion of distillates to
gasoline and adding a limited amount of refinery equipment. This also
assumes no major changes in the crude oils available to U.S. refineries.

   <u>Refinery Energy Considerations</u> Published studies (Lawrence <u>et al.</u>,
1980) demonstrate that a modern fuel refinery will consume less energy
as diesel fuel production is increased. Most of this energy reduction
occurs by virtue of reduced catalytic cracking operations. As a typical
modern refinery shifts from a maximum gasoline production mode to a
maximum distillate mode, the energy output in product as a percent of
energy input in raw material will increase from about 92 percent to
about 93 percent.

   <u>Product Cost and Pricing</u> The cost of crude oil is the major factor
determining the cost of producing diesel fuel. Typically, the crude
oil raw material cost may represent 85 to 90 percent of the cost of
producing diesel fuel. Although a minor part of the total, the other
significant element is the refining cost. In an absolute sense, future
diesel fuel prices will be largely determined by crude oil costs. With
the uncertainties of crude oil availability and producer pricing
policies, it is impossible to forecast the impact of raw material costs
on diesel fuel prices.
   From 1967 to 1976, diesel fuel sold for 4 to 6 cents less per gallon
than regular grade gasoline. Over the past 3 years, the prices of the
two have been nearing equality. The principal question is, What will
happen to the price of diesel fuel as demand for it increases and the
demand for gasoline decreases?
   In the face of declining gasoline demand and increasing distillate
demand, the petroleum companies surveyed predict a narrowing of the
differential between gasoline and distillate prices. If the price of

diesel fuel does not increase as production shifts from gasoline to diesel, refining revenues and net refining profit will decrease because the income from diesel fuel will be lower and refinery operating costs will not have decreased by the same percent. The refiner will attempt to maintain the rate of return on investment and operations by increasing the price of diesel fuel. Assuming free-market conditions, some refiners predict that sometime between 1985 and 1990 the price of a gallon of No. 2 diesel fuel will be the same as the price of a gallon of unleaded regular gasoline. During the same period, kerosene jet fuel prices are expected to rise above the price of gasoline. To maintain refinery profit margins, the refiner will shift a larger proportion of total refining costs to middle distillate production. This will result in a reduction of the current operating cost advantage of the diesel passenger car relative to the gasoline car. Demand for lower sulfur and aromatic content will cause a further cost disadvantage to diesel fuel.

2

AIR POLLUTANT AND NOISE EMISSIONS

PARTICULATE EMISSIONS

Light-duty diesels emit from 30 to 100 times more particulate than comparable catalyst-equipped vehicles powered by gasoline. Some of the particles emitted are known to be biologically active; all may contribute to air pollution. Light-duty gasoline vehicles produced for the 1979 model year emitted particulates at an average of 0.008 g/mi, based on Environmental Protection Agency (EPA) data. For the same model year, particulate emissions from light-duty diesels ranged from 0.23 to 0.84 g/mi. The comparatively high particulate emission rate is the principal emission disadvantage of the diesel engine. Table 2.1 provides data on total particulate emissions from 1979 model year certificatation diesel vehicles.

Fundamentals of Diesel Particulates

Diesel exhaust particulate consists of a solid fraction, with a carbon/hydrogen molar ratio varying from 8:1 to 11:1, a soluble organic fraction, sulfates, and trace elements. The particles themselves are carbonaceous solid chain aggregates on which organic compounds are adsorbed. Most particles are the result of incomplete combustion of fuel hydrocarbons; some are contributed by the lubricating oil; and others may be contributed by fuel carbon components. The heaviest portions of the extractables from particulate have properties very similar to engine lubricating oil.

At combustion temperatures above 500°C, the small agglomerated chains of particles are composed principally of carbon-hydrogen spheres with diameters ranging from 100 to 800 angstroms (Å) (Vuk et al., 1976; Carpenter and Johnson, 1979). As temperatures decrease below 500°C, however, the resulting particles become coated with adsorbed and condensed species in the form of high molecular weight organic compounds including:

- unburned hydrocarbons;
- oxygenated hydrocarbons (ketones, esters, ethers, organic acids);

- polynuclear aromatic hydrocarbons; and
- inorganic species such as sulfur dioxide, nitrogen dioxide,

and sulfuric acid (sulfate) (Khatri et al., 1978; Vuk et al., 1976).

Thus, to describe particulate content adequately, the sample collection temperature and the means by which that temperature was reached must be specified. Accordingly, diesel particulate is generally defined as any material that is collected, at a temperature of 52°C or less, on a filtering medium after dilution of the raw exhaust gases. Water that condenses on the filter is not considered to be particulate.

A diagram of basic diesel exhaust particulates is provided in Figure 2.1. The size distribution of diesel automobile particulates is compared with that of particulates emitted from catalyst-equipped gasoline-powered automobiles in Figure 2.2. As is shown in the latter figure, the diesel particulate volume distribution peaks at about 0.15 μm, and the gasoline engine particulate peaks at 0.02 μm. These dimensions are somewhat misleading because they represent equivalent spherical aerodynamic diameters, even though diesel particles occur as agglomerated chains. Photomicrographs show that diesel particulates have a very light, fluffy structure, with a density of about 0.07 g/cm^3 (Vuk et al., 1976; Khatri et al., 1978; Carpenter and Johnson, 1979). Because of these features, particulate emissions tend to load up and plug any static filtering device that might be used in the exhaust stream.

Combustion and the Particulate Formation Process

Particulate formation begins when fuel is injected into the combustion chamber and continues during and after the dilution of the exhaust in the atmosphere. The first steps of the process occur rapidly--within microseconds or milliseconds. The last steps may take hours or even days.

The ignition delay period, the time interval between injection and ignition of the fuel, is significant in the formation of particulates because a variety of chemical and physical processes take place during this period. These processes include: heating, vaporization, decomposition, and mixing. As the fuel is heated and exposed to increasing amounts of heated air, it begins to undergo significant preignition reactions. The chemical processes accelerate rapidly until the fuel is ignited.

Pyrolysis of the fuel, which also begins during the ignition delay period, accelerates as the temperature rises. Some of the fuel is quickly vaporized and burned; the remainder undergoes a complex series of pyrolytic reactions before combustion occurs. Although most of the fuel is burned, a small, but significant, percentage is cracked, polymerized, and partially oxidized into an array of compounds. Most of these substances undergo further oxidation before leaving the combustion chamber.

Experimental evidence obtained by using a wide variety of combustion systems (premixed and diffusion flames, stirred reactors, etc.), and a variety of fuels, indicates that particulate is formed whenever the local carbon/oxygen ratio exceeds 0.5 (Wagner, 1978). Because this is twice the amount of oxygen required to oxidize all the carbon atoms to carbon monoxide (the thermodynamically favored product for fuel-rich combustion), it is apparent that chemical kinetics play an important role in the particulate formation process.

The mechanism by which combustion-generated particulate is formed in the vapor phase is depicted in Figure 2.3. It is thought that each of these events occurs within the diesel engine cylinder; however, there also is some evidence that particulate can be formed from individual fuel droplets through liquid phase polymerization and carbonization (Johnson and Anderson, 1961). With properly designed and adjusted fuel injection equipment, the amount of particulate formed through this mechanism is thought to be negligible (with the possible exception of operation at very light load). Concurrent with the vapor phase formation process outlined in Figure 2.3, particulate destruction occurs in flame regions where the oxidizing species are present.

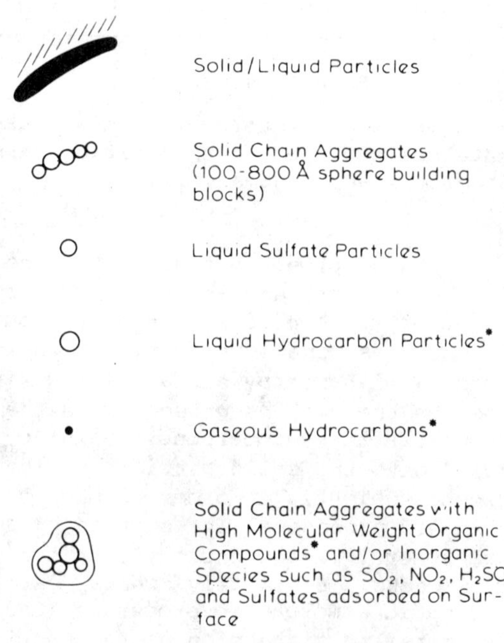

FIGURE 2.1 Schematic of basic diesel particles. SOURCE: Lipkea et al., 1978.

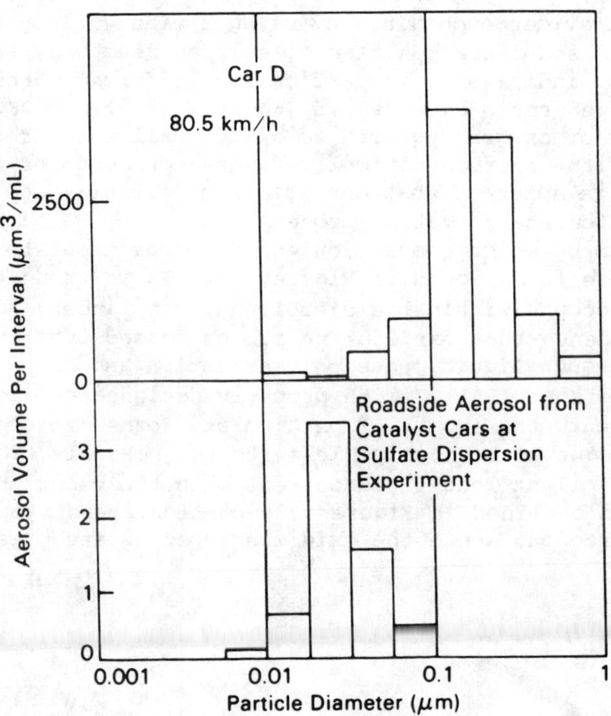

FIGURE 2.2 Comparison of aerosol from a diesel car
and from catalyst-equipped spark ignition cars.
SOURCE: Groblieki and Begeman, 1979.

After combustion, exhaust products are emitted and cooled in the
exhaust manifold, in the muffler, and in the associated pipes. Some
reactions continue, but physical processes seem to dominate. The size
distributions of the solid carbonaceous particles change as they
interact with each other; with nearby surfaces; and with other gaseous,
liquid, and solid species that may be present. As the temperature of
the exhaust drops, some of the hydrocarbons, sulfuric acid, and water
may condense and adsorb onto the solid particles. These processes
occur within 0.05 second, which is an order of magnitude longer than
the time for "in-cylinder" processes. The emitted exhaust cools as it
is diluted in the atmosphere. Reactions may continue over periods of
minutes, days, or even weeks. Ultimately exhaust particulate is
disposed of through dynamic atmospheric processes (Cadle, 1971). (See
Appendix B for a more detailed discussion of the particulate formation
process.)

Other Sources of Particulate

Although most particulate emissions are formed through incomplete
combustion of fuel hydrocarbons, recent studies show that up to 25

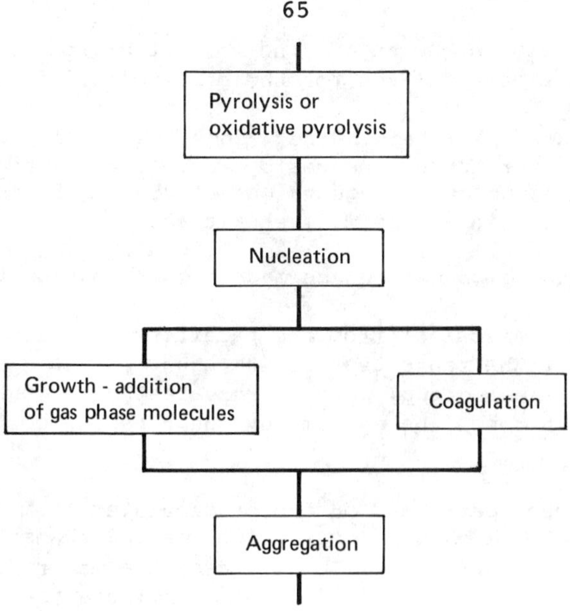

FIGURE 2.3 The mechanism of particulate formation in combustion systems.

percent may be contributed by the engine oil and that fuel components other than hydrocarbons may also contribute.

In one study, General Motors determined the contribution of engine lubricating oil by using a 1978, 5.7-liter light-duty diesel engine. An organic compound (n-octacosane), containing carbon 14, was added as a tracer for the engine lubricating oil (Mayer <u>et al</u>., 1980). The experiments proceeded in two phases: first, the oil contribution to the diesel particulate material was determined; and second, the material was extracted with a solution of 80 percent benzene and 20 percent ethanol, by volume. Both the amount of extractable material and the contribution of oil to this material were measured.

As is illustrated in Figure 2.4, the study showed that engine oil can be a significant source (1.5 to 25 percent by mass) of the particulate emissions produced by a light-duty diesel engine, and a major source (16 to 80 percent by mass) of the extractable organic portion of these emissions. The greatest contribution from the oil was observed at the highest test speed. The data further showed that all of the oil contribution to the particulate was extractable and, hence, appeared in the soluble organic fraction.

General Motors also carried out a fuel tracer study to determine whether, and to what extent, the various constituents of the fuel (i.e., paraffins, aromatics, or olefins) contribute to particulate emissions (Burley and Rosebrock, 1979). The method selected was to:

• blend radioactive carbon 14 tagged fuel constituents, one at a time, with a No. 2 diesel fuel;

- run the fuels in the engine and collect all particulates; and
- analyze the particulates for the presence of carbon 14.

A four-cylinder 2.1-liter indirect injection Opel engine, modified so that only cylinder number one was operating, was used during the study. Tests were run at the equivalent of 25- and 55-mph road loads. The results, as seen in Figure 2.5, show that:

- each of the compounds chosen made a contribution to particulates;
- particulate emissions from the paraffins were independent of boiling point over the range tested, which was 174°to 317°C; and
- aromatics, as a group, were the greatest contributors to particulates, with ring labeled 1-methyl naphthalene contributing the most.

Extractions were performed on the particulates from these tests, and both the insoluble carbonaceous fractions and the soluble fractions were tested for radioactivity. In all cases the major (80 percent) portion of the carbon 14 was found in the insoluble fraction. The remainder (20 percent) went to the soluble fraction.

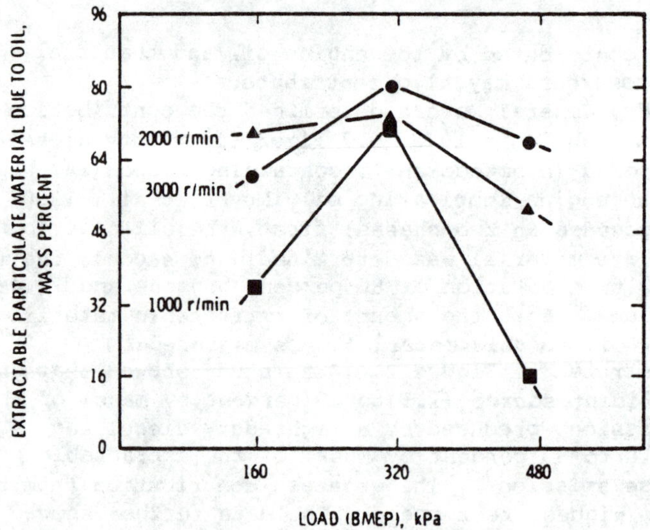

FIGURE 2.4 Contribution of engine oil to extractable organic particulate emissions.
SOURCE: Mayer et al., 1980.

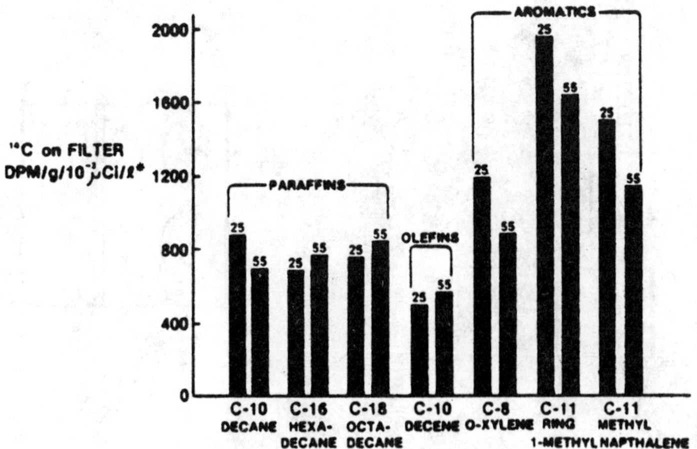

FIGURE 2.5 Contribution of various elements of
diesel fuel to particulate emissions.
SOURCE: Mayer et al., 1980.

Particulate Measurement and Characterization

Particulates are measured and characterized in three categories:
physical qualities (mass and size), chemical composition, and
biological character. Techniques for particulate measurement and
characterization range from simple smokemeter opacity readings to
analyses using dilution tunnels. Most techniques require lengthy
sample collection periods because the rate of emission of individual
species is likely to be low.

The physical conditions under which particulate measurements are
made are critical because the emitted species are unstable and may be
altered through loss to surfaces, change in size distribution (through
collisions), and chemical interactions among other species in the
exhaust at any time during sampling, storage, or examination. Agglomeration with like particles and liquid droplets, heterogeneous chemical
reactions, sedimentation, etc., are typical of some of the processes
that can occur (Lipkea et al., 1978).

Particulate Emission Rates

The objective of particulate measurement is to determine the amount of
particulate being emitted in the atmosphere. The most basic information
is normally obtained on a mass basis, for example grams per mile for the
total vehicle, grams per brake horsepower-hour for the engine, grams
per kilogram of fuel, and milligrams per cubic meter of the exhaust
volume. Smokemeters and dilution tunnels are used to measure particulate emission rates.

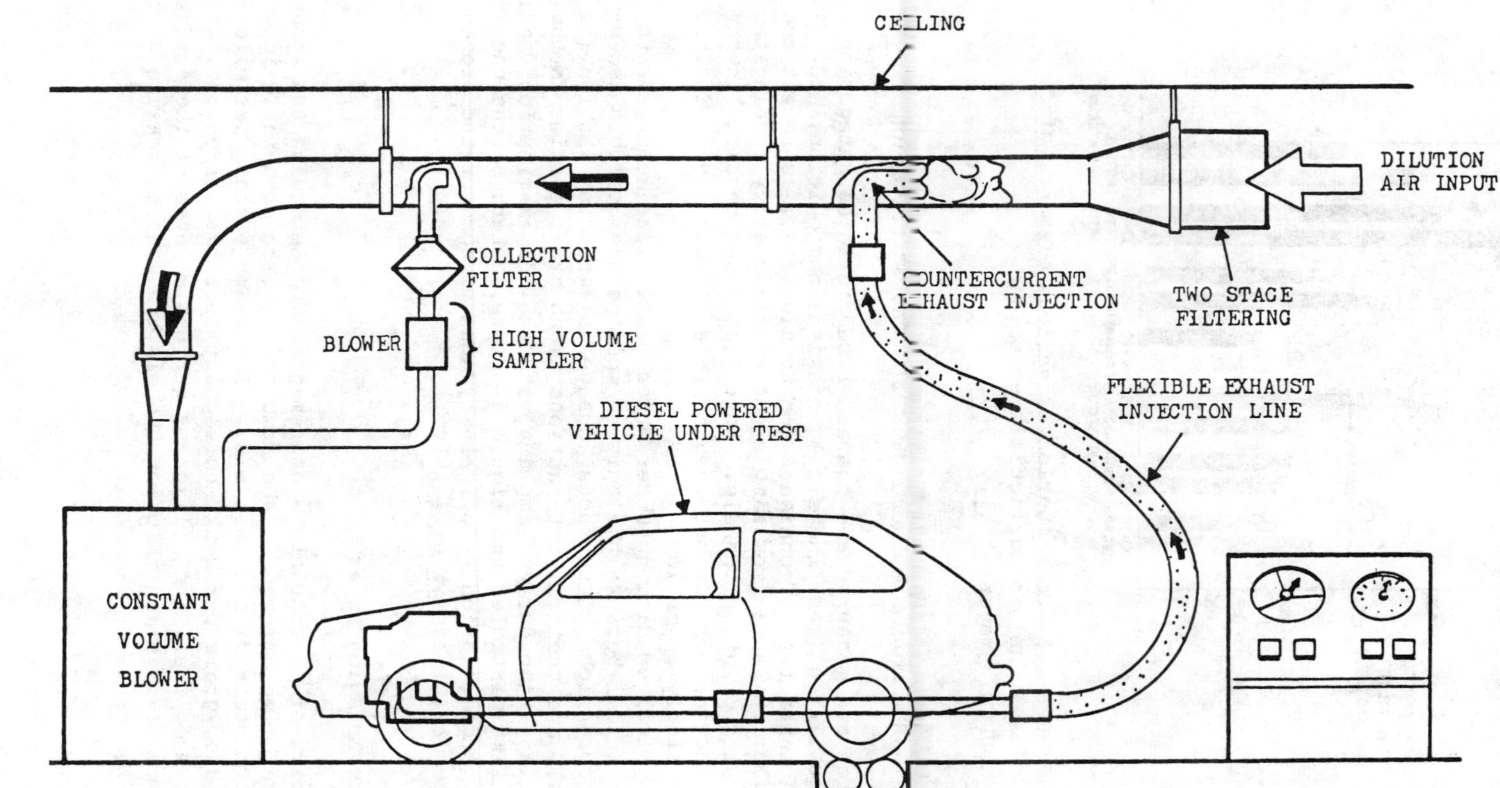

FIGURE 2.6 Typical light-duty chassis dynamometer full exhaust dilution tunnel test setup.
SOURCE: Smaby et al., 1979.

<u>Smokemeters</u> Smokemeters measure the relative quantity of light that passes through the exhaust or the relative reflectance of particulate collected on filter paper. They are generally used as approximate indicators of engine solid particulate emissions.

Some researchers have developed means for correlating smoke density and particulate mass concentrations (e.g., Vuk <u>et</u> <u>al</u>., 1976); however, these correlations are limited. They are not consistent at light-load engine conditions and may not be consistent from engine to engine or fuel to fuel. Hence, smokemeters are now only used to approximate particulate mass concentrations when other more sophisticated sampling equipment, such as a full dilution tunnel, is not available.

<u>Dilution Tunnel Sampling Systems</u> Dilution tunnel systems are used to determine the diesel exhaust contribution to the total particulate content in the atmosphere. Gaseous emissions can be accurately measured in undiluted exhaust, but particulate emissions, because they continue to undergo physical and chemical processes in the atmosphere, must be diluted with ambient air before sampling if accurate data are to be obtained. To that end, dilution tunnels permit exhaust sampling under conditions that simulate the actual processes that the particulate undergoes once it is emitted from the tail pipe.

The dilution tunnels used in the United States are typically 8 to 18 inches in diameter and 10 to 35 feet long. Within the tunnel, part or all of the engine exhaust is diluted with a stream of filtered ambient air. Thus, as in actual road driving, the exhaust is immediately diluted with a relatively large portion of turbulent air. In the atmosphere, the dilution ratio (ratio of exhaust pollutant concentration in the exhaust to its concentration in the diluted condition) changes continuously; in the tunnel, it is constant for a given condition. To allow time for thorough mixing, probes for obtaining samples are inserted into the tunnel 10 to 20 diameters downstream from the point at which the exhaust and air are mixed. Particulate matter is collected by using the sampling system that is appropriate to the type of analysis desired. Figure 2.6 illustrates the typical setup of a full exhaust dilution tunnel for a light-duty chassis dynamometer.

The variability of dilution ratio, tunnel temperature and humidity, tunnel length and diameter, and volume flow rate (which determines the elapsed time from the point of injection to the sampling point) provide flexibility in dilution tunnel sampling. Of the variables, the volume flow rate and dilution ratio are the most important. The volume flow rate is important because it, in effect, determines the residence time for the chemical and physical processes in the tunnel including: nucleation; particle growth by agglomeration; and reactions between the liquid, vapor, and solid phases present. The dilution ratio is important because it is the basis of the simulation.

The sampled exhaust from a dilution tunnel is relatively homogeneous if good mixing has occurred before the sampling. Although this is not the case in the ambient atmosphere—the dilution ratio changes continuously during actual vehicle operation—the data indicate that the tunnel provides a good simulation of short-term conditions. Figure 2.7 compares the history of atmospheric dilution of exhaust gases with

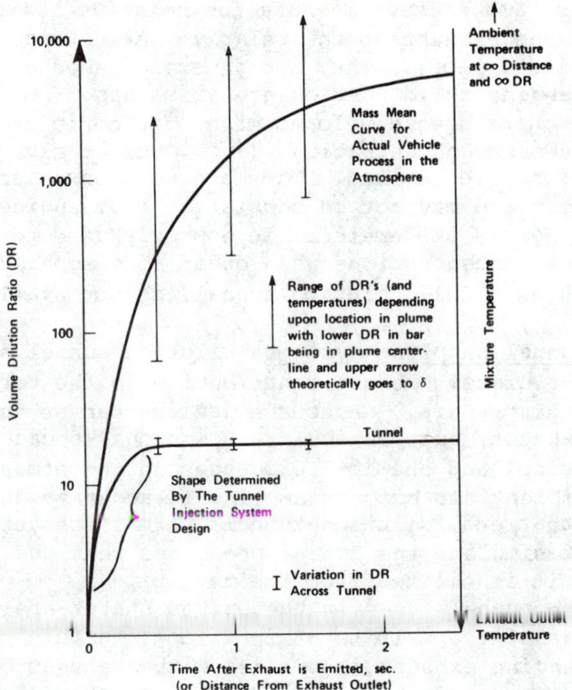

FIGURE 2.7 The atmosphere dilution process versus the dilution tunnel dilution process.
SOURCE: Frisch _et al._, 1978.

dilution achieved in a typical tunnel (Frisch _et al._, 1978). (In the figure, the curve representing atmospheric dilution is an average as mean dilution curve.)

Collection and Extraction of the Soluble Organic Fraction for Chemical and Biological Characterization

Studies have shown that the extractable organic components of diesel particulate emissions include compounds that may cause health and environmental hazards. For this reason, the chemical and biological characterization and accurate measurement of the soluble organic fraction are important. It is also important that any emission control technology under consideration be assessed with respect to its effects on the soluble organic fraction. A brief description of the procedures and methods used to characterize the soluble organic fraction is presented below.

Extraction Methods The soluble organic fraction can be extracted from a 47-mm filter in a micro Soxhlet apparatus by using various solvents (Smaby et al., 1979). The filters contain only enough of the fraction for mass characterization. Larger masses for more extensive chemical and biological characterization are now usually collected on one or more 20- by 20-inch ultrahigh volume samplers (Killough et al., 1979). Typically, the soluble organic fraction constitutes 20 to 40 percent of the total particulate mass. The actual mass can vary from 5 to 90 percent, depending on the individual operating mode of the engine (speed and load).

The mass and types of compounds of the soluble organic fraction are a function of the solvent used. It is often suggested that the solvent used to extract the fraction should simulate body fluids. Although this might be possible when more is known about the health effects of the particles, the discussion here addresses the need to identify the chemicals bound to the particles as a function of engine variables, as well as the reactions they may have undergone when they were acted on by aftertreatment control systems.

Because diesel exhaust particulates are mixtures of both polar and nonpolar components, extraction of these components requires different solvents; no solvent will extract both. Any one solvent or solvent system is, therefore, a compromise. A variety of solvents have been used, individually and in combination, in diesel particulate extraction studies (Smaby et al., 1979; Funkenbusch et al., 1979; CRC, 1980; Williams and Chock, 1979), but the most commonly used extractant is methylene chloride (dichloromethane). This solvent has as good an extraction efficiency as benzene, can be obtained in a very pure form, is easily redistilled, and has a relatively high vapor pressure.

Soxhlet and sonication methods are both used for the extraction of organics from diesel particulates. Use of Soxhlet apparatus is the most common technique, although sonication shows several good characteristics (CRC, 1980).

The organics associated with diesel particulates are usually complex mixtures. A preliminary fractionation scheme is used, therefore, to group similar types of compounds before other methods are used for final separation and identification of individual compounds. The scheme most frequently used was originally developed to identify organics contained in airborne particulates and was subsequently modified for use on diesel particulates (Hoffman and Wynder, 1968; Huisingh et al., 1978).

The scheme illustrated in Figure 2.8 consists of liquid-liquid extractions of the sample; it uses an aqueous base to remove the strongly acidic components, and acid to remove the basic components. The neutral portion of the sample is fractionated, on the basis of polarity, by column chromatography. This results in seven fractions generally labeled as basics, acidics, paraffins, aromatics, transitionals, oxygenates, and ether insolubles. Some researchers also include hexane insolubles in the list (Funkenbusch et al., 1979).

<u>Measuring Biological Activity</u> The Ames salmonella/microsomal test is the most popular short-term bioassay (Ames et al., 1975) for measuring the activity of the soluble organic fraction and its subfractions. With this test a quantitative dose-response curve showing the mutageniccity of a sample compound can be obtained. In its principal form, the

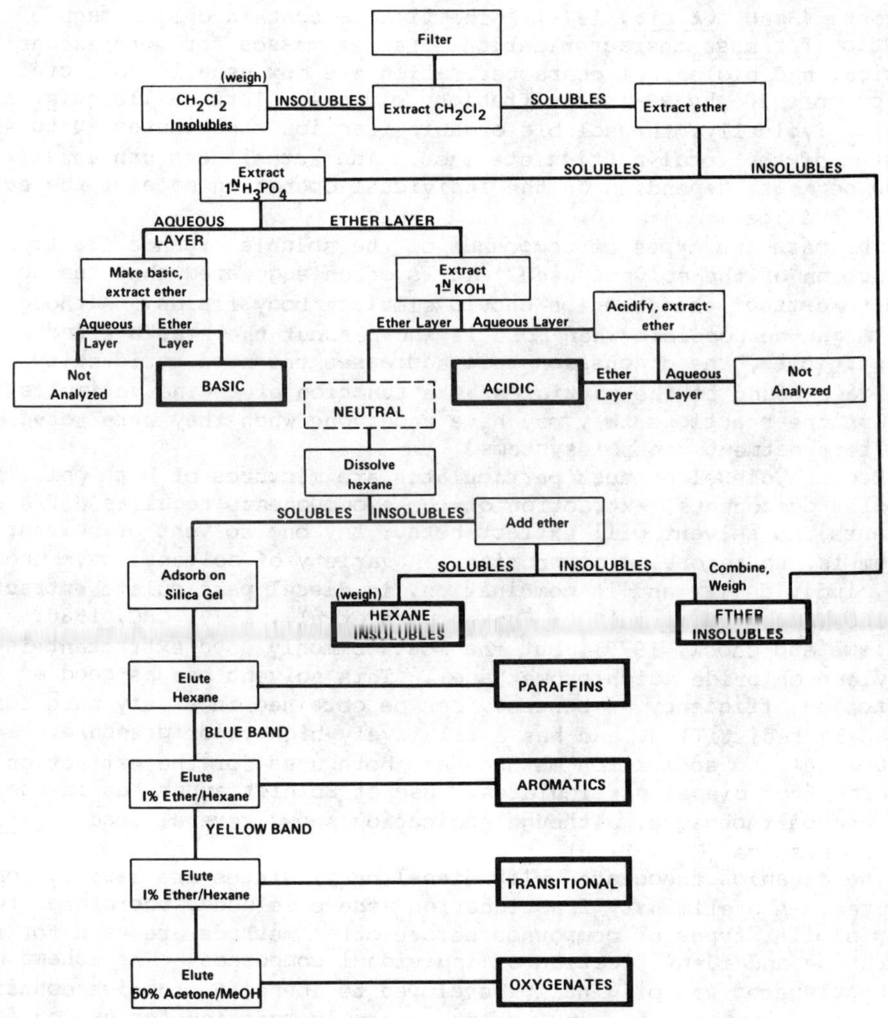

FIGURE 2.8 Scheme for extraction of the soluble organic fraction of diesel particulates.
SOURCE: Funkenbusch et al., 1979.

Ames test uses a mutant strain of Salmonella typhimurium that is incapable of producing histidine. Mutagenicity is defined as the ability of a tested compound to revert--back mutate--this bacterium to its wild state, where it regains its ability to produce histidine. (A further technical description of the Ames test is provided in Appendix C.)

Although the Ames test is widely accepted, it must be remembered that the mutagenicity of compounds is defined only in the context of the test system. Results obtained from the Ames test, using a mutant prokaryotic organism, cannot be extrapolated with any degree of confidence to a higher eukaryotic organism. The Ames test by itself, therefore, has only a limited ability to classify compounds with respect to human health hazards. The test is useful in diesel engine research because it helps engineers and chemists to determine the effect of variables and control systems on the biological activity (mutagenic) of compounds or classes of compounds in the soluble organic fraction. With the many thousands of compounds in the soluble organic fraction, chemical identification studies would be never-ending if engineers were not able to make use of the Ames test.

Using a number of engines and operating conditions, EPA has carried out extensive Ames tests on the seven fractions mentioned above. The tests resulted in the following relative biological activity: transitionals > oxygenates > acidics > and ether insolubles, basics, aromatics, paraffins (Zweidinger et al., 1978a,b). In addition, the EPA Ames test data indicate that the diesel soluble organic fraction contains direct-acting mutagens.

Chemical Identification Tests There is a real need to tie together Ames-type bioassay work with chemical identification studies to determine the chemical identities of the direct-acting mutagens (Wei et al., 1980).

Extensive research to fractionate extracts through the use of high-performance liquid chromatography (HPLC) is underway. If successful, these results may permit the estimation of mutagenic activity and provide a starting point for characterization studies (Zweidinger et al., 1978a,b). For one such study, HPLC was used to obtain low-resolution chromatograms of the extracts on silica gel employing a hexane to methylene chloride gradient. Using fluorescence detection, four major peak areas were obtained. The eluate of the first set of peaks contained the paraffins and aromatic fractions; the second and third sets contained the transitional fraction; and elution of the fourth set contained most of the oxygenate fraction. Results from 16 different extractions, representing several different vehicles, fuels, blends and modes of operation, have shown a high correlation (r_2 value of 0.983) when the fluorescence area of the third peak was plotted against Ames specific activity (revertants per microgram of extract) (Zweidinger et al., 1978b).

The use of this HPLC technique as a rapid screening and development method in engine research work to quantify the various fractions shows much promise: It is being investigated in round-robin experiments by a number of laboratories (CRC, 1980). A more complete description of the subfractions of the soluble organic fraction is provided in Table 2.1.

Emission Characteristics of Uncontrolled Vehicles

Diesel and spark ignition vehicle emission data obtained prior to 1980 provide a basis for discussing emission control technology for the 1985 to 1990 period. Particulate data are presented below; regulated and unregulated gaseous emission data are provided in the section on gaseous emissions.

TABLE 2.1 Components of the Soluble Organic Fraction

Fraction	Components of Fraction	Percent of Fraction

Acidic	Aromatic or aliphatic
	Acidic functional groups
	Phenolic and carboxylic acids
	  Benzoic acid and substituted[a]
	    benzoic acid
	  Napthoic acid and substituted[a]
	    napthoic acid
	  Phthalic acid and substituted[a]
	    phthalic acid
	  Phenathroic acid and substituted[a]
	    phenathroic acid
	  Phenols
	    α and β napthols
	    Higher fused-ring phenols
	    Cresol
	    Phenol
		3–15

Basic	Aromatic or aliphatic
	Basic functional groups
	Amines
	  Benzacridines
	  Dibenzacridines
	  2-aminofluorene
	  Pyridine
	  Aniline
		<1–2

Paraffin	Aliphatics, normal and branched
	Numerous isomers
	From unburned fuel and/or lubricant
	  C_nH_{2n+2}
	  n-paraffins predominate
	  Undecane
	  Nonadecane
	  Hexane
	  Cyclohexane
	  Cyclohexene
	  Pentane
	  Heptane
	  Octane
		34–65

Aromatic	From unburned fuel, partial
	  combustion, and recombination of
	  combustion products; from
	  lubricants
	Single ring compounds
	  Benzene
	  Toluene
	  Ethyl benzene
	  Styrene
		3–14

TABLE 2.1 (continued)

Fraction	Components of Fraction	Percent of Fraction

	Polynuclear aromatics
	  Anthracene
	  Phenanthrene
	  Fluoranthrene
	  Naphthalene
	  Phenanthrene derivatives
	  Pyrene
	  Benzo(a)pyrene
	  Substituted BaP's
	    6-nitro
	    1-nitro
	    3-nitro
	    6,8-diol
	    9,10-epoxide
	  Benzo(e)pyrene
	  Benz(a)anthracene
	  Benzo(ghi)pyrene
	  Perylene and 3-nitro-perylene	

Oxygenated	Polar functional groups but not
	  acidic or basic
	Aldehydes, ketone, or alcohols
	Aromatic phenols and quinones
	  Anthrone
	  Benzanthrone (known mutagen)
	  Alkylated 9-flourenones
	  Dibenzofuran
	  Hydroquinone
	  x-naphthoquinone
	  9,10-anthroquinone
	  9,10-phenanthronequinone
	  Higher fused-ring quinones
		7-15

Transitional	Aliphatic and aromatic Carbonyl functional groups Ketones, aldehydes, esters, ethers	1-6
Insoluble	Aliphatic and aromatic Hydroxyl and carbonyl groups High molecular weight organic species Inorganic compounds Glass fibers from filters	6-25

[a]Substituent groups are alkyl and alkoxy.

SOURCE: Funkenbusch *et al.*, 1979; Andon *et al.*, 1979.

TABLE 2.2 Particulate Emission Data for 1979 Light-Duty Diesel Certification Vehicles

Vehicle	Particulate	
	g/mi	g/km
Typical gasoline-powered vehicle (with catalyst)	0.008	0.005
VW Rabbit	0.23	0.14
Peugeot 504	0.29	0.18
VW Dasher	0.32	0.20
IHC Scout (Nissan)	0.32-0.47	0.20-0.29
Mercedes 300 SD	0.45	0.28
Mercedes 240D	0.53	0.33
Chevrolet Pickup (Oldsmobile)	0.59	0.37
Dodge Pickup (Mitsubishi)	0.61	0.38
Oldsmobile 260	0.73-1.02	0.45-0.63
Mercedes 300D	0.83	0.52
Oldsmobile 350	0.84	0.52

SOURCE: EPA, 1980d.

Total particulate emission data from 1979 certification vehicles are presented in Table 2.2; fuel-specific data on the soluble organic fraction, from a Mercedes 240D and a Volkswagen Rabbit, are provided in Table 2.3. As is illustrated in this table, the weight percent of the soluble organic fraction of the particulate varies with driving cycle. This subject is further discussed in the section on federal particulate and gaseous emission measurement procedures. Because the other components of diesel particulate have been discussed, only sulfate and visible smoke emissions are discussed below.

Sulfate Emissions

In humid atmospheres, sulfates can hydrate to form sulfuric acid and acid sulfate mists, which can cause human health and environmental hazards and destroy materials. Hence, sulfate emissions from diesel engines are of concern. Sulfates appear in diesel particulates as either acid "droplets" or more complex forms. There may be some free sulfur, but most of the sulfur will combine with carbon, hydrogen,

TABLE 2.3 Organic Soluble Content by Weight Percent and Major Elements of Particulate

| Fuel | Weight Percent Organic Solubles in Particulate by Operating Schedule[a] | | | | | | | |
	Cold FTP	Hot FTP	FET	NYCC	Idle	50 km/hr	85 km/hr	Mean Percentage
				Mercedes 240D				
EM-238-F	10.7	11.5	6.4	11.4	9.3	10.2	7.7	9.4
EM-239-F	9.7	9.8	7.6	7.9	6.0	9.0	8.6	8.3
EM-240-F	12.2	11.9	11.7	25.0	20.1	12.7	9.8	14.3
EM-241-F	8.3	7.0	4.7	3.1	4.1	4.7	6.0	5.4
EM-242-F	9.5	8.5	2.3	3.9	6.5	7.3	4.9	6.2
Mean percentage	10.1	9.7	6.5	10.4	9.2	8.8	7.4	8.7
				VW Rabbit Diesel				
EM-238-F	12.2	9.8	9.1	14.4	10.0	14.9	14.0	12.4
EM-239-F	11.6	14.4	14.3	27.5	18.4	22.7	12.7	16.8
EM-240-F	13.2	16.4	15.3	18.7	19.6	15.7	13.4	15.9
EM-241-F	11.8	18.4	15.4	18.0	16.3	15.4	21.6	16.5
EM-242-F	13.5	13.7	12.9	11.6	11.4	16.3	7.4	11.4
Mean percentage	12.5	14.5	13.4	18.0	13.1	17.0	13.8	14.6

TABLE 2.3 (continued)

| Fuel | Weight Percent Element(s) in Organic Solubles | | | | | |
	Carbon	Hydrogen	Nitrogen	Sulfur	Oxygen	CHNSO
			Mercedes 240D			
EM-238-F	82.8	12.4	0.10	0.40	4.2	99.9
EM-239-F	83.5	12.2	0.08	0.36	3.8	100.0
EM-240-F	83.2	12.4	0.10	0.39	3.7	99.8
EM-241-F	83.9	12.2	0.08	0.36	3.4	100.0
EM-242-F	83.7	12.4	0.13	0.41	3.3	99.9
Mean values	83.4	12.3	0.10	0.38	3.7	99.9
			VW Rabbit Diesel			
EM-238-F	83.9	12.7	0.12	0.37	2.9	100.0
EM-239-F	84.2	12.1	0.08	0.41	3.2	100.0
EM-240-F	83.7	12.8	0.21	0.35	2.9	99.9
EM-241-F	84.2	12.4	0.16	0.43	2.7	99.9
EM-242-F	83.8	12.6	0.11	0.38	3.0	99.9
Mean values	84.0	12.5	0.14	0.39	2.9	99.9

[a]Averages used where possible.

SOURCE: Hare and Baines, 1979.

oxygen, or nitrogen to form compounds such as hydrogen sulfide, alkyl sulfates, hetrocyclic compounds, sulfur dioxide, or sulfur trioxide. The relative amounts of individual species will depend on factors such as the amount of sulfur in the fuel, the mode of engine operation, the combustion temperature, the exhaust temperature, the dilution factor, and the overall particulate level. In the absence of a catalyst in diesel operation, the surface area of the diesel particulate, relative humidity, nitric oxide and nitrogen dioxide level, the air/fuel ratio, and the number of diesel particles may play an important part in sulfate formation (Khatri _et al._, 1978).

Table 2.4 provides data on sulfur emissions as compared with fuel-sulfur levels and total particulate emissions. The data were obtained using a naturally aspirated, direct injection, V-8, 1973 Caterpillar 3150 diesel engine (a medium-duty engine). A further discussion of sulfate emissions is provided in the section on effect of fuel properties.

Visible Smoke Emissions

Visible smoke is an indication of a high exhaust particulate level. The smoke emissions of light-duty diesels, at conditions of full-load acceleration, are not federally regulated, because the FTP is a relatively low-speed, light-load cycle, with engine speeds ranging from 1,300 to 1,800 rpm at 10 to 25 percent of full load.

Diesel engines generate the highest particulate concentrations and visible smoke under full-throttle operation and during cold starting conditions. Heavy-duty diesel vehicles have been regulated for worst case visible smoke since 1970. Heavy-duty diesel smoke standards for 1970 and 1974 are shown in Table 2.5.

The federal smoke cycle test contains both an acceleration mode and a full-load lug down mode from rated speed to peak torque speed. The "a" factor standard shown in the table is based on a calculation of the smoke opacity from the 15 highest 0.5-second increments of smoke from the acceleration mode; the "b" factor is calculated from the 5 highest 0.5-second increments of smoke from the lug down mode; and the "c" factor is calculated from the 3 highest 0.5-second increments from both modes—typically the acceleration mode.

Manufacturers set their own internal product specifications for other than FTP conditions. Generally, a Robert Bosch smoke number of 3 is the maximum level permitted for full-load conditions. This number correlates to an 8 percent opacity when the Green and Wallace (1980) correlation is employed. Generally, smoke values above 4 percent opacity are visible to the human eye.

Hare and Baines (1979) have recently made smokemeter measurements on a Mercedes 240D and a Volkswagen Rabbit by using an EPA type smoke-meter during the first 505 seconds (the "transient phase") of the FTP, from start of both the cold and the hot cycle. A summary of the data is provided in Table 2.6. The measurements were made on the plumes emitted from 2-inch (51-mm) O.D. exhaust pipes. The fuels used in the tests are described in Table 2.24 (section on effect of fuel proper-ties). Starting peaks were generally much higher during cold starts

TABLE 2.4 Sulfate and Total Particulate Emissions[a]

Mode No.	Test No.	Emissions Sulfate, mg/m$_3$ act.	Particulate mg/m$_3$ act.	Sulfur in Fuel by Mass, %	Sulfate = Mass/ Tot. Part. Ratio
2	10	0.80	30.9	0.31	0.0326
	11	0.92	29.8	0.31	0.0309
	12	0.31	29.3	0.31	0.0277
3	73	0.92	27.8	0.31	0.0329
	74	0.37	32.4	0.31	0.0268
	75	0.74	35.7	0.31	0.0209
4	37	1.80	23.9	0.31	0.0752
	40	1.87	30.8	0.31	0.0685
	42	1.77	29.8	0.31	0.0594
	45	1.41	21.9	0.31	0.0643
5	24	1.51	28.4	0.31	0.0530
	25	1.99	27.8	0.31	0.0727
	47	2.34	23.7	0.31	0.0987
6	50	1.00	72.1	0.35	0.0138
	52	0.97	64.5	0.35	0.0151
	55	1.04	73.3	0.35	0.0141
	57	1.30	39.4	0.35	0.0145
8	58	1.08	79.3	0.34	0.0136
	80	0.97	78.2	0.34	0.0111
	67	1.26	93.3	0.34	0.0135
	69	1.13	86.0	0.34	0.0132
9	62	1.06	32.3	0.34	0.0327
	64	1.09	33.0	0.34	0.0329
	69	1.08	34.0	0.34	0.0318
11	78	1.23	26.0	0.31	0.0491
	86	1.33	28.1	0.31	0.0472
	87	1.18	26.2	0.31	0.0451
12	88	0.84	36.2	0.31	0.0231
	89	0.71	33.5	0.31	0.0230

[a]Amoco No. 2 diesel fuel with sulfur contents ranging from 0.23 to 0.35 by mass used throughout these tests.

SOURCE: Khatri et al., 1978).

TABLE 2.5   New Federal Heavy-Duty Diesel Smoke Standards

Year	Smoke Factor, % Opacity		
	a (accel.)	b (lug)	c (peak)
1970	40	20	--
1974	20	15	50

SOURCE: Federal Register, 1972.

for the Volkswagen Rabbit, but this was true only for the "minimum quality" No. 2 (EM-241-F) fuel in the Mercedes 240D. In addition, the Volkswagen Rabbit was highest in all the first, or "cold," idle smoke levels for all fuels. The cold idle average (after start) was by far the highest for the Volkswagen Rabbit (72 percent versus next highest at 4.2 percent opacity) when using the minimum-quality fuel. It appears that the relatively low cetane number of this fuel (42) made cold-engine

TABLE 2.6   Summary of Smoke Data by Vehicle and Fuel

Condition	Average Smoke, PHS %, by Fuel[a]				
	238	239	240	241	242
Mercedes 240D					
Cold-start peak	22.0	36.0	39.0	59.0	21.0
Cold idle avg. (after start)	4.2	2.6	2.3	2.2	3.4
1st accel. peak	18.0	15.0	15.0	14.0	18.0
Idle at 125 sec., avg.	2.0	1.4	1.4	2.4	2.1
Accel. at 164 sec., peak	6.6	13.0	7.2	9.8	12.0
Hot-start peak	26.0	38.0	23.0	29.0	22.0
Hot idle avg. (after start)	2.2	1.4	1.4	2.3	2.3
Hot 1st accel. peak	7.0	7.8	5.8	11.0	7.5
Hot idle at 125 sec., avg.	1.8	1.7	1.3	2.2	1.8
Hot accel. at 164 sec., peak	5.4	6.2	4.6	5.2	6.2
VW Rabbit					
Cold-start peak	71.0	79.0	58.0	89.0	85.0
Cold idle avg. (after start)	0.5	3.0	1.8	72.0	3.5
1st accel. peak	16.0	5.8	6.8	41.0	23.0
Idle at 125 sec., avg.	0.3	0.5	0.4	0.4	0.4
Accel. at 164 sec., peak	18.0	22.0	10.0	27.0	22.0
Hot-start peak	41.0	37.0	28.0	48.0	34.0
Hot idle avg. (after start)	0.4	0.4	0.2	0.2	0.3
Hot 1st accel. peak	3.5	3.0	2.9	4.2	2.2
Hot idle at 125 sec., avg.	0.4	0.4	0.2	0.2	0.5
Hot accel. at 164 sec., peak	31.0	23.0	13.0	28.0	17.0

[a]See Table 2.24 (section on noise emissions and control) for properties of fuels.

SOURCE: Hare and Baines, 1979.

operation marginal in the Volkswagen Rabbit (Hare and Baines, 1979). The fuels differed widely in smoke, but generally the highest smoke levels resulted with the minimum quality No. 2 fuel, and the lowest resulted when No. 1 fuel was used. The Rabbit produced more smoke during a start than the Mercedes, but the Mercedes produced more smoke during acceleration peaks and idling.

Specifying properties of diesel fuels with limits as outlined in the section on effect of fuel properties will, in general, reduce the visible smoke from diesels. Further, it is clear that "worst case" (full-load acceleration conditions, cold start, and full-load lug) smoke limits must be developed to control visible smoke. This is likely to be necessary in order to gain public acceptance of the light-duty diesel.

Interpreting Particulate Data

One of the important considerations in interpreting NO_x and particulate data is whether diesel-powered vehicles are influenced by the test inertia weight and, in the case of light-duty trucks, whether there is an effect of test horsepower. Light-duty trucks are typically tested at higher road-load horsepower and inertia weight settings than light-duty vehicles. This fact is particularly important in determining the technological feasibility of an individual standard. Appendix D provides data on the effects of inertia weight and road load horsepower on NO_x and particulate emissions.

The data show that NO_x and particulate are both influenced by vehicle inertia weight. On the average, it appears that particulate decreases by approximately 25 percent, and NO_x decreases by about 30 percent, for each 1,000-pound reduction in inertia weight. These effects make it more difficult to achieve 1.0-g/mi NO_x and less than 0.5 g/mi particulate levels for the over 3,000-pound inertia-weight cars.

In addition to inertia weight and road load, fuel components and test cycle will also influence particulate data. These variables are discussed in the section on effect of fuel properties and in the section on federal particulate and gaseous emission measurement procedures, respectively.

Deterioration Factors and Particulate Emissions

Diesel particulate deterioration factors play a role in the technological feasibility of testing a given standard. General Motors was the only manufacturer to submit data for the EPA particulate and NO_x hearings held in April and May 1979. These limited data are provided in Table 2.7 (EPA, 1980c). General Motors concluded that the particulate deterioration factors for light-duty diesels will range from 1.2 to 1.4; EPA believes that the range will be 1.0 to 1.1 because the average diesel hydrocarbon deterioration factor for 1979 certification vehicles was 1.06. As shown in recent General Motors particulate data, Figures 2.9 and 2.10, deterioration factors may increase as more complex

TABLE 2.7 General Motors Particulate Deterioration Factors

Car/Displacement	Test	Particulate DF
80 Olds Delta 88, 5.7-liter engine	50°K AMA	1.03
80 Olds Delta 88, 5.7-liter engine + EGR	27.6°K AMA	0.66
78 Opel, 2.1-liter engine + EGR	50°K AMA	1.26
76 Opel, 2.1-liter engine	Trap development baseline tests	1.53

SOURCE: EPA, 1980c.

control systems are added to the diesel engine. Vehicles used to obtain these data were equipped with mechanical exhaust gas recirculation systems, which probably explains the higher deterioration factors.

Equipment and Research Needs for Particulate Sampling and Characterization

The basic EPA light-duty constant-volume sampler (CVS) certification dilution system will continue to be the main method for particulate measurements. However, the cost and complexity of that system, particularly for larger engines and vehicles, indicate that minitunnels (less than 8 inches in diameter) and alternate dilution systems (such as two-stage dilution and constant-mass tunnels) are needed, particularly for routine development tests. In addition, there is a need for theoretical and empirical correlations between undiluted exhaust mass particulate or smokemeter measurements and diluted tunnel measurements using the basic theory outlined in references such as Plee and MacDonald (1980).

The Coordinating Research Council is planning to support research on a constant mass system in which a fraction of the dilution tunnel flow, downstream from the air inlet, is drawn off as supply air to the engine, after which the total exhaust flow is mixed back with the remaining tunnel air and sampled downstream. The constant-mass system would be cheaper and simpler than the present EPA system because a simple fan that produces a constant mass flow rate through the tunnel is substituted for the expensive heat exchangers and CVS Roots blower.

These systems require refinement and additional development to provide research laboratories with the ability to engineer advanced engines and emission control systems efficiently. The EPA certification systems also need refinements, particularly for heavy-duty vehicles, because the larger engines make the CVS systems large, costly, and overly complex.

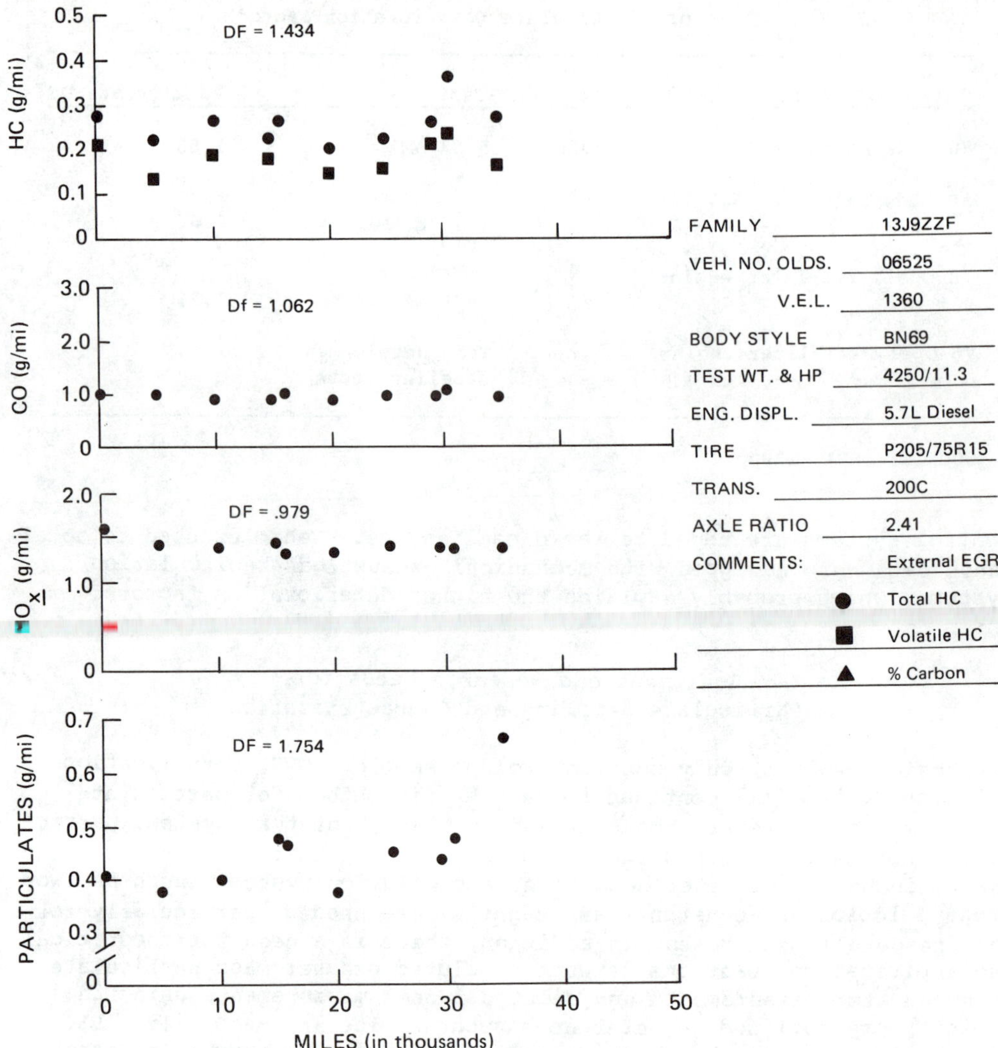

FIGURE 2.9 Diesel certification durability--car. SOURCE: GM, 1980.

GASEOUS EMISSIONS

Although high particulate emission levels are the most obvious concern
stemming from the projected increase in the use of diesel engines in
the light-duty fleet, they are not the only area of concern. Many of
the gaseous species emitted from the light-duty diesel can damage human
health and the environment. Current federal emission standards apply
only to carbon monoxide, oxides of nitrogen (NO_x), and total hydrocarbons. There is no differentiation between gaseous hydrocarbons and
particulate hydrocarbons, and no regulation on sulfurous emissions.
Because of the consequences that some of the unregulated emissions in

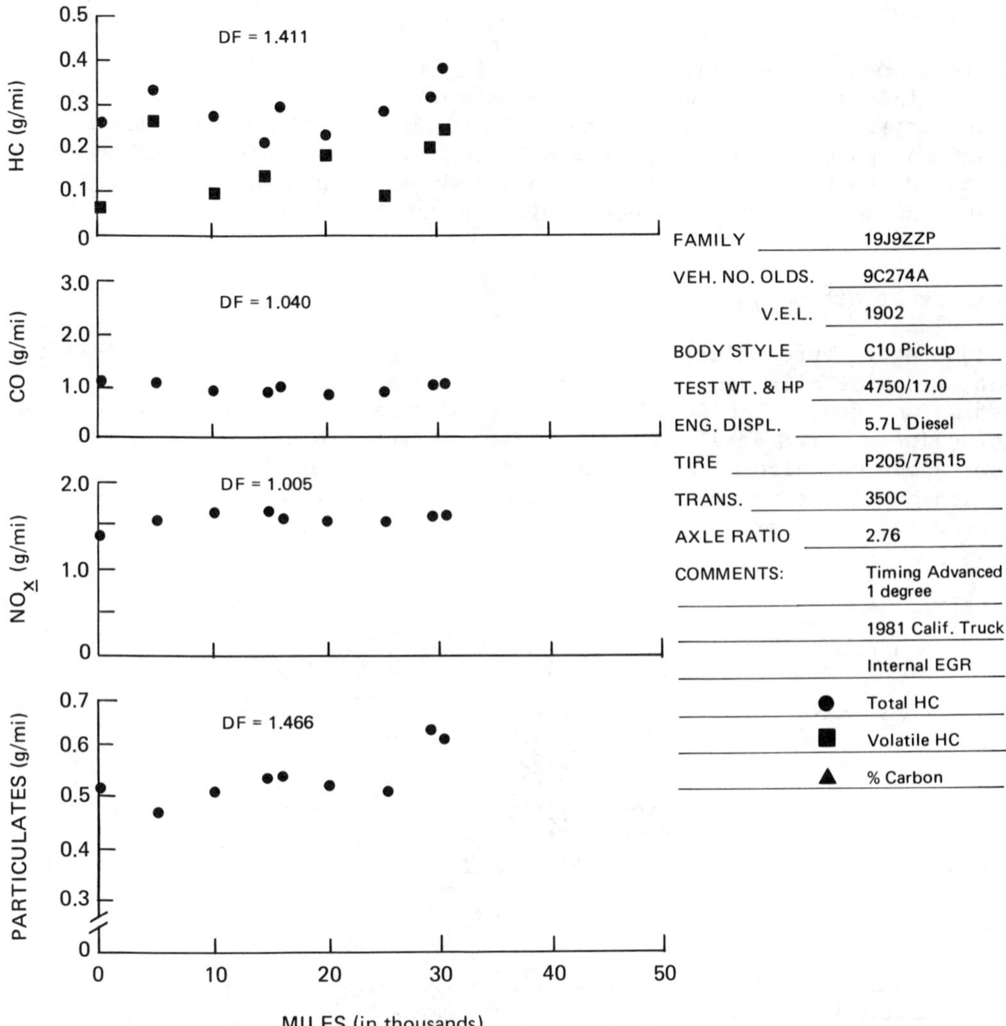

FIGURE 2.10 Diesel certification durability--truck. SOURCE: GM, 1980.

diesel exhaust may have on human health and the environment, these
emissions are discussed below.

Regulated Emissions

Carbon monoxide, oxides of nitrogen, and hydrocarbon emission levels
are all regulated by the federal government. Table 1.6 contains
certification data on these emissions for the 1979 model year.

Carbon Monoxide

Carbon monoxide emissions are caused by incomplete combustion. As is
shown in Figure 2.11, carbon monoxide emissions from light-duty diesels
are comparable to, or lower than, the minimum levels emitted from
gasoline-powered vehicles. For all diesels, the levels of carbon
monoxide emitted are well below mandated emission limits; hence, it is
not considered to be a gaseous emission problem.

Oxides of Nitrogen

High-temperature oxidation of molecular nitrogen leads to the formation
of oxides of nitrogen (NO_x). As is shown in Figure 2.12, NO_x
emissions from light-duty diesels are comparable to those from
gasoline-powered vehicles. The effect of vehicle weight on NO_x
emissions can also be ascertained from this figure. Diesel NO_x
emission levels are comparable to, or below, minimum emission levels
for gasoline-powered vehicles when both vehicles weigh less than 2,500

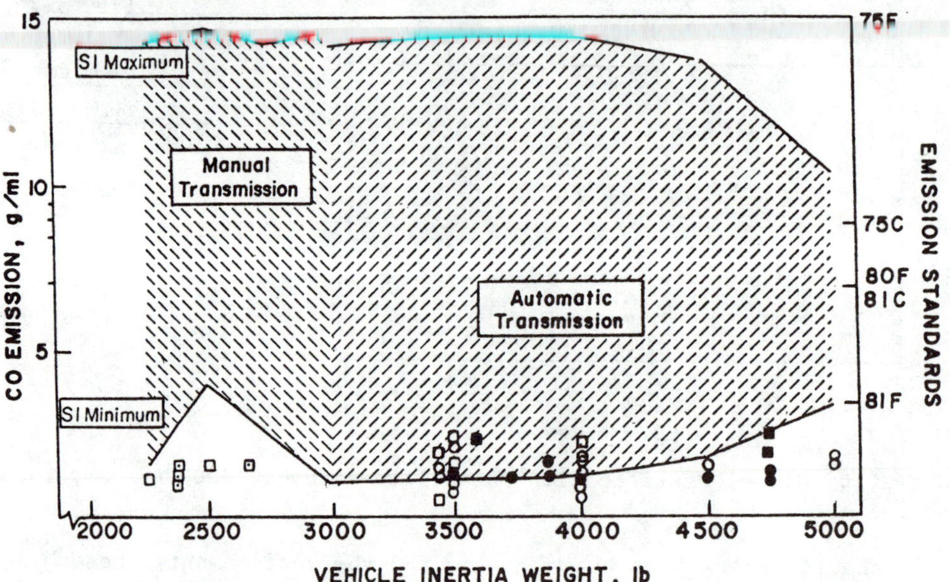

FIGURE 2.11 CO emissions of certification vehicles as a function of
vehicle inertia weight, in comparison to the emission standards (C,
California; F, Federal). Model year 1978 and 1979 certification
light-duty diesels in comparison with 1979 spark ignition light-duty
vehicles. Closed symbols, light-duty diesels with EGR, squares,
vehicles with manual transmission; and circles, vehicles with automatic
transmission. (See Appendix A for a tabulation of the certification
data upon which this figure is partially based.) SOURCE: Adapted from
Roessler et al., 1980; EPA, 1980c.

pounds. At weights of 4,000 pounds or more, NO_x emissions from the diesel are comparable to, or greater than, the maximum levels for gasoline-powered vehicles. However, as illustrated by the data for the vehicles in the 3,500-pound class, the effects of engine and vehicle parameters other than vehicle weight play a more significant role in determining NO_x emissions from the uncontrolled diesel. Overall, the data in Figure 2.12 show that very few of the 1978, 1979, or 1980 model year certification diesels had NO_x emission levels below the future 1.0-g/mi standard. Many of the 1980 light-duty diesels incorporated exhaust gas recirculation (EGR), but this system was generally used for hydrocarbon rather than for NO_x control.

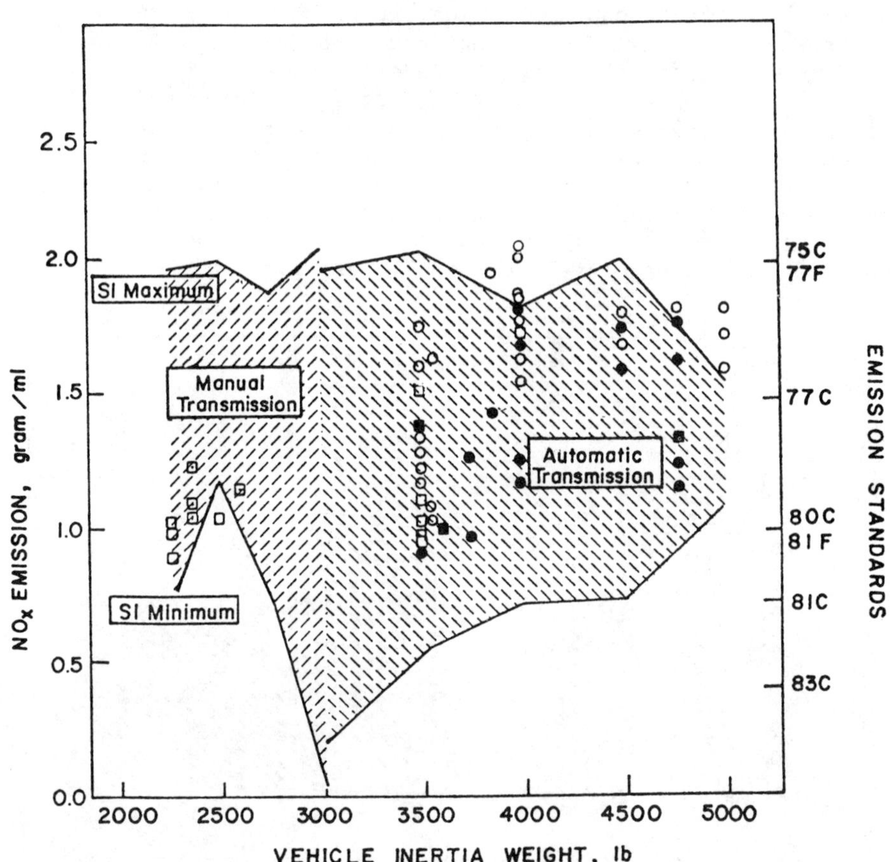

FIGURE 2.12 NO_x emissions from certification vehicles as a function of vehicle inertia weight in comparison to the emission standards. (See Appendix A for a tabulation of the certification data upon which this table is partially based.) SOURCE: Adapted from Roessler et al., 1980; EPA, 1980c.

Hydrocarbons

   <u>Exhaust Emissions</u> Tail pipe emissions of hydrocarbons are due to
incomplete combustion, mid-air or lean-limit quench, and pyrolysis of
the lubricating oil. Wall-quench and crevice area hydrocarbon emissions
are low because of the droplet-burning nature of the combustion process
(Wade, 1980). As is illustrated in Figure 2.13, exhaust hydrocarbon
emissions from the diesel generally fall within the range of values
recorded for certification gasoline-powered vehicles. However, for
vehicles weighing 3,500 pounds or more, seven Mercedes diesels exhibited
exhaust emissions lower than the minimum levels reported for spark
ignition engines, and one Peugeot and four General Motors diesels
emitted hydrocarbons at levels greater than the maximum gasoline-powered
engine values. The Mercedes vehicles are prechamber diesels, Peugeot
and Volkswagen use swirl chamber engines, and General Motors uses a
modified prechamber/swirl chamber. Fewer than half of the 1978 and
1979 light-duty diesels meet the future 0.41-g/mi standard. Of the
1980 light-duty diesels, all of the passenger cars, but few of the
trucks, meet the 0.41 standard. Many used EGR and fuel injection
timing modifications to lower hydrocarbon emissions while keeping NO_x
within the 2.0 g/mi 1977 to 1980 federal standards. This control
technique will probably not be successful when the 1.0 g/mi NO_x
standard (or the waiver level between 1.0 and 1.5 g/mi) is imposed in

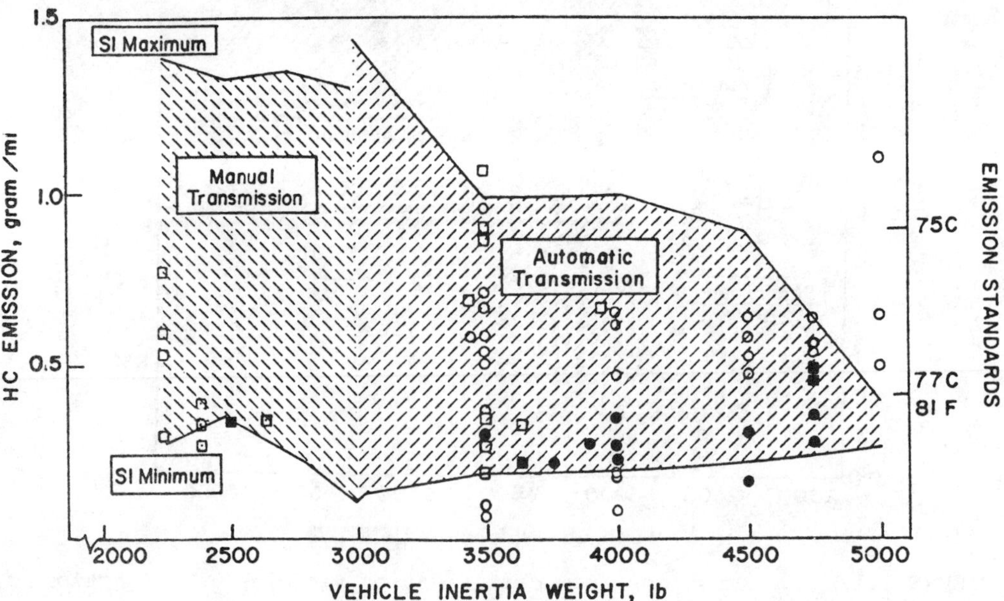

FIGURE 2.13 Hydrocarbon emissions from certification vehicles as a
function of vehicle inertia weight in comparison with the emission
standards. (See Appendix A for a tabulation of the certification data
upon which this table is partially based.) SOURCE: Adapted from
Roessler <u>et al.</u>, 1980; EPA, 1980c.

1981, because the EGR system will be needed primarily for NO_x rather than for hydrocarbon control. It is expected that future-generation diesel passenger cars will have to employ hydrocarbon emission controls, especially considering the NO_x/hydrocarbon trade-offs characteristic of most control technologies.

Evaporative Emissions Because diesel fuel has a very low vapor pressure, evaporative hydrocarbon emissions from diesels are negligible. This fact is emphasized by the exclusion of diesel-powered vehicles from the EPA evaporative emission certification procedure. Evaporative emissions from some gasoline-powered vehicles approach the current federal standard of 6.0 g per test (Federal Register, 1979)--a level that researchers at General Motors calculate to be the equivalent of 0.65 g/mi (GM, 1978). Gasoline-powered vehicles produced for the 1981 model year and beyond are required to meet an evaporative emission standard of 2.0 g per test, and no crankcase emissions are allowed. This evaporative emission rate is the equivalent of 0.22 g/mi, according to the General Motors calculations. General Motors researchers also calculate that a gasoline-powered vehicle averaging 13.4 mpg will have a refueling loss equivalent to 0.4 g/mi (and, presumably, linearly less for cars with better fuel economy). The refueling loss from light-duty diesels would be very low. The EPA certification procedure does not credit the diesel for negligible evaporative emissions and very low refueling losses, even though, if ambient air quality is the goal, it might be expected that such credit (for example, the 0.16-g/mi hydrocarbon credit for diesels in California, effectively increasing the allowable exhaust hydrocarbon standard) would be justifiable. It should be noted that the exhaust hydrocarbon distribution for the light-duty diesel is different from that for the gasoline-powered vehicle, as will be discussed in the subsection on noncriteria hydrocarbons.

In-Use Emission Deterioration

It is generally believed that light-duty diesel engines will not have emission rate deteriorations over the vehicle lifetime that are significantly lower than those of gasoline engines (Drexle, 1979). This expected lower in-use deterioration was demonstrated (Hergenrother, 1979) when emission levels from 66 diesel-powered taxicabs were compared with an equal number of spark ignition engine taxis. All vehicles were operated for 120,000 miles. The results of this comparison are shown in Figure 2.14. The average emissions over the lifetime of the taxis were 3 to 6 times lower for the diesels. The gasoline-powered taxicabs were equipped with EGR and exhaust catalysts. The high deterioration of the gasoline-powered vehicles confirmed an earlier study by Gibbs and co-workers (1978) of 56 catalyst-equipped light-duty gasoline-powered vehicles. This study included both fleet and privately owned cars with mileage accumulation of up to 60,000 miles; it showed generally good agreement with the taxi study, except that the carbon monoxide deterioration occurred sooner and the NO_x deterioration was not as

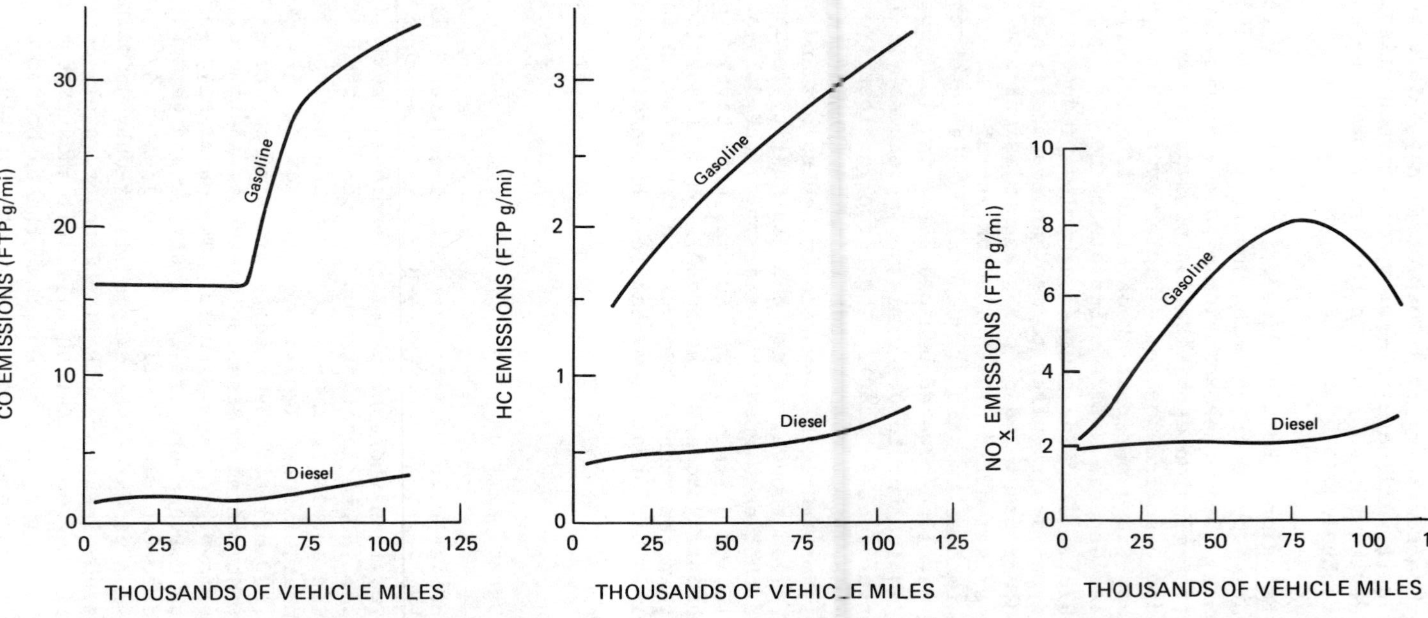

FIGURE 2.14 Emission deterioration with mileage accumulation: light-duty diesels versus light-duty gasoline-powered vehicles. SOURCE: Hergenrother, 1979.

significant over the first 50,000 miles. These results indicate another advantage of the diesel as an automotive power plant: an insignificant level of exhaust emission deterioration in comparison with the gasoline-powered vehicle, resulting in lower emissions over the vehicle lifetime.

Unregulated Emissions

Prior to the 1979 model year, automobile manufacturers were required to measure unregulated pollutants to ensure that emission control devices were not causing increased emissions of other pollutants (40 CFR 86.078-5 and 86.408-78). This policy was expanded and extended by the Clean Air Act Amendments of 1977 (Sections 202(a)(4)(A), and 206(a)(3)(A)), as clarified by EPA Advisory Circular 76 (EPA, 1978). The latest report outlining analytical procedures for characterizing unregulated emissions from diesel vehicles is by Smith and co-workers (1980). Brief descriptions of the 10 measurement procedures developed are provided in Appendix E.

Sulfurous Species

Sulfur-containing species are emitted in the exhaust of internal combustion engines because sulfur exists as an impurity in the fuel. Most of this sulfur is emitted as sulfur dioxide: Less than 5 percent is emitted as sulfates (sulfuric acid and sulfate salts), and even smaller amounts are emitted as hydrogen sulfide, carbonyl sulfide, and other organic sulfides.

<u>Sulfur Dioxide</u> Sulfur dioxide is an eye and respiratory irritant. It is of concern as an exhaust pollutant because it may undergo further oxidation in the atmosphere to form sulfuric acid, which in turn is the most significant cause of acid rain (EPA, 1979c; Cronan and Schofield, 1979).

As is shown in Figure 2.15, EPA researchers have demonstrated a linear relationship between the mass percent of sulfur in fuel and the sulfur dioxide emission rate for light-duty diesels (Braddock and Bradow, 1975; Braddock and Gabele, 1977). The ASTM upper limit for No. 1 diesel fuel is 0.2 percent by mass fuel sulfur, and for No. 2 fuel it is 0.5 percent by mass, although higher values have occasionally been measured. The average diesel fuel in the United States contains approximately 0.23 percent by mass fuel sulfur. The average for gasoline is only about 0.03 percent by mass.

The sulfur concentration in diesel fuel is expected to increase in the future because of the deteriorating quality of crude oil. Therefore, sulfur dioxide emissions from the diesel are expected to increase from the present average of about 450 mg/mi. Gasoline-powered cars emit about 100 mg/mi. (Note that the difference in sulfur dioxide emission rates on a milligram per mile basis is not as great as might be expected from the relative fuel-sulfur levels because of the difference in fuel economy between the light-duty diesel and light-duty gasoline-powered vehicle.) This difference will probably increase in the future because refiners must control the sulfur level in gasoline to protect their catalytic cracking units, a constraint not present in the refining of diesel fuel. The light-duty diesel sulfur dioxide linear curve fit of Figure 2.15 is extrapolated to much lower fuel-sulfur levels in Figure 2.16 and compared to the sulfur dioxide emission levels of catalyst-equipped and noncatalyst light-duty gasoline-powered vehicles. This figure demonstrates that if diesel fuel were constrained to the same fuel-sulfur levels as those required for gasoline, the sulfur dioxide emission rate from diesels would be reduced to levels comparable to those from the light-duty gasoline-powered vehicle.

In the past, sulfur dioxide emissions from passenger cars have been of little concern because of the low levels emitted in comparison with stationary sources, especially those burning coal or residual fuel oil. With increased dieselization this situation may change. Assuming that total passenger car travel per year is about 10^{12} miles (EPA, 1975) and that diesel fuel sulfur does not increase from 0.23 percent by mass (yielding a conservative estimate), a 25 percent dieselization level in

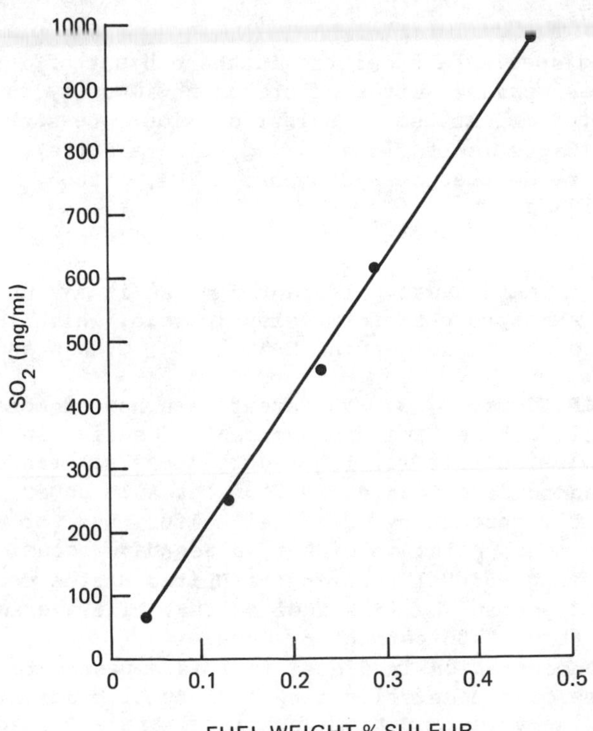

FIGURE 2.15 Linear relationship between sulfur dioxide emissions from the light-duty diesel and fuel-sulfur concentration. SOURCE: Braddock and Bradow, 1975.

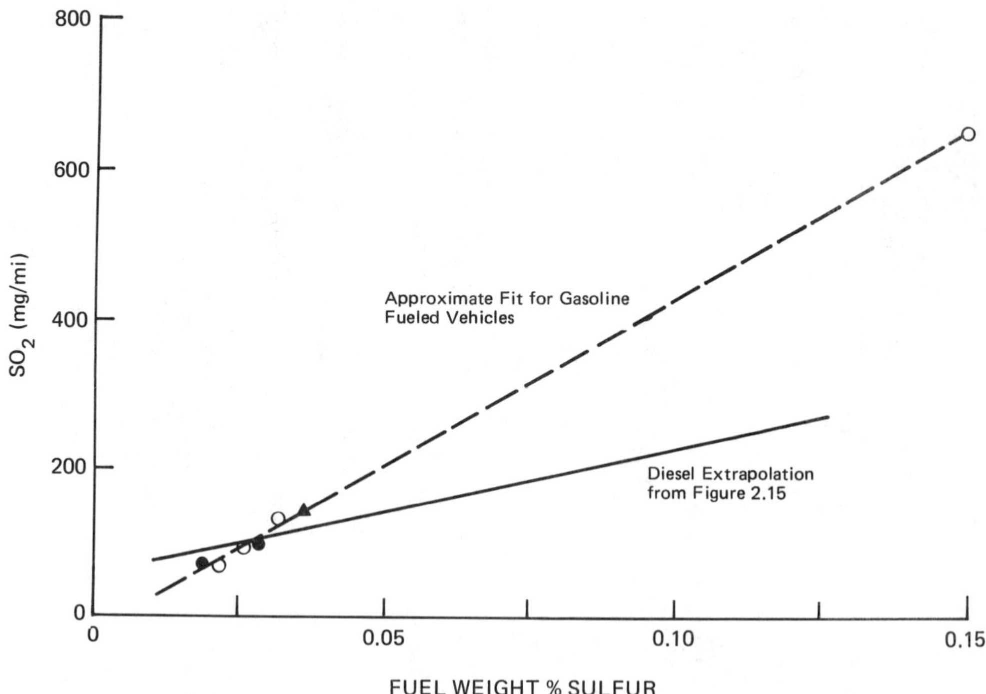

FIGURE 2.16 Extrapolation of the curve fit of Figure 2.15, demonstrating that light-duty diesel sulfur dioxide emissions would be comparable to those of gasoline-powered vehicles if diesel fuel were constrained to the same fuel-sulfur level as that required for gasoline. Closed symbols, catalyst-equipped spark ignition engine. Open symbols, noncatalyst spark ignition engine. SOURCE: Data obtained from Cadle et al., 1979; Braddock and Bradow, 1975.

the passenger car fleet would result in an increase of about 87.5 x 10^3 tonnes (metric) of sulfur dioxide emitted annually by passenger cars, or roughly a 48 percent increase compared to the sulfur dioxide emissions from the 1970 passenger car fleet (practically no diesels) (Cavender et al., 1973).

Other Sulfurous Species Sulfates were discussed in the section on particulate emissions because they are in the particulate state and are influenced by fuel-sulfur levels. Besides sulfur dioxide and sulfates, hydrogen sulfide, carbonyl sulfide, methyl sulfide, ethyl sulfide, and methyl disulfide are the only specific sulfurous species that have been measured in automotive exhaust. Table 2.8 compares the emission rates of hydrogen sulfide, carbonyl sulfide, and total organic sulfides from various light-duty vehicles. Only carbonyl sulfide measurements have been reported for the light-duty diesel. However, because of the oxidizing characteristic of the diesel combustion process, it is not expected that exhaust levels of these other sulfurous species would be significant.

TABLE 2.8 Average Unregulated Emissions for Various Types of Light-Duty Vehicles and Comparison to Values Observed During Rich Malfunction Operations (in mg/mi)[a]

		Non-catalyst S.I.	Catalyst Spark Ignition			Rich Malfunction Maxima[b]			
	Diesel		Oxidation	3-Way	Dual	Non-catalyst	Oxidation	3-Way	Dual
H2S	–	0.06[c]	0.04	0.18	0.01	0.8	1.1	9.0	1.6
COS	0.55[c]	0.18[c]	0.19	0.34	0.05	–	0.7[c]	5.0	1.1
Organic sulfides[d]	–	0.30[c]	0.35	0.42	0.02	0.6[c]	1.9	5.2	1.3
NH_3	1.5	7.2	3.9	40.6	5.9	80.5	91.1	512.1	407.3
HCN[e]	1.8	8.6	2.6	3.8	1.5	15.3	10.0	107.7	180.8
N_2O	–	4.8[c]	33.8	23.0	32.1	6.4[c]	98.2	44.3	141.5
DMNA	0.001	–	0.09	–	–	–	–	–	–
Organic amines	–	0.04[c]	0.20	0.01	0.03	0.32[c]	0.16	0.08	0.06
Formaldehyde	22.3	25.8	2.8	0.3	2.1	–	9.7	2.7	2.0
Total carbonyls	44.7	46.5	12.8	2.2	4.1	198.8[c]	29.6	10.3	4.7
Total carbonyls, %HC	13.8	1.8	–	0.6	1.3	–	–	–	–
Benzene	7.7	64.4[c]	15.9	22.3	5.6	370.3[c]	165.2[c]	160.0[c]	201.9
NMHC, %HC	97.1	93.4[c]	83.8	81.4	64.8	–	–	–	–
NRHC, %HC	6.8	18.5[c]	25.8	35.5	43.9	–	–	–	–

[a]See Appendix F for detailed information and references.
[b]Ford (1979b) reports maximum rates for catalyst-equipped vehicle (without specifying the type of catalyst) of: H_2S-13, COS-100, NH_3-548, N_2O 167, formaldehyde-15, and benzene-193 mg/mi. Other maxima are taken from Urban and Garbe (1979, 1980), Urban (1980a,b), and Keirns and Holt (1978).
[c]Only one vehicle.
[d]Includes carbonyl sulfide.
[e]May include other nitriles.

Nitrogenous Species

Nitric oxide is generally recognized as one of the most important combustion generated air pollutants. More than 95 percent of the nationwide emissions of nitric oxide are attributable to combustion sources (Cavender et al., 1973). Traditionally, nitric oxide has been thought to be the only significant nitrogen-containing air pollutant that is emitted from combustion systems, including diesel- and gasoline-powered engines. Recent research has shown that other potentially important nitrogenous species may be present in the exhaust of auto-motive engines (see Matthews and Sawyer (1979) for a review of emissions of noncriteria nitrogenous species from combustion systems).

Nitrogen Dioxide Although nitrogen dioxide is indirectly regulated as part of the total NO_x standard, it is of growing concern because of recent evidence that it is emitted directly from combustion systems. In the past, the only significant source of nitrogen dioxide was thought to be atmospheric oxidation of nitric oxide. However, recent investigations have measured nitrogen dioxide directly in products of combustion. This finding is significant because emission of nitrogen dioxide rather than nitric oxide will affect local nitrogen dioxide concentrations and may shorten the induction period for photochemical smog and acid rain formation. Nitrogen dioxide is a significant atmospheric pollutant because:

   • it decreases visibility;
   • it is an important intermediary in the photochemical smog formation process;
   • it is a direct-acting irritant contributing directly to property and plant damage;
   • it reacts in the atmosphere to form aerosols, peroxyacyl nitrates, nitric acid, nitrous acid, nitroolefins, and the highly carcinogenic nitrosamines (Matthews, 1980a); and
   • it reacts with benzo(a)pyrene or other polynuclear aromatics to form mutagenic and carcinogenic products (Stuart, 1980).

Concern over direct nitrogen dioxide emissions from the diesel was raised by Braddock and Bradow (1975), who reported that the emission rate of this species is higher for the light-duty diesel than for comparable gasoline-powered vehicles, especially under low-speed driving conditions. They found that the nitrogen dioxide/NO_x ratio for the light-duty diesel was as high as 27.6 percent at idle (2.1 ppm nitrogen dioxide), 16.3 percent at 30 mph (5.7 ppm), and 9.1 percent at 50 mph (5.0 ppm). Unfortunately, only few Federal Test Procedure (FTP) data are available because of NO_x reactions and nitrogen dioxide wall adsorption in the sampling bag that occur during the course of the driving cycle. Cadle and co-workers (1979) avoided this difficulty by using dual chemiluminescent analyzers, and they continuously sampled from the tail pipe. For an experimental 5.7-liter indirect injection diesel, they found nitrogen dioxide production rates of 0.08 g/mi (11.27 percent nitrogen dioxide/NO_x) with high EGR and 0.22 g/mi

(17.4 percent NO_2/NO_x) with low EGR. This compares to a survey of over 250 catalyst-equipped gasoline-powered vehicles, which showed average precatalyst nitrogen dioxide/NO_x ratios of 12.4 percent and postcatalyst ratios of 5.6 percent. This comparison indicates that the light-duty diesel exhibits a higher direct nitrogen dioxide emission rate than the catalyst-equipped gasoline-powered engine. However, a prototype gasoline-powered vehicle equipped with a dual-bed catalyst emitted nitrogen dioxide at the rate of 0.14 g/mi with a nitrogen dioxide/NO_x fraction of over 25 percent.

Hilliard and Wheeler (1977, 1979) also found that a platinum catalyst decreased nitrogen dioxide emissions for rich and stoichiometric operation, but increased them by up to a factor of 4 for lean mixtures, and that carbon monoxide inhibited the catalytic oxidation of nitric oxide to nitrogen dioxide. Therefore, although the current generation of light-duty diesels appear to have a higher nitrogen dioxide emission rate than most gasoline-powered light-duty vehicles, the nitrogen dioxide emission rate of future gasoline vehicles may surpass that of the diesel. Control technologies for the diesel may increase its nitrogen dioxide emissions to even higher levels. This is an area that appears to merit further investigation.

Assuming that both gasoline- and diesel-powered light-duty vehicles meet a 1.0-g/mi NO_x standard and that the nitrogen dioxide/NO_x fraction is 5 percent for gasoline and 20 percent for diesel light-duty vehicles, a 25 percent dieselization level would result in an increase of 3×10^4 tonnes (metric)/year of nitrogen dioxide emissions, predominantly in urban areas, if the number of passenger car travel miles per year is 10^{12}.

Other Nitrogenous Species Hydrogen cyanide, ammonia, nitrous oxide, nitrosamines, and various secondary and tertiary amines, nitroparaffins, nitroolefins, and nitriles have been measured in automotive exhaust. Recently, Matthews and Sawyer (1979) reviewed the emission of unregulated nitrogenous species from combustion systems, and Matthews (1980a,b,c) discussed the significance of these species from the atmospheric chemistry and public health perspectives. The average emission rates of hydrogen cyanide, ammonia, nitrous oxide, the organic amines, and dimethylnitrosamine for various light-duty vehicles are presented in Table 2.8. The available data indicate that diesels emit much lower levels of ammonia than gasoline-powered light-duty vehicles and that hydrogen cyanide emission levels are comparable to, or lower than, those from other light-duty vehicles. Nitrosamines are typically measured as dimethylnitrosamine, but the methodology for quantifying nitrosamines has not been fully developed. It is thought that nitrosamine formation in engine crankcases may be more of a problem than exhaust emissions because of the longer residence times available for crankcase reactions. The results of an investigation of diethylnitrosamine levels in crankcase gases of heavy-duty diesels by Hare and Baines (1977) were inconclusive. Emission levels ranging from 27 to 170 parts per trillion were measured using one technique, but they were not confirmed using other measurement methods. Similarly Stone and co-workers (1980) found no evidence of nitrosamines in diesel exhaust or blow-by, but Goff and co-workers (1980) detected up to 136 mg/hr in the crankcase emissions of heavy-duty diesels. Considering the potential significance of these species, further research is warranted.

Many other nitrogenous species have been measured in light-duty gasoline-powered vehicle exhaust but not in that of light-duty diesels. Nitrous oxide is the most significant of these species, and its measurement in diesel exhaust is suggested. Also, it has been shown that exhaust oxidation catalysts promote the formation of nitromethane (Seizinger, 1975) and nitrous oxide (see Table 2.8). These species may become significant if oxidation catalysts are used as a control technique for hydrocarbon emissions from light-duty diesels.

Oxygenates

Automotive exhaust oxygenates are known to participate in the formation of photochemical smog. Some oxygenates are also irritants and odorants. The oxygenates can be generally categorized as carbonyls, phenols, and other noncarbonyls.

<u>Carbonyls</u> The carbonyls of interest are the low molecular weight aldehydes and aliphatic ketones. Formaldehyde is of particular significance because it is the predominant carbonyl and because it has been identified as a potential carcinogen. The National Research Council recently recommended that formaldehyde exposure be reduced to the lowest practical level. The volatile aldehydes are eye and respiratory tract irritants. In diesel engines, aldehydes and other oxygenates are thought to be formed during precombustion or very lean region reactions. Because these species are already partially oxidized, they have a higher specific photochemical reactivity than pure hydrocarbons (Kerrebrock and Kolb, 1978; Braddock and Gabele, 1977).

Table 2.9 presents the average individual carbonyl emission rates and carbonyl distributions for diesels and gasoline-powered vehicles (with and without catalysts). The individual carbonyl emission rates from diesels are similar to those of noncatalyst light-duty gasoline vehicles, but they are higher than those of catalyst-equipped vehicles. There are some differences in the carbonyl distributions, but formaldehyde is the major component and thus accounts for more than 20 percent of the total carbonyls. Of special interest is the fact that the carbonyls account for about 10 percent of the hydrocarbon emissions from diesel-powered passenger cars and, at most, only a few percent of hydrocarbon emissions from gasoline-powered vehicles. Table 2.8 presents much of the same data but includes additional comparisons with the maximum values exhibited by gasoline-powered vehicles under rich malfunction conditions.

It is important to note that the measurement method used for total gaseous hydrocarbons has a very low sensitivity to carbonyls (Colket <u>et al.</u>, 1974). This does not result in a significant error for gasoline-powered vehicles because carbonyls are such a low fraction of the

TABLE 2.9 Average Individual Carbonyl Emission Rates and Distributions[a]

Vehicle Type	Formaldehyde	Acetaldehyde	Isobutyraldehyde	Crotonaldehyde	Hexanaldehyde	Benzaldehyde	Acetone[b]	Methyl Ethyl Ketone	Total Carbonyls[c]
				Emission Rate, mg-mi					
Diesel	18.4	6.9	6.8	3.9	0.7	1.2	9.6	0.0	47.4
Noncatalyst	25.8	6.9	0.0	0.5	0.2	4.0	1.5	0.3	46.5
Oxidation	2.8	0.6	1.1	6.5	0.2	0.6	0.7	0.2	12.8
Dual	2.1	0.1	0.1	0.0	0.1	0.4	0.1	0.1	3.0
3-way	0.3	0.3	0.0	0.0	0.2	0.2	0.1	0.1	1.2
				Distribution, Mass % of Carbonyls[c]					
Diesel	38.8	14.6	14.3	8.2	1.5	2.5	20.3	0.0	10.9
Noncatalyst	55.5	14.8	0.0	1.1	0.4	8.6	3.2	0.6	3.8
Oxidation	21.9	4.7	8.6	50.8	1.3	4.8	5.7	1.3	2.8
Dual	71.4	3.0	3.4	0.7	1.7	14.1	2.7	3.7	1.6
3-way	27.6	23.6	0.0	0.0	14.6	17.9	4.9	10.6	0.6

[a]Results may not coincide exactly with Table 2.8, because more extensive data were available for that table. (See Appendix F.)
[b]May include acrolein (acrylaldehyde) and propanal.
[c]Total carbonyls column in the distribution section is in percent of total hydrocarbons calculated under the assumption of zero sensitivity of flame ionization detectors (FID) to carbonyls. Rows may not sum to 100% (neglecting last column) due to round-off errors.

exhaust. However, because carbonyls account for 10 percent of the hydrocarbons in diesel exhaust, total hydrocarbon measurements will indicate a significantly lower value for carbonyls than is the case.

   <u>Phenols and Other Noncarbonyls</u> Phenols are odorants and irritants. There are few data on phenol levels in diesel exhaust. Hare and Baines (1979) report that phenol levels under steady state vehicle operating conditions are a factor of 100 to 1,000 times lower than aldehyde levels for driving cycle operation. Gaseous phenol levels were comparable to, or lower than, particulate phenol when No. 2 diesel fuel with properties at the national average is burned (although this result is strongly dependent on the fuel properties). Therefore, gaseous phenol emissions are not considered to be significant. Other noncarbonyls measured in gasoline-powered vehicle exhaust include methanol, ethanol, nitromethane, methyl formate, etc. There are no reports on attempts to measure these species in diesel exhaust. For the gasoline-powered vehicle, noncarbonyls do not appear to be significant when compared with the carbonyls, especially formaldehyde. However, it may be surmised that the noncarbonyl oxygenate levels are higher in diesel exhaust than in gasoline engine exhaust because the carbonyl levels are known to be higher. This appears to be an area that merits further investigation.

Noncriteria Hydrocarbons

Although hydrocarbons are regulated species, some hydrocarbon emissions merit further discussion. Figure 2.17 illustrates the molecular weight distribution of hydrocarbon emissions from diesels. The figure also depicts the nature of the hydrocarbon double-counting problem that will be discussed in the section on federal particulate and gaseous emission measurement sensitivities. The hydrocarbon distribution in diesel and gasoline engine exhaust is different. As a result, there are also differences in the photochemical reactivity of the two types of exhaust. The benzene emission rates are also different. Both of these topics are discussed below.

   <u>Hydrocarbon Distribution and Reactivity</u> The hydrocarbon distribution is of interest because some hydrocarbons are not considered to be very photochemically reactive. Table 2.10, which forms the basis for the following discussion, presents the averaged individual emission rates and hydrocarbon distributions for various categories of light-duty vehicles.
   The easiest method of weighting exhaust hydrocarbons to account for reactivity is to assign methane a relative reactivity of zero and assume that the remaining hydrocarbons are equally reactive. This is a portion of the logic behind the nonmethane hydrocarbon standard that was implemented in California in 1980. Table 2.10 lists the nonmethane hydrocarbon percentages (100 percent minus percent methane) for the diesel in comparison with other light-duty vehicles. From the perspective of photochemical reactivity, it is best if the nonmethane hydro-

carbon percentage is as low a fraction as possible. The methane
fraction of diesel hydrocarbons is extremely low--much less than that
of the catalyst-equipped gasoline-powered light-duty vehicle. This is
partially because oxidation catalysts do not oxidize paraffins as
easily as they do other hydrocarbons. It is also due to the higher
methane emission rate of the gasoline engine without the catalyst.
Therefore, according to the simple nonmethane hydrocarbon system,
diesel exhaust is more reactive than light-duty gasoline engine exhaust
for equal amounts of total hydrocarbons. Also, the low methane fraction
of diesel hydrocarbons will make it more difficult for light-duty
diesels to meet the California nonmethane hydrocarbon emission standard.

A slightly more sophisticated reactivity scale would subtract all
of the hydrocarbons that are not considered to be significantly photochemically reactive--methane, ethane, propane, acetylene, and benzene.
This scheme, and the other more sophisticated reactivity schemes, would
be much more expensive as regulatory standards because of the difficulty
of quantifying the hydrocarbon distribution. Table 2.10 provides
nonreactive hydrocarbon percentages for several light-duty vehicles.
It is best that the nonreactive hydrocarbon fraction be as high a

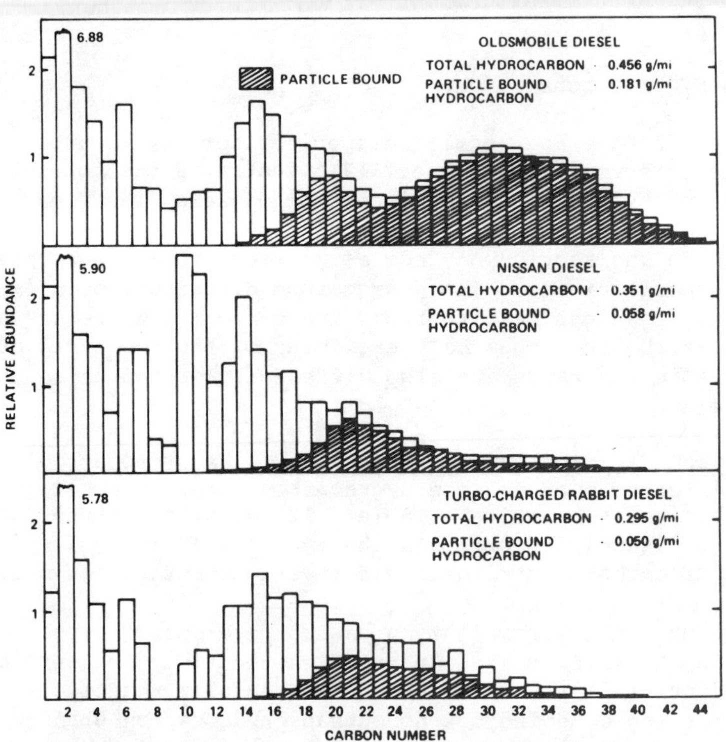

FIGURE 2.17 Molecular weight distribution of
gaseous and particle bound hydrocarbons in light-
duty diesel exhaust. SOURCE: Black and High, 1979.

TABLE 2.10   Average Individual Hydrocarbon Emission Rates and Distributions[a]

Vehicle Type	Nonreactive					Reactive			
	Methane	Ethane	Acetylene	Propane	Benzene	NRHC	Toluene	Ethylene	Propylene
Emission Rate, mg/mi									
Diesel	13.2	1.7	7.7	0.4	7.7	30.6	3.9	43.9	13.8
Noncatalyst[b]	77.3	11.3	56.4	8.1	64.4	217.5	96.6	133.6	61.2
Oxidation	71.0	19.7	4.0	2.4	15.9	113.0	35.6	30.0	14.7
Dual	67.1	5.4	3.4	1.1	5.6	82.6	11.6	5.8	2.6
3-way	39.0	5.5	5.8	1.7	22.3	74.3	18.7	16.9	9.1
Distribution, Mass % Total Hydrocarbons									
Diesel	2.9	0.4	1.7	0.1	1.7	6.8	0.9	9.4	3.0
Noncatalyst	6.6	1.0	4.8	0.7	5.5	18.5	8.2	11.4	5.2
Oxidation	16.2	4.5	0.9	0.5	3.6	25.8	8.1	6.8	3.4
Dual	35.7	2.9	1.8	0.6	3.0	43.9	6.2	3.1	1.4
3-way	18.6	2.6	2.8	0.8	10.6	35.5	8.9	8.1	4.3

[a]See Appendix F for more detailed information and references.
[b]Only one vehicle.

percentage as possible. The diesel exhibits a nonreactive hydrocarbon value less than half that of its noncatalyst gasoline-powered counterpart and 3 to 6 times less than that of the catalyst-equipped vehicles. Thus, the diesel would fare even less well on a nonreactive hydrocarbon basis than it would on a nonmethane hydrocarbon standard.

More sophisticated reactivity rating schemes assign a relative reactivity to each individual hydrocarbon. The fact that there are at least six of these scales, each yielding different reactivity results, indicates that they are arbitrarily defined. Jackson (1978) compared the exhaust reactivities of various gasoline-powered vehicles but did not include diesels. Jackson also developed a linear regression analysis that allows determination of reactivities based only on the concentrations of methane, ethane, ethylene, propylene, and acetylene. This technique provides good agreement with reactivities calculated from more extensive hydrocarbon analyses. His linear regression equations were used to calculate the relative reactivities presented in Table 2.11. This technique, like the nonmethane hydrocarbon method, indicates that diesel exhaust is about as reactive as exhaust from a

TABLE 2.11 Reactivity of Light-Duty Vehicle Exhaust

	NO_2 Formation Scale		Altshuller Scale	
	This Work	Jackson (1978)	This Work	Jackson (1978)
		Reactivity per Gram		
Noncatalyst	0.0421	0.0487	0.0434	0.0520
Oxidation	0.0363	0.0339	0.0369	0.0340
Dual	0.0224	0.0262	0.0214	0.0304
3-way	0.0375	0.0416	0.0378	0.0398
Diesel	0.0414	–	0.0419	–
		Relative Reactivity		
Noncatalyst	100.0	100.0	100.0	100.0
Oxidation	86.2	69.6	85.0	65.4
Dual	53.2	53.8	49.3	58.5
3-way	89.1	85.4	87.1	76.5
Diesel	98.3	–	96.5	–

noncatalyst gasoline vehicle but more reactive than catalyst-equipped gasoline vehicle exhaust. Fairly good agreement with Jackson's data for gasoline-powered vehicles is exhibited.

It should be noted that the nonmethane hydrocarbon, nonreactive hydrocarbon, and relative reactivity scales do not account for the very high reactivities of aldehydes. Aldehydes do not contribute significantly to the reactivity of gasoline-powered vehicle exhaust, because they account for less than 3 percent of the total hydrocarbons. Because carbonyls account for about 10 percent of the total hydrocarbons in diesel exhaust, they will magnify the photochemical reactivity that should be assigned to the diesel. Thus, all of the scales mentioned underestimate the reactivity of diesel hydrocarbon emissions.

Benzene

Benzene has been the focus of much recent interest, primarily because it has been identified as a carcinogen. Inhalation of gaseous benzene results in decreased red and white cell levels and depressed platelet counts. Exposure to high levels of benzene has been shown to cause leukemia. A benzene standard for gasoline has been implemented. This standard is primarily aimed at reducing exposure due to evaporative losses, refueling losses, and spills.

As shown in Table 2.10, the benzene emission rate from the diesel is comparable to, or lower than, emissions from catalyst-equipped and noncatalyst light-duty gasoline-powered vehicles. However, it is currently believed that benzene exposure from automotive exhaust is not significant in comparison to other sources (Sigsby, 1980). Thus, the exhibited diesel advantage does not appear to be important.

FEDERAL PARTICULATE AND GASEOUS EMISSION MEASUREMENT PROCEDURES

With the implementation of the 0.6-g/mi particulate standard for the 1982 model year, and the more stringent 0.2 g/mi standard for the 1985 model year, federal emission measurement procedures will have important consequences for light-duty diesel certification. Although the Clean Air Act specifies that particulate standards apply to all motor vehicles, they only have consequences for diesels because the particulate emissions from gasoline-powered vehicles already fall well below the required limit for 1985 model year vehicles. If light-duty diesels are to meet the 1985 federal particulate limits, new emission control systems will have to be added. (Particulate control devices are discussed in Chapter 3 of this report.) Because of the importance of the particulate and hydrocarbon emission test procedures that the Environmental Protection Agency (EPA) intends to use for certification of 1982 model year light-duty diesels (EPA, 1979d, 1980a), they are reviewed and evaluated below.

Test Procedures

Federal regulations (EPA, 1980d) specify that particulate and regulated gaseous emissions must be measured simultaneously during the Federal Test Procedure (FTP) driving cycle.* Under this procedure, a dilution tunnel with a positive displacement pump-constant volume sampler (PDP-CVS) or a critical flow venturi-constant volume sampler (CFV-CVS) must be used to dilute the total exhaust stream. A schematic of the PDP-CVS system is provided in Figure 2.18. For hydrocarbon analysis, diesel vehicles require a heated (375° ± 20°F (191° ± 11°C) wall temperature) flame ionization detector (HFID) sample, which must be taken directly from the diluted exhaust through a heated probe in the dilution tunnel. The flow rate through the PDP-CVS must be sufficient to maintain the diluted exhaust stream (from which the particulate and gaseous hydrocarbon sample flow is taken) at a temperature of 125°F (52°C) or less. The regulation specifies:

> The transfer of heat from the vehicle exhaust gas shall be minimized between the point where it leaves the vehicle tailpipe(s) and the point where it enters the dilution tunnel airstream. To accomplish this, a short length (not more than 12 feet (365 cm)) of uninsulated, or not more than 20 feet (610 cm) of insulated Smooth Stainless Steel tubing from the tailpipe to the dilution tunnel is required. This tubing shall have a maximum inside diameter of 4.0 inches (10.2 cm). The dilution tunnel shall be at least 8.0 inches (20.3 cm) in diameter and sized to permit development of turbulent flow (Reynolds' No. >4000) and complete mixing of the exhaust and dilution air between the mixing orifice and each of the two sample probes (i.e., the particulate probe and the heated HC sample probe).

In addition, the regulation recommends "that uniform mixing be demonstrated by the user," although the procedure and limits for achieving uniform mixing are not specified. Furthermore, it states:

> The particulate sample probe shall have a minimum inside diameter of 0.5 inch (1.27 cm) and be installed facing upstream at a point where the dilution air and exhaust are well mixed (i.e., near the tunnel centerline approximately 10 tunnel diameters downstream from the point where the exhaust enters the dilution tunnel). The total hydrocarbon probe shall have a minimum inside diameter of 0.19 inch (0.457 cm) and be installed facing upstream at a point approximately 10 tunnel diameters downstream from the point where the exhaust enters the dilution tunnel.

---

*See Figure 1.6 for a further description of the FTP cycle.

Fluorocarbon-coated glass-fiber filters or fluorocarbon based (membrane) filters are required for particulate collection. The particulate filter must have a minimum diameter of 47 mm (37 mm stain area). Larger diameter filters are also acceptable and may be desirable for reducing the pressure drop across the filter when testing vehicles that produce large amounts of particulate. The recommended minimum loading on the 47-mm filter is 2 mg (equivalent mass loadings/ stain area is recommended for larger filters). The combination of the minimum loading and the 125°F maximum mixture temperature requirements places tight limits on the overall procedure in some labs when the ambient temperature is high or when larger vehicles with low particulate emissions are being tested.

During each phase of the cycle, dilute exhaust is simultaneously sampled by paired "primary" and "back-up" test filters. The back-up filter is located 3 to 4 inches downstream from the primary filter holder. The ratio of the primary filter weight to the primary plus back-up weights must be greater than 0.95 to use the primary filter only. If the weight is less than 0.95, both filters must be used to determine the particulate emissions.

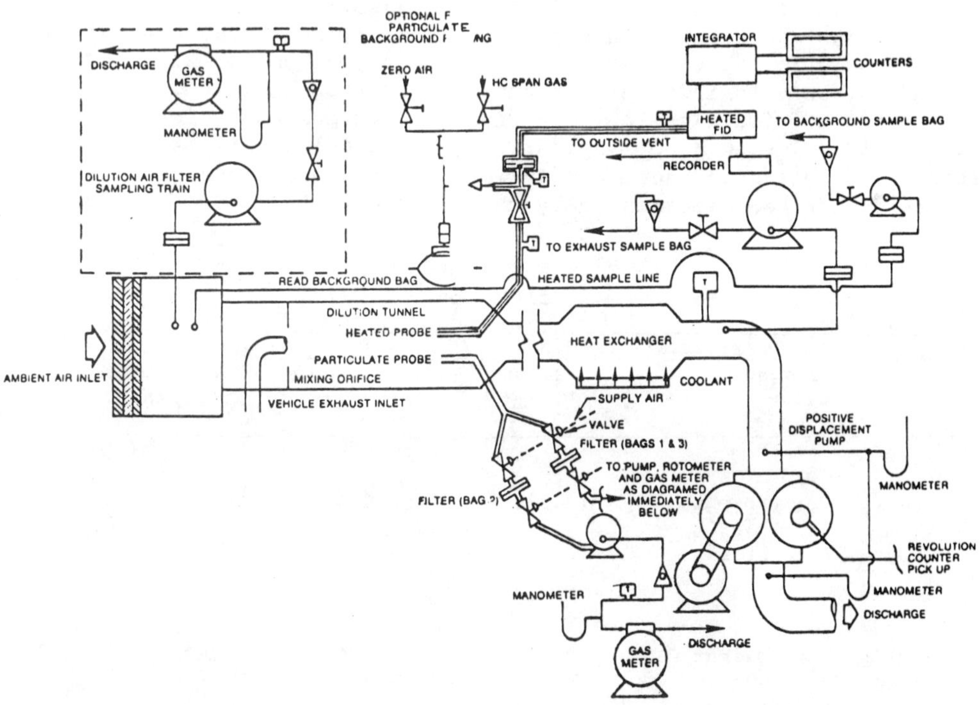

FIGURE 2.18 Gaseous and particulate emissions sampling system (PDP-CVS) (for diesel vehicles only). SOURCE: EPA 1980d.

Repeatability and Measurement Sensitivities

In general, the EPA certification procedure is a satisfactory method
for the mass characterization of particulates needed for the 1982 and
1985 model year standards. Below is a discussion of the state-of-
the-art in the measurements, including areas that could be improved in
the future.

Table 2.12 shows collection efficiency data reported by Toyota. The
data were obtained by using teflon-coated glass-fiber filters (Pallflex
TX 40HIZOWW) of the type required for certification testing. Per EPA
specifications, two filters in series were used to collect the particulate. The data from Toyota show that the 95 percent collection
efficiency is achievable and that for a 95 percent or higher collection
efficiency, only one filter is required. Data in the laboratory
generally show that particulate measurements are about as repeatable as
carbon monoxide and NO_x measurements and better than hydrocarbon
measurements.

Hydrocarbon Measurement

Current federal measurement procedures for gas-phase pollutants, except
hydrocarbons, are the same for diesel- and gasoline-powered vehicles.
Vehicle exhaust is diluted in a dilution tunnel, filtered, and collected
in sample bags for subsequent determination of carbon monoxide, NO_x,
and hydrocarbon emissions for the gasoline-powered vehicles, and carbon
monoxide and NO_x for diesels. For diesels, hydrocarbons are measured
continuously by using a hot (375° ± 20°F) particulate filter and a
heated flame ionization detector (HFID). The hydrocarbon concentration
data are integrated electronically to determine total hydrocarbons.

TABLE 2.12 Diesel Particulate Collection Efficiency

Driving Cycle	Particulate Loading, mg		Collection Efficiency, %
	Primary Filter	Back-up Filter	
		Test 1	
Cold transient	3.339	0.114	96.7
Stabilized	2.107	0.104	95.3
		Test 2	
Cold transient	2.927	0.086	97.1
Stablilized	1.863	0.097	95.1

Filter: teflon-coated glass fiber filter (Pallflex TX40HI20WW).

SOURCE: Toyota, 1980.

This method is more accurate for diesels because it does not allow adsorption of hydrocarbons onto the walls of the bags.

Bag and electronic integration methods were recently compared in a series of experiments that used gasoline-fueled vehicles. Six tests, using continuous electronic integration, yielded an averaged 73.5 ± 3.0 ppm carbon; bag integration yielded 72.4 ± 2.8 ppm carbon (Black and High, 1979).

Collecting particulate at the tunnel dilution temperature, and hydrocarbons at 375°F, measures the lighter portion of the hydrocarbons on the particulate filter and again in the HFID. The original purpose of using HFIDs on diesel exhaust was to obtain an accurate and repeatable measurement of the hydrocarbons. Now, with particulate measurements, the hydrocarbons are being counted twice.

EPA, in its summary and analysis of this point, offered the following interpretation:

General Motors' comment emphasized that the temperature specifications are the problem. However, the real issue centers around EPA's intent to measure both total hydrocarbons and total suspended particulates. The temperature specifications are naturally the result of this intent, and not the problem.

The intent to measure both total hydrocarbon and total suspended particulate material results from the fact that each is regulated for a different reason. Hydrocarbon emissions standards are intended to reduce the atmospheric photo-oxidant smog level; standards for total suspended particulates are intended to reduce the respiratory health hazard associated with these fine materials. Therefore, the proposed measurement practice as specified in 86.110-81(b) should be continued until it can be conclusively proven that particle-bound hydrocarbons do not leave the particle while suspended in the atmosphere, and hence do not first constitute a respiratory health hazard and later contribute to the atmospheric photo-oxidant level.

EPA further states:

It must be emphasized that since the standards have been based on a baseline allowing some double-counting, and because the standards are technology-based, the double-counting does not affect the stringency of the standards. Halting the double-counting would necessitate a new baseline, a new technological review, and a lower set of standards. (EPA, 1979d)

It takes either high temperature or a powerful solvent to remove the adsorbed hydrocarbons bound to the solid chain aggregate particles. General Motors recently carried out a series of studies in which the weight of filters containing diesel particles was monitored for up to 7 months. These data from these studies are shown in Table 2.13. The average weight loss of the particulate on the filters was only 0.40

percent. The extractable portion of the particulate in the samples
ranged from 15 to 45 percent, which is representative of the wide range
of sampling and operating conditions used to generate them (Williams
and Chock, 1979). In an experiment where diesel fuel was added to a
clean filter, General Motors showed that it took 40 days to evaporate
94 percent of the fuel (Williams and Chock, 1979). These data confirm
that the vapor pressure of hydrocarbons is dramatically reduced when
they are adsorbed on carbon (Pupp et al., 1974; Commins, 1962).

Even if the hydrocarbons adsorbed onto the particles were removed
in the atmosphere and took part in the smog reactions, there would be
no need to count them twice, because if they participated in the smog
reactions, they would not be available to contribute to the respiratory
health hazard associated with fine materials.

Some argue that the standards have been based on a baseline
allowing double-counting: This argument cannot really be applied to
hydrocarbons. The federal 0.41-g/mi hydrocarbon standard was promul-
gated on the basis of data from gasoline-powered vehicles. It would
seem, therefore, that a test procedure for hydrocarbons from diesels
should be developed to give numbers equivalent to those for
gasoline-powered vehicles.

An equivalent procedure for diesels would measure the gaseous
hydrocarbons behind the first particulate filter or use an identical
particulate filter at the tunnel dilution temperature in the HFID
sampling line before heating. This would in effect use the filter

TABLE 2.13 Evaporation of Hydrocarbons From Diesel Particulate Aged
on Filters (Room Temperature Storage)

Days After Collection	Particulate Mass on Filter, mg	Percent of Particulate Mass Evaporated	Percent of Particulate Mass Extractable[a]
80	17.7	0.29	28
80	19.5	2.05	37
80	18.6	0.84	36
80	39.2	0.02	16
80	32.4	0.19	17
190	52.8	-0.10	15
190	38.7	-0.07	17
190	37.4	-0.05	22
190	42.6	-0.41	18
190	29.1	-0.97	41
210	16.8	0.35	29
210	17.5	0.10	30
210	34.8	0.80	42
210	34.5	0.58	45
Avg. percent evaporated		0.40	

[a]Determined 24 hours after sample collection on paired filter.

SOURCE: Williams and Chock, 1979.

media to determine which hydrocarbons are gaseous and which are associated with particulates. It should also improve hydrocarbon repeatability because the gaseous hydrocarbons would be the lighter components, and they could easily be kept in the gaseous phase with heated lines. At a temperature of 375°F, the higher molecular weight fraction of the hydrocarbons is being adsorbed and desorbed from the lines. Looking at the gaseous hydrocarbons at that temperature would probably, therefore, contribute to the variability in the hydrocarbon measurement.

Table 2.14 shows some recent General Motors durability-vehicle emission data (GM, 1980). The hydrocarbons were measured using the standard EPA hot filter and 375°F line, following a particulate filter at the dilution tunnel temperature. As the data show, the hydrocarbon level is approximately 40 percent higher when the EPA hydrocarbon measurement procedure is used than when the measurement is made from behind the particulate filter at the dilution tunnel temperature. It would seem that the General Motors approach to hydrocarbon measurements would be correct for comparing diesel and gasoline vehicles and that diesel vehicles should be given credit for the fact that evaporative emissions are extremely low. California, for instance, allows a 0.16 g/mi hydrocarbon credit to diesels because of this effect. These suggested changes in hydrocarbon measurement procedures are technically sound and would allow industry to concentrate on particulate control to the maximum extent possible. As will be discussed in Chapter 3 of this report, the change would allow better optimization of engines for both particulate and NO_x control.

Mixture Temperature Control

In the certification procedures EPA has specified that a temperature of 125°F should not be exceeded in the dilution tunnel sampling zone. A simple maximum temperature limitation does not in itself provide mixture temperature control throughout the complete FTP driving cycle for different-sized cars operating at various emission levels. It would be more appropriate to integrate the mixture temperature throughout the complete cycle, holding to an average of 90° ± 10°F, for instance. This would allow excursions above and below a mean temperature, but it would ensure that different-sized vehicles follow a similar mixture temperature and dilution ration profile throughout the cycle and would thus provide more comparable data.

Furthermore, this approach would put less weight on brief temperature excursions. The mean temperature should be set low enough that no excursions above 140°F will occur; at temperatures above 140°F, desorption of hydrocarbons from the filter and nitrogen dioxide/ hydrocarbon reactions are possible. In addition to the mean temperature, a temperature maximum of 140°F, for instance, might also be set to prevent higher peak temperatures during accelerations. The mean and maximum temperatures could be set lower if higher average dilution ratios were necessary because of nitrogen dioxide, temperature, or other dilution ratio effects on the chemical or biological character of the particulates.

TABLE 2.14 General Motor 1980, 5.7-Liter Durability-Vehicle Emission Data (in g/mi)

	Car No.						
	1	2	1902	1360	02218	0439	04351
Hydrocarbon (measured after filter at dilution tunnel temperature)	0.187	0.195	0.192	0.184	0.202	0.161	0.196
Particulate	0.406	0.509	--	--	--	--	--
Hydrocarbon (after hot filter) (EPA test procedure)	0.263	0.289	0.287	0.262	0.255	0.208	0.228
Number of measurements	--	--	9	11	3	4	4
Percent increase of hydrocarbons	41	48	49	42	26	29	16

SOURCE: GM, 1980.

FTP mass particulate data show little effect of dilution ratio and mixture temperature, although carefully controlled experiments generally produce slight reductions of total particulate at increased temperatures and dilution ratios, due to decreases in the soluble organic fraction of the particuate emissions. Appendix E shows these data.

Test Cycle Effects

The EPA certification procedure uses an urban driving cycle (the FTP). Emissions of both regulated and unregulated pollutants are a function of the driving cycle of the vehicle. Table 2.15 shows emission data for a Mercedes 240D for four different driving cycles and two steady state conditions. The data show that the urban New York cycle, which is more stop-and-go oriented than the FTP, results in higher particulate, organic, aldehyde, and benz(a)pyrene emission levels than the other cycles. It is at these low-speed, light-load conditions that the diesel has the greatest fuel economy advantage over the gasoline engine.

ODOR AND IRRITANTS

Vehicular exhaust odors and irritants are generally considered to be nuisance pollutants. They can, however, affect human well-being by eliciting unpleasant sensations, by triggering harmful reflexes or other physiological reactions, and by modifying olfactory functions. Unfavorable responses are said to include nausea, headache, and coughing; upsetting of sleep, digestion, and appetite; irritation of eyes, nose, and throat; and destruction of the sense of well-being and

TABLE 2.15 Effect of Driving Cycle on Emissions[a]

| | g/mi | | | | mg/mi | | mg/mi | |
Driving Cycle	HC	CO	NO_x	Parti-culates	Organ-ics	Sul-fates	Alde-hydes	BaP
FTP	0.19	0.92	1.26	0.53	58.3	13.2	27.1	0.62
FET[b]	0.10	0.56	1.09	0.34	21.8	20.8	13.5	0.12
Sulfate test	0.14	0.63	1.35	0.42	32.3	16.1	13.3	0.12
New York City	0.43	1.79	1.88	1.09	124	20.9	130	3.06
31 mph	0.13	0.43	0.76	0.28	28.6	6.4	11.6	0.23
53 mph	0.13	0.58	1.35	0.32	24.6	15.7	7.9	0.13

Vehicle: Mercedes 240D.

[a] No. 2 emission test fuel.
[b] Fuel economy test.

SOURCE: Hare and Baines, 1979.

enjoyment of food, home, and external environment (e.g., National Research Council, 1979; Moulton, 1980; Cain and Garcia-Medina, 1980). Odor and irritants must, therefore, be considered in any treatment of emissions from vehicle power plants.

This section deals with the measurement and characterization of light-duty vehicle odor and irritant emissions and with their environmental consequences. The engine control technologies being introduced to limit other exhaust pollutants are discussed here only as they relate to the problems of odor and irritation from light-duty diesels. Possible controls for the latter are discussed in Chapter 3.

Structure and Function of the Sensory System

The National Research Council has recently published a detailed discussion of the olfactory system, including its anatomy and physiology (National Research Council, 1979). That work is highlighted here in order to clarify the distinction between the separate, yet related, responses to odors and irritants.

What is commonly called the "sense of smell" is a function of two different organs in the nose. One of these organs, the olfactory epithelium, is a yellow pigmented area of a few square centimeters located in the uppermost part of the nose, remote from the main respiratory airstream. This area contains millions of bipolar receptor cells that connect directly to the olfactory bulbs of the brain. During quiet breathing, only 3 percent of the odorous molecules that enter the nose reach and contact this area. The other organ consists of the free endings of the trigeminal nerve distributed throughout the nasal cavity. Odorants cause irritation, tickling, or burning by stimulating the trigeminal receptors. In many practical situations involving smell, the distinction between the two senses is overlooked; they are not always easy to separate, because most odorants stimulate both systems. However, the two organs are connected to different regions of the brain, and their effects are different.

The human olfactory system can discriminate among many thousands of different substances and can detect many of them in extremely low concentrations. Odors convey information about their sources and elicit a wide variety of emotional and physical effects. The human memory for odors is retained over long periods--often over a lifetime.

Most odorous matter discharged into the atmosphere consists of a complex mixture of many components. Human sensory responses to the individual components of such mixtures vary over wide ranges, from component to component and, to some extent, from person to person. Many atmospheric contaminants are odorless, or very nearly so; others are readily detectable in minute concentrations. Unfortunately, the sensory response to complex mixtures usually cannot be predicted on the basis of responses to individual compounds because of masking, counteraction, and synergistic effects.

The magnitude of the human sensory response to odor (the perceived odor intensity) decreases as the concentration of the odorant decreases. However, perceived odor intensity and odorant concentration are by no means directly proportional. Unfortunately, when odorous air is diluted with odor-free air, the perceived odor decreases less sharply than the concentration; for example, a 10-fold reduction in the concentration of amyl butyrate in air is needed to reduce its perceived odor intensity by half. Furthermore, odorants do not all respond in the same way. In general, however, the typical logarithmic organoleptic response behavior is followed. This also applies to the response to irritants.

Hedonic judgment (pleasantness versus unpleasantness) and odor quality (character) also can be used to quantify odors, even though the physiology and psychology of these effects are not clear at this time.

Odor Measurement

Both sensory and analytical methods have been applied to exhaust odor measurements. Sensory methods use simple intensity category scales, a reference-sample scale, or signal detection theory; all require the use of human subjects as judges. In the United States, concern over possible adverse health effects from the inhalation of diesel exhaust has halted all such studies involving direct human exposure. The analytic system in use for diesel odor evaluations measures the exhaust concentration of oxygenated species. These methods are described in detail elsewhere (Dravnieks, 1978; National Research Council, 1979; Levins, 1979; CRC, 1979) and are only summarized here.

Odor Intensity and Character

Exhaust is usually diluted with nonodorous air before sensory evaluations are carried out. The dilution is designed to replicate the odor experienced by a pedestrian at a curb close to the tail pipe. A dilution ratio of 100:1 is typically chosen for a diesel bus, and 700:1 for a truck with a vertical exhaust stack. A ratio of 100:1 is typically used for passenger cars.

During a typical diesel odor evaluation, the exhaust is diluted by a factor of 100 and piped to sniffing ports or funnels. A set of bottles containing 28 diesel odor reference chemicals are used (Turk, 1967). Four sets, with four bottles in each, are used to characterize the intensities of smoky-burnt, oily, pungent-acid, and aldehyde-aromatic notes, on a 1 to 4 scale. A set of 12 bottles, termed the D-scale, is used to characterize the overall diesel odor intensity.

In another evaluation, exhaust is diluted by a factor of 600 and supplied to a walk-in odor test room for evaluation. Odor intensity is evaluated on a total intensity of aroma (TIA) scale, which has seven ratings ranging from 0 (not detectable) to 3 (strong). The intensity of smoky-burnt and oily kerosene notes is rated separately on a similar scale. The presence or absence of eye, nose, and throat irritation is also recorded. It has generally been found that the smoky-burnt fraction is the primary contributor to the TIA.

The relationship between the D and TIA diesel odor intensity scales has been investigated in a limited study by the Coordinating Research

Council, Inc. (CRC, 1979). The resulting comparison is provided in
Figure 2.19. When there is a need to put odor measurements on a common
basis, this particular relation has been used.

It is important to remember that the intensity of diesel odor is an
exponential function of odorant concentration, as is the case for most
sensory responses--the odor intensity increases or decreases as a
logarithmic function of concentration. This dependence is shown in
Figure 2.20: A 10-fold reduction in odorant concentration only gives
rise to a reduction of one TIA unit in intensity.

<u>Hedonic Tone</u> The overall pleasantness or unpleasantness of engine
exhaust has been evaluated by using various hedonic scales. Springer
(1974a) used a five-position pictorial hedonic scale ranging from
'pleasant' to 'unbearable' for conducting public opinion surveys.
Ricardo & Co., Ltd. (Such, 1978) uses a four-point "nastiness index"
for rating overall odor impact, which also has been related to the TIA
scale (Hames <u>et al</u>., 1980). These techniques are valuable for odor
evaluations where trained panels are not available; however, they
remain coarse measurements.

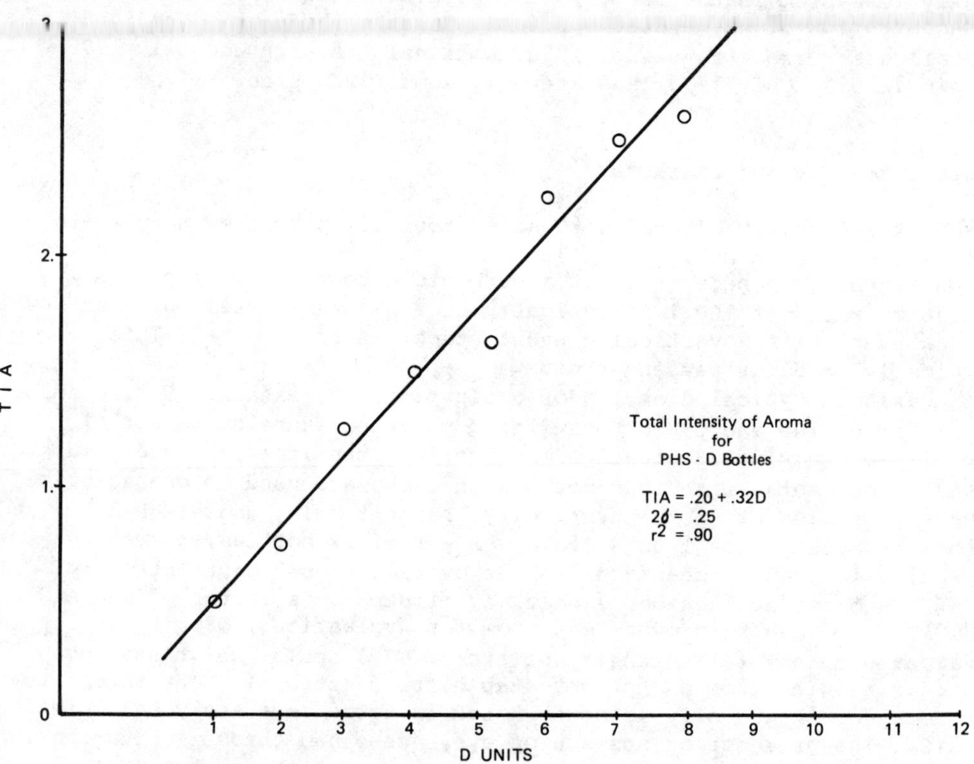

FIGURE 2.19   Comparison of D and TIA diesel odor intensity scales.
SOURCE: CRC, 1979.

<u>Odor Threshold</u> Another approach to assessing odor intensity is to determine the dilution ratio at which the odor reaches the threshold of perceptibility. If experimental conditions are controlled, if panelists are appropriately screened and selected, and if presentation protocols are strictly adhered to, reasonably reproducible values of dilution thresholds can be obtained. Even with these controls, different systems with different flow rates and sample presentation characteristics will result in different numerical values; however, calibrations relating one method to another can generally be developed (Dravnieks, 1978). Threshold dilution ratio measurements have been successfully related to other odor measurement methods (Daudel <u>et al</u>., 1979; Hsieh <u>et al</u>., 1979a).

Threshold dilution ratio measurements present the lowest concentrations of diesel exhaust to the panelists of any of the sensory measurement systems. Thus, if sensory measurements are required, this method may be preferred because it minimizes human exposure to the exhaust.

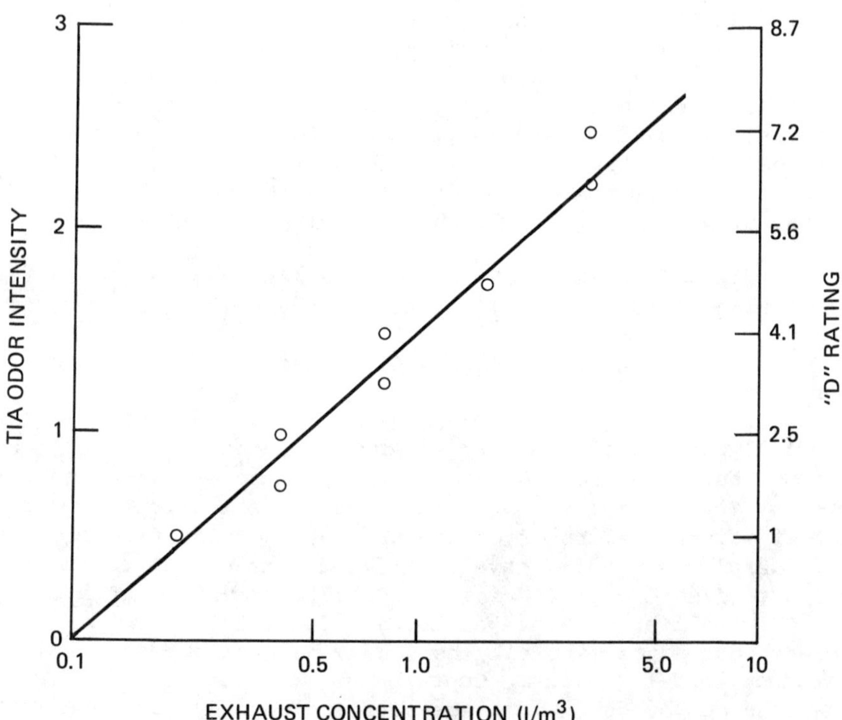

FIGURE 2.20   Dose-response dependence of diesel exhaust odor. Data represent the odor emitted by an engine operated at mid-speed and -load using No. 1 diesel fuel.
SOURCE: Levins, 1969-1971.

__Differential Olfactometry__ Degobert (1980) has recently applied MacLeod's (1972) olfactometric concept to the study of diesel odors. MacLeod's method is to compare an unknown odor (exhaust gas) simultaneously with a standard odor. Each odor is presented in one nostril of the same sensory evaluator. The principle of detection is that the two halves of the olfactory system exhibit reciprocal inhibition, so that odor is perceived only in the nostril presented with the more intense odor. By changing the concentration of the standard odor, the unknown odor can be bracketed and balanced. With respect to accuracy and repeatability, this test method is reportedly better than other sensory evaluations ($\pm$10 to 20 percent versus 100 to 200 percent). However, once again, concern about direct exposure to undiluted diesel exhaust will limit the use of this technique.

Analytical Measurements

To satisfy the need for objective measurement of diesel exhaust odors, an analytical method called the Diesel Odor Analysis System (DOAS) has been developed (Levins, 1973; Levins __et al.__, 1974). The method uses a group-analysis approach, based on the results of odor-chemical studies. The DOAS involves: collection of a sample by passing heated hot-filtered undiluted exhaust gas over a Chromosorb 102 adsorbent, elution of that sample with cyclohexane to yield the total organic extract (TOE), and analysis of the eluate by using silica gel liquid chromatography with ultraviolet absorption detection. The chromatograph separates the eluate, or TOE, into three primary fractions:

- a paraffinic fraction (LCP) containing little or no odorants;
- an aromatic fraction (LCA) containing the oily-kerosene odor species; and
- a polar or oxygenate fraction (LCO), mostly containing the smoky-burnt odor species.

DOAS essentially only measures species above a carbon number of 5 and below a carbon number of 14. Species with lower molecular weights are not trapped by the Chromosorb 102, and those above a carbon number of 14 are not efficiently eluted from the trap by the specified procedures.

Chemical analysis studies have identified some of the compounds contributing to diesel odor (Levins, 1969–1971; Dravnieks __et al.__, 1969–1970; O'Donnell and Dravnieks, 1970). Contributors to oily-kerosene odor notes were alkylbenzenes, naphthalene derivatives, indans, indenes, and tetralins. Contributors to smoky-burnt odor notes were oxygenated compounds. Smokiness carriers were hydroxy and methoxy indanones. Methylphenols and methoxyphenols also assisted in this note. Burnt character was associated with furans and alkylbenzaldehydes. Aliphatic aldehydes up to octanal were found, probably contributing some aldehydic character. Aliphatic acids up to nonanoic were found, and one foul-odor species containing sulfur, trimethylthiophene, was found also.

It was estimated that some 200 distinct chemicals may contribute to the oily-kerosene odor and 2,000 to the smoky-burnt odor. Thus it is important to measure the odorants as a group, as is done in the DOAS, to reflect odor and character accurately (Levins, 1969, 1971).

The sensory studies showed that the smoky-burnt fraction is the primary contributor to the total diesel exhaust odor intensity; it is, therefore, expected that the LCO should correlate with the TIA scale. As shown in Figure 2.21, the correlation found by Levins and co-workers (1974) was strong. The correlation represented by these data is:

$$TIA = 1.0 + 1.0 \ \log_{10} LCO \ (mg/m^3)$$

where r^2 is 0.996, and 2σ is 0.32. The ±0.32 TIA 95 percent confidence limits are better than normally observed (0.4) in odor observation. A number of studies have shown that the DOAS instrument

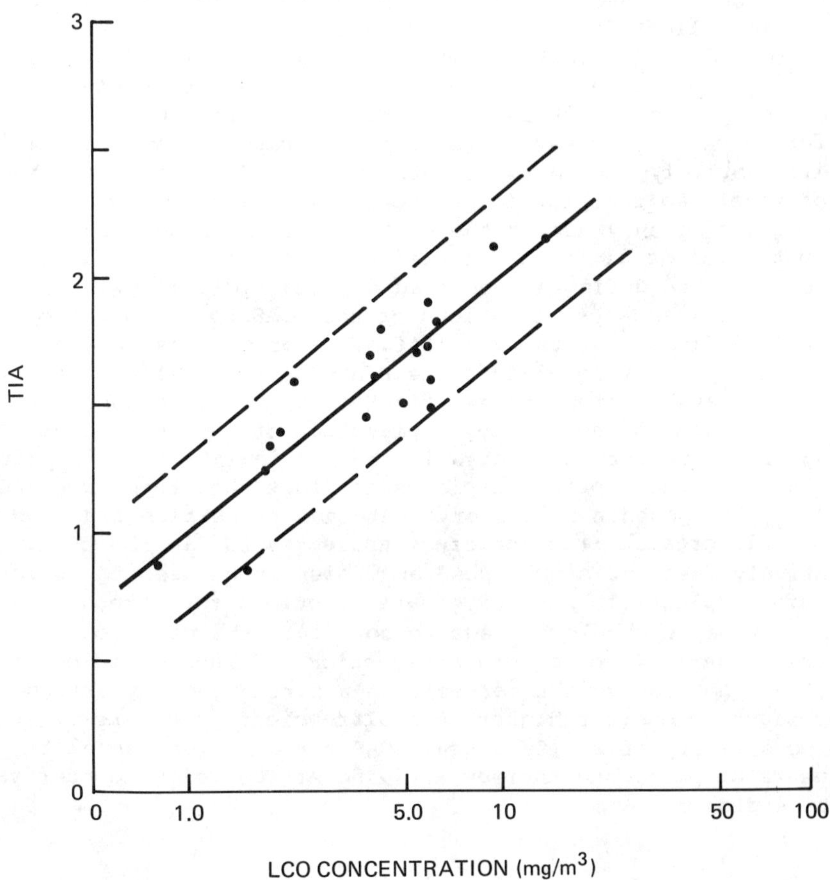

FIGURE 2.21　Summary correlation of exhaust odor intensity and LCO.　SOURCE: Levins et al., 1974.

gives a good correlation with odor intensity, particularly for light-duty diesels (Springer, 1974b; Levins, 1973; Daudel et al., 1979; Such, 1978); other studies show a more scattered correlation, particularly for heavy-duty vehicles (Springer, 1977a, 1979).

Studies conducted by CRC showed that it is possible to obtain precise measurements of diesel exhaust odor intensity by using the DOAS (CRC, 1979; Hames et al., 1980). The smoky-burnt oxygenate group (LCO) measurement gave an interlaboratory precision of about 10 percent, with a TIA precision of 0.15 units. However, these studies and others conducted at Drexel University (Cernansky et al., 1978), SGRD Ltd. (Reading and Greeves, 1980), and elsewhere have indicated that the details of the sampling and collection conditions need to be more carefully defined if problems with odor compound breakthrough and recovery are to be avoided and better reproducibility in measurements is to be achieved.

The main advantage of the DOAS system is that it is relatively simple and does not depend on human judgment or human exposure to exhaust pollutants. It can provide reliable odor measurements during steady state operating conditions; however, the sampling requirements (5 minutes to collect 25 liters of exhaust) are not well suited for measurements under transient conditions. Nonetheless, DOAS odor measurements of acceleration and deceleration transients can be made if sufficient sample volume is collected rapidly during the transient and stored for subsequent passage through the sample trap at the specified rate. Alternatively, the measurements can be made if the transient is run repetitively so that sufficient volume is collected, with odor sampling occurring only during the portions of interest. The DOAS also has the potential of giving a weighted average odor sample if the sample is collected during a specified driving cycle (e.g., the FTP cycle). However, these applications of the DOAS to transient operations are not well defined at this time. Also, major changes in fuels and diesel engine type may invalidate the DOAS-TIA correlation and thereby necessitate a revalidation and verification by using sensory methods.

An interesting and noteworthy application of the DOAS is as a screening procedure for oxygenates in the particulate organic extract (Perez, 1979). The approach exploits the fact that the low-molecular-weight ($<C_{14}$) components of the oxygenate and transition fractions of the soluble organic fraction are measured as LCO in the DOAS. Thus, the relatively fast and simple DOAS analysis can be used to isolate particulate samples with high oxygenate fractions for more detailed analysis. Attempts should be made to correlate the measured LCO with the soluble organic fraction characteristics (including biological activity) and to improve the correlation and relevance by extending the DOAS technique to monitor higher-molecular-weight oxygenates.

The possibility of modifying the DOAS for dilution tunnel measurements of odor, and thereby arriving at FTP weighted odor values, should be explored.

Irritant Measurements

There are no reported studies on the chemical composition of the
species giving rise to the irritation associated with diesel exhaust.
Thus, there is no measurement available to quantify irritancy. However,
as the portion of diesels in the vehicle fleet grows, it may be the
irritation associated with diesel exhaust that becomes more important
than odor, in terms of public acceptance and potential health effects.
Although the instantaneous effects of diesel odor disappear when the
source is removed, the irritation from diesel exhaust can linger.
Thus, a program to measure systematically the sensory irritation of
diesel exhaust and to find the chemicals primarily responsible for it
is indicated. Such a program would indicate which species or classes
of species must be measured to quantify irritation.

The species known to be present in diesel exhaust, and also known
to act as irritants, include: nitrogen dioxide, sulfuric acid,
formaldehyde, acrolein, and phenol. Emission levels of these compounds
were discussed in the section on gaseous emissions. The concentrations
of aldehydes in diesel exhaust are sometimes greater than the levels
where irritancy effects are known to occur. Levels above 0.1 ppm for
acrolein and 3 ppm for formaldehyde have been observed (Levins, 1979).
Thus, it is expected that aldehydes will be pinpointed as the primary
contributors to exhaust irritancy. The fact that they are not measured
by the DOAS may contribute to the variability in the DOAS-sensory
impact correlations.

Uncontrolled Vehicle Emissions

Odor Characteristics of Engine/Vehicle Systems

There are two sources of odors in diesel engines (National Research
Council, 1979): unburned fuel and its thermal breakdown products,
principally hydrocarbons and some nitrogen compounds; and products of
incomplete combustion, which result from wall or bulk gas-quenching.
Incomplete combustion results in partial oxidation of hydrocarbons.
The first source is associated with the fuel-rich spray core and later
phases of injection. The second source is associated with preignition
mixing during early injection phases (Reading and Greeves, 1980).
The work of Barnes (1968), Reading and Greeves (1980), Hsieh and
co-workers (1979a,b), and Cernansky and co-workers (1980) established
that high concentrations of aromatics (LCA) and oxygenates (LCO) are
formed in preignition reaction zones and are subsequently consumed in
high-temperature flame zones. Thus, production of odorants is inherent
in the diesel combustion process, but the ultimate exhaust levels
depend on the dynamics of the combustion and mixing processes.
Odorants can survive if they are transported to combustion chamber
regions too lean or too cool to permit complete combustion.
The intensity of odors emitted by diesel engines operated at steady
conditions has been generally thought to be greatest at idle and
full-load conditions. Some researchers believe that engine speed is

not an important parameter (Rounds and Pearsall, 1957; Monaghan et al., 1974; Reading and Greeves, 1980); however, engine studies by Ingram (1978) and passenger car studies by Springer and Stahman (1975) and Springer (1979) showed opposite trends and thus indicated that odor may not be a predictable function of power. It was thought that the high odor levels at no load, which can be accentuated at high speeds by misfire, could be minimized by advancing the injection timing at light loads to prevent misfire conditions. Related to this, Reading and Greeves (1980) found lower odor with decreases in ignition delay. However, the study of Springer and Baines (1977) tested production engines and found that no consistent effect resulted from variable injection timing or from changing the injector design.

In another study of Ricardo & Co., Ltd., Lesley and French (1976) reported that odor increased with engine speed but decreased with increasing load in a four-cylinder direct injection engine. Other engine variables had lesser effects. Marshall and Seizinger (1978) found odor level increases at either high or low loads for mid- and high-speed ranges, and both the Lesley study and the Marshall study found that an oxidation catalyst in the exhaust reduced odor levels. Reading and Greeves (1980) found no consistent speed and load trends with two direct injection and two indirect injection engines; and Ingram (1978), Petrow and co-workers (1978), and Petrow and Oxynandry (1979) found little effect on odor intensity except during extreme operating conditions when tests using small direct and indirect injection engines were conducted.

These results indicate that each system has characteristic odor emission levels. Changes in operating variables (speed, load, injection angle, etc.) over normal operating ranges do not significantly affect odor emissions. However, under extreme conditions, where the combustion process is significantly affected, odor also is affected. In general, direct injection engines are more odorous than indirect injection engines. However, there is considerable latitude in the reported characteristic emission levels, with the best direct injection engines having lower odor emissions than the worst indirect injection engines.

The only other engine factor with a substantial effect on odor levels at steady state operation is the uncontrolled fuel volume remaining in the injector after each injection (Ford et al., 1970; Ingram, 1978; Such, 1978; Reading and Greeves, 1980). This sac volume fuel, because it is poorly atomized and mixed with air, increases the partially burned hydrocarbons. Both odor and hydrocarbons were reduced somewhat when this uncontrolled fuel volume was decreased.

Fuel composition also affects odor intensity. Ford and co-workers (1970) found reduced odor when the 10 percent distillation temperature was increased. Hills and Schleyerbach (1977) found similar results with increases in the 5 and 95 percent distillation points, but Hardenberg (1972) reported exactly the opposite effect. Hills and Schleyerbach also found increased odor when the fuel aromatic content increased, and Such (1978) found lower odor using a fuel lower in boiling range, aromatics, and cetane number. Ingram (1978) found that improving ignition quality reduced odor intensity. Hsieh and co-workers

(1980) found only small effects. The results of a recent study of fuel composition effects (Hare, 1979) show odor increasing with use of a minimum-quality fuel (see Tables 2.16 and 2.17). The properties of the fuels used in this study are shown in Table 2.24. The problems with the apparently contradictory data can be appreciated if the close interrelationship and dependencies of fuel properties are recognized.

Transient operation may result in the highest odor intensities. Springer and Statiman (1975), Springer (1979), Hare (1979), and Such (1978) report much higher odor levels for acceleration and deceleration transients than for steady state conditions. This behavior is also shown in Tables 2.16 and 2.17. Public surveys have found that at a D-2 level, objections to odor are significant and that at a D-3 level, 77 percent of the people surveyed found the odor objectionable (Hare and Springer, 1971; Hare et al., 1974). Thus, streetside odor emissions could become a serious problem.

Volkswagen has measured the odor from four of their research diesel-powered vehicles (Wiedemann and Hofbauer, 1978; Wiedemann and Schmidt, 1979a,b). Figure 2.22 compares the odor from these automobiles. No difference between naturally aspirated and turbocharged engines was noted; however, when exhaust gas recirculation (EGR) was introduced, odor intensity increased substantially. An influence of fuel on odor also was noted: U.S. No. 2 diesel fuel produced more intense odor than did European fuel. Volkswagen did not note the differences in the properties of the two fuels, but European fuels

TABLE 2.16 Summary of Odor and Corresponding Emission Data, Mercedes 240D

Fuel	Average Data Values by Engine rpm/Load						
	1,800/2	1,800/50	1,800/100	3,000/2	3,000/50	3,000/100	Idle
			Odor Panel "D" Rating				
EM-238-F	2.0	1.9	2.4	2.3	2.0	2.8	2.0
EM-239-F	2.2	2.0	2.6	2.0	2.2	2.9	2.3
EM-240-F	2.1	1.9	2.0	2.2	1.8	2.6	2.2
EM-241-F	2.2	1.8	2.2	2.2	2.0	2.6	2.2
EM-242-F	2.0	1.9	2.4	2.0	2.3	2.7	2.2
			DOAS-TIA Rating				
EM-238-F	1.5	1.5	1.5	1.4	1.4	1.6	1.4
EM-239-F	1.4	1.4	1.6	1.5	1.6	1.8	1.6
EM-240-F	1.5	1.3	1.3	1.2	1.4	1.6	1.0
EM-241-F	1.2	1.2	1.4	1.2	1.3	1.6	1.0
EM-242-F	1.2	1.4	1.5	1.2	1.3	1.6	1.1

Fuel	Odor panel "D" Rating by Transient Operating Condition			
	Idle-Acceleration	Acceleration	Deceleration	Cold Start
EM-238-F	2.8	2.5	4.2	4.0
EM-239-F	3.0	2.9	2.8	2.8
EM-240-F	2.6	2.4	4.4	2.8
EM-241-F	2.8	2.6	4.4	4.0
EM-242-F	2.5	2.5	4.0	4.1

SOURCE: Hare, 1979.

TABLE 2.17 Summary of Odor and Corresponding Emission Data, Volkswagen Rabbit

Fuel	Average Data Values by Engine rpm/Load						
	1,800/2	1,800/50	1,800/100	3,000/2	3,000/50	3,000/100	Idle
			Odor Panel "D" Rating				
EM-238-F	2.7	2.5	3.0	2.4	2.7	3.0	2.9
EM-239-F	2.6	2.4	3.4	2.2	2.6	3.1	3.4
EM-240-F	2.6	2.6	3.1	3.1	3.0	3.0	3.0
EM-241-F	2.8	2.5	3.5	2.8	3.2	3.3	3.4
EM-242-F	2.6	2.6	2.8	2.2	2.8	2.9	2.8
			DOAS-TIA Rating				
EM-238-F	1.8	2.0	2.2	1.5	2.1	2.1	1.7
EM-239-F	1.7	1.9	2.2	1.5	2.1	2.0	1.8
EM-240-F	1.6	1.8	2.0	1.7	2.0	2.0	1.9
EM-241-F	2.2	2.2	2.4	1.9	2.4	2.3	2.2
EM-242-F	1.5	1.6	2.0	1.4	1.9	1.9	1.4

Fuel	Odor panel "D" Rating by Transient Operating Condition			
	Idle-Acceleration	Acceleration	Deceleration	Cold Start
EM-238-F	3.6	3.7	3.0	4.1
EM-239-F	3.2	3.6	2.8	5.0
EM-240-F	3.3	3.3	3.3	4.0
EM-241-F	3.9	4.4	3.7	5.4
EM-242-F	3.4	3.8	3.0	4.8

SOURCE: Hare, 1979.

typically have higher cetane numbers than U.S. fuels (55 versus 46).
With its lower cetane rating, the U.S. fuel would lead to a greater
ignition delay and to higher hydrocarbon and odor emissions.

Irritant Emissions

As was indicated previously, irritancy levels are not measured directly,
because no quantitative measurement is available. However, information
is available about some of the known and suspected contributing compounds, including nitrogen dioxide, sulfuric acid, aldehydes, phenols,
and oxygenates. The sources and emission characteristics of these
compounds were discussed in the section on gaseous emissions. In
general, diesel engines have higher emissions of these compounds than
gasoline-powered engines.
The impact of these factors on any given community or driving
situation is unclear. Unlike odor intensity, which is a short-term
transient response, irritation from exhaust gas can linger for a
considerable period of time. Based on the anticipated significance of
odor emissions, diesel exhaust gas irritancy might be expected to
become a significant problem as well.

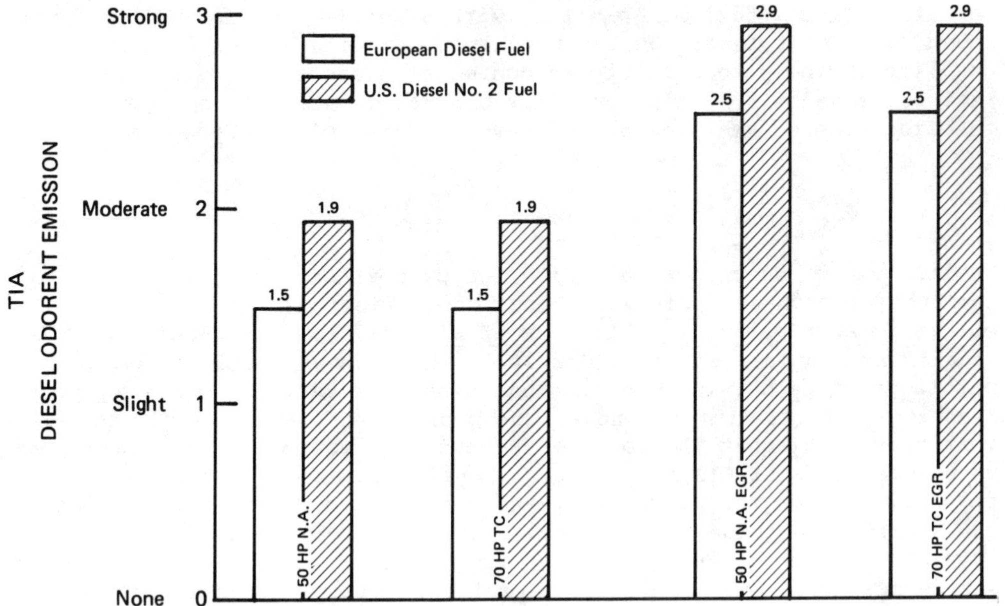

FIGURE 2.22 Diesel odor emission averaged over a city driving cycle influenced by fuel composition and EGR. Odor values may vary by +0.25 units. SOURCE: Wiedemann and Hafbauer, 1978.

Gasoline Engine Comparisons

As in diesel engines, the cause of exhaust odors from gasoline engines is incomplete combustion of fuel (Gruden, 1972). Without catalytic converters, partially burned hydrocarbons are more highly concentrated in gasoline exhaust than in diesel exhaust.

Catalytic converters promote the oxidation of unburned hydrocarbons and partial-combustion products. However, when the temperature is too low and the gas flow too rapid, the catalysts themselves may promote the formation of additional partial-oxidation products. Under the reducing conditions that may exist during deceleration, hydrogen sulfide and probably hydrogen cyanide may be generated in catalytic converters. Both gases have highly unpleasant odors. Thus, odors may be created both in the engine and in the exhaust system.

The DOAS method has not been calibrated for, and, therefore, is not applicable to, quantifying gasoline engine exhaust odors. Sensory measurements using approaches similar to those for diesel exhaust odors must be employed, even though odor intensities and odor notes are different.

In a comparison between several diesel-powered and gasoline-powered passenger cars, the odor intensities of exhausts (at a 100-fold dilution) were 1.1 to 3.0 on the D-scale for the gasoline engine, averaged over several operational conditions; for the diesels they were 3.2 to 5.3 (Springer, 1979). One engine, the Texaco TCCS, could be operated on either diesel fuel or gasoline; with gasoline, the intensities were about 30 percent lower (on the D-scale) than with diesel fuel. A gasoline engine with a catalytic converter produced odor intensity ratings close to 1. Thus, properly operating catalyst- equipped gasoline-powered vehicles do not have an odor emission problem.

Environmental Impact

Public opinion surveys to evaluate the perceptions of, effects of, and community response to diesel exhaust odor emissions have been carried out (Hare and Springer, 1971; Hare et al., 1974). The results of these studies are summarized in Figure 2.23, which shows the fraction of participants perceiving diesel exhaust odor at various levels and the fraction of participants finding the odor objectionable to the point that they think that the odor is "so bad that someone should take steps to reduce it." At D-2 and D-3 odor levels, 56 and 77 percent, respectively, of the people surveyed found the odor objectionable.

Considering these survey results and the typical odor emission levels reported earlier (D ratings ranging from 2 to 3 for steady state engine operation and from 3 to 5 for transient operating conditions are shown in Tables 2.16 and 2.17), diesel odor emissions could become a

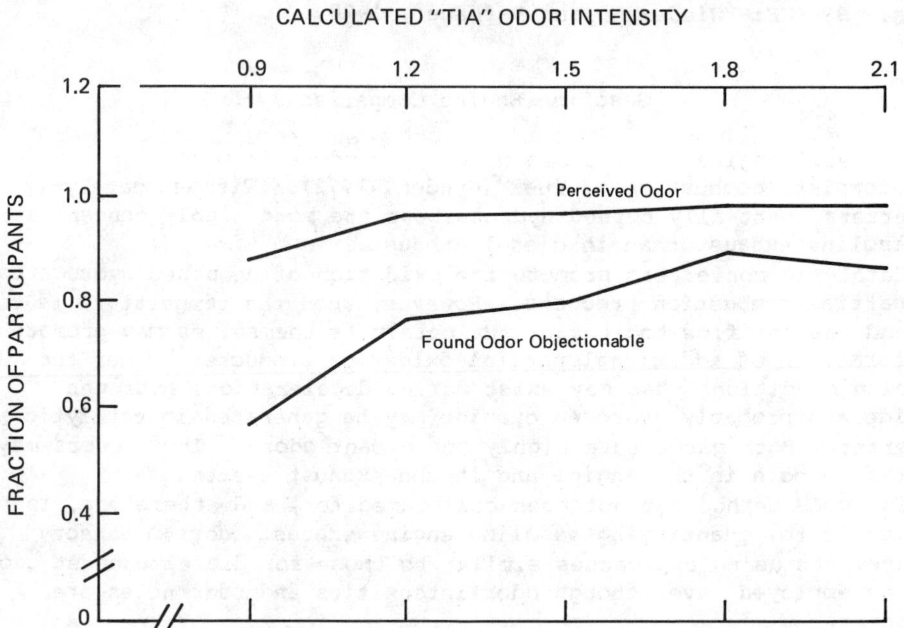

CALCULATED "TIA" ODOR INTENSITY

NOMINAL "D" RATING OF ODOR PRESENTED

FIGURE 2.23 Fractions of participants who perceived or objected to odors. SOURCE: Springer, 1974a.

serious problem when large numbers of diesels are operating. This is true not only for pedestrians at streetside but also for occupants of vehicles following or passing diesel-powered vehicles. Today, because there are few diesel-powered passenger cars on the road, there are few encounters in which exhaust fumes cause a nuisance; thus public complaints are limited. However, as the fraction of diesel-powered vehicles in the fleet increases, the number of such occurrences will also increase, and public complaints will probably increase proportionately.

A recent study examining the impacts of light-duty-vehicle dieselization included an analysis of potential increased time-averaged ambient odor levels (Forrest et al., 1979, 1980). Modal steady state odor emission factors were applied to projected modal traffic conditions, weighted according to the mix of diesel vehicles, and corrected to the time-averaged dilution ratio observed between a receptor and the vehicle emission source. Instantaneous or transitory odor concentrations at receptor locations were not considered. Diesel penetration in the light- and heavy-duty-truck categories was considered, but no ambient contributions to odor levels were assumed. Calculations were made for urban freeway, urban street canyon, and enclosed parking garage situations for levels of light-duty-vehicle dieselization up to 25 percent. In each case, Forrest and co-workers concluded that no detectable effects on ambient odor levels would be found. However, recognizing that odor levels may be strongly influenced by short-duration localized concentrations and that in any random event a high fraction of diesel vehicles may occur, the case for a 100 percent diesel population was investigated as well. This investigation yielded time-averaged ambient D numbers of 1.5 for the urban freeway, 2.6 for the street canyon, and 1.8 and 4.0 for off-peak activity and overload conditions, respectively, in the enclosed parking garage.

These results indicate that localized time-averaged high-level odor concentrations that will generally be regarded as unpleasant or worse by a large segment of the public are likely to occur. The calculations also indicate that with a nominal 25 percent diesel population, time-averaged ambient diesel odor will not be a significant problem. Although Forrest and co-workers concluded that a diesel odor problem is not likely to develop and that odor should not be regarded as a constraint on dieselization, these conclusions are arguable. The worst case calculation indicates potential problems with time-averaged ambient odor levels in both street canyon and parking garage situations. Consideration of the instantaneous odor levels at the receptor locations will exacerbate these problems. Also, transient operations and lower temperatures, where odor emissions are significantly higher, were not taken into account. As can be seen from the data in Tables 2.16 and 2.17, transient operations can increase odor levels at a 100:1 dilution ratio (typical streetside value) by 1 to 2 D numbers.

These factors suggest that diesel odor may become a serious problem when large numbers of diesels are operating. It is particularly important to realize that although average ambient odor concentrations may be low, it is the transitory high concentrations occurring during the exhaust plume mixing and transport process that elicit the greatest negative responses.

Summary and Conclusions

The physiological responses to odorants and irritants are separate, yet related. The magnitude or intensity of odor results from stimulation of the olfactory nerve, which transmits precise information with a relatively short reaction delay. Irritation results from stimulation of the trigeminal nerve, which transmits imprecise information. Its reaction delay is relatively long, and it leaves durable impressions concerning nose, throat, and eye irritation. Odorants and irritants, as organoleptic stimuli, follow a typical logarithmic response behavior. This means that a 10-fold reduction in concentration or level is needed to reduce the perceived intensity by half.

Both sensory and analytical methods have been applied to exhaust odor measurements. The sensory methods require the use of human subjects as judges and have been all but halted in the United States because of concern over possible adverse health effects. The analytical system in use for diesel odor evaluation lumps and measures the exhaust concentration of the 200 to 2,000 contributing oxygenated compounds and has shown good correlation with sensory methods for indirect injection light-duty diesels. The method should be developed to provide better precision, to allow for dilution tunnel and transient operation measurements, and to provide a screening procedure for particle-bound oxygenates.

Quantitative measurements of diesel exhaust irritancy are not available. A program to measure sensory irritation systematically and identify the chemicals primarily responsible is needed.

Production of odorants is inherent in the diesel combustion process. Odorants are formed by partial oxidation in preignition reaction zones and are subsequently consumed in high temperature flame zones. The ultimate exhaust levels depend on the dynamics of the combustion and mixing processes. Irritants probably behave similarly and are also formed directly in quench regions.

Gasoline engines typically have imperceptible-to-slight odor intensities. Diesel engines have slight-to-strong odor intensities. Each engine system has a characteristic odor emission level; this level does not have a predictable behavior, nor is it significantly affected over the normal operating ranges of the engine. In general, direct injection engines are more odorous than indirect injection engines. Transient and cold-start/low-temperature operation can result in significantly higher odor emission levels.

Widespread dieselization of light-duty vehicles (up to 25 percent) is expected to result in a significant diesel odor problem. Although overall ambient odor levels may not become significant, localized or transient high-level odor concentrations that will be regarded as unpleasant or worse by a large segment of the public are likely to occur. Because diesels have higher emission levels of several known irritants, diesel exhaust gas irritancy may be expected to become a significant problem as well. In contrast to odor intensity, which is a short duration transient response, irritation from exhaust gas can linger for a considerable period of time.

Control technologies that reduce regulated pollutants generally tend to exacerbate odor and irritants. The use of EGR for NO_x control has been found to increase odor intensity significantly. Combustion and injection modifications for particulate control also result in increased odor. Furthermore, catalysts and trapping systems can enhance odorants and irritants during low-temperature, high-flow, off-design operation. However, fuel modifications for particulate control (tighter specification limits on cetane number and composition) will reduce odor, but poorer quality fuels will have a negative effect. For all these reasons, the problems with diesel odor and exhaust irritation could become much worse in the future, as the number of diesel-powered vehicles increases, unless specific attention is given to their characterization and control.

NOISE EMISSIONS AND CONTROL

As with vehicular odor and irritants, noise from light-duty diesels is considered a nuisance pollutant. Although the noise emitted from a diesel-powered vehicle is usually well below the level at which permanent physiological damage occurs, it may affect human well-being and auditory functions. Unfavorable responses include: headache, upsetting of sleep, destruction of the sense of well-being, annoyance, fatigue, and depression (Hornig, 1968; Bragdon, 1970; Kryter, 1970; EPA, 1972; Bugliarello et al., 1976; Eldred, 1977; Harris, 1979). In addition, noise emissions may well affect customer acceptance and marketability of the diesel-powered passenger car (Cohen, 1980).

The anatomy and physiology, including perception and response, of the auditory system are well understood and have been extensively reported (e.g., Davis, 1957; EPA, 1972; Harris, 1979). No attempt is made here to summarize that body of knowledge and literature. Community impact and control technologies are considered, but this section primarily deals with the measurement and characterization of vehicular noise emissions.

Causes and Sources of Emissions

The sources, characteristics, and control of diesel engine and vehicle noise have been treated extensively in a number of recent reviews (e.g., Priede, 1980; Russell, 1979; Roessler et al., 1980) and in the technical literature (e.g., SAE, 1975, 1979; Morrison et al., 1980). Consequently, only a brief review is presented here.

Three major forms of noise are emitted by diesel engines: combustion noise, mechanical noise, and structural resonance noise. In light-duty indirect injection engines, the combustion process and mechanical parts are the predominant sources of noise. The magnitude of noise from each of these sources is roughly equivalent (Russell, 1973).

Combustion noise is generated by random fluctuations in the nonisotropic cylinder pressure field. Its magnitude depends upon the rate of heat release and pressure rise, and the peak pressure within the cylinder; these in turn depend upon: compression ratio, turbulence, air inlet temperature, fuel injection characteristics, cetane number, and fuel composition. The alteration of any of these parameters for the purpose of noise control generally results in small noise reductions, but they are limited by offsetting penalties of increased fuel consumption and exhaust emissions. Combustion noise is particularly pronounced during a long injection delay, such as occurs when idling or during a cold start, making "diesel knock" an attribute of the engines (Oetting and Papez, 1979).

Mechanical noise results from abrupt discontinuous forces acting on the working parts of the engine. Typically, the parts respond to changes in the applied transient inertial forces by accelerating across the clearances between moving parts and striking other engine components. The main source of mechanical noise is the rattle of piston components.

Structural noises are caused by vibrational resonances in engine parts that have been excited by combustion, mechanical, or other response noise sources. They are generally small and can be eliminated in a straightforward manner. The piston and connecting rod assembly, for example, vibrate longitudinally as a forced oscillator. The engine block, crankcase, and crankshaft respond with bending and torsional modes. These vibrations are then transmitted to the air by the oil pan and light structural covers, such as the rocker covers, the timing gear cover, and side panels.

Noise Measurements

Because human perception is subjective, an objective measure of sound levels is used. The sound pressure energy is measured with respect to a specific reference. Ten times the common logarithm of this ratio is defined as the sound pressure level (SPL) in decibels (dB). The reference level is frequently chosen as the faintest sound that the average person can hear--the so-called "threshold of hearing." The international standard for this level is 20 µPa (Harris, 1979). When considering separate noise sources, the sound energy levels are added to yield the overall SPL, not the decibels. Thus, two sources, each of which emits 90 dB, together emit 93 dB.

In addition to intensity level, a sound has a frequency. The human ear is able to perceive sounds between 20 and 20,000 Hz, with maximum sensitivity in the range of 800 to 5,000 Hz (Davis, 1957). To correct the sound pressure level for the frequency dependence of the ear, a weighting network is often applied. The most common weighting network is the "A" rating, which yields an A-weighted decibel scale (dBA)(Lipscomb and Taylor, 1978; Harris, 1979).

Even at the same dBA levels, some sounds seem more annoying than others. Although attempts have been made to quantify annoyance, these methods still remain somewhat subjective. One important source of annoyance is the masking of speech, which occurs primarily in the frequency range of 800 to 1,900 Hz (Webster and Cluff, 1974). Descriptors of long-term average intensities of noise exposure and annoyance have been developed in an effort to quantify the overall exposure and annoyance that a person experiences. These descriptors are the equivalent sound level (L_{eq}) and the day-night sound level (L_{dn}). The L_{eq} is the level of a steady sound which, in a given period of time, would contain the same noise energy as the time-varying sound level that is being described or considered. To give greater weight to the annoyance caused by a sound intrusion at night, the equivalent sound level for noise occurring between 10 p.m. and 7 a.m. must be augmented by 10 dB before being combined with the equivalent sound level for the period 7 a.m. to 10 p.m. For a sound that is continuous over the 24 hours, L_{dn} exceeds L_{eq} by 6 dB. Some noise intrusions, like those from transportation systems, occur less often at night than by day, and the difference between L_{dn} and L_{eq} may be small. Although, for simplicity, L_{eq} and L_{dn} are widely used, they fail to identify brief but intrusive sounds. Thus, better measures of exposure and annoyance are still required.

Noise measurements of engine and vehicle systems are relatively straightforward, with sound levels normally being measured and reported on the dBa scale. A number of standardized test procedures and practices are in use for regulatory purposes and for comparative testing under a variety of engine and vehicle conditions (ANSI, 1971; NBS, 1977; EPA, 1979a; SAE, 1980). These comparative procedures include simulations of, among others, the following conditions:

* urban acceleration (fixed acceleration rate from a standing start);
* full-throttle acceleration (from a fixed starting speed); and
* maximum sound level (attainment of rated speed at a fixed point).

Typically, the sound level is measured by a microphone located at a fixed distance from the center line of the vehicle path. Other tests to measure exhaust, engine, and interior noise levels are also prescribed. However, in many cases the tests and procedures are modified to suit the equipment and facilities available.

Measurements of the frequency distribution of the noise and vibration provide more quantitative information and are routinely made for research purposes. Such measurements are essential for tracing the sources, transmission paths, and radiating surfaces of the sound energy. Also, they can provide insight into the subjective response to the noise of different engine systems. For example, it is well recognized that impulsive noise characteristics, as evident in diesel idling, may not necessarily be adequately assessed on the dBA scale (Morrison and Challen, 1979). Conventional measurements are proving to be inadequate for research, manufacturing, and community impact evaluations (Cohen, 1980). Thus, additional work on measurement methods, equipment, procedures, and human response is required.

Characteristics of Engine/Vehicle Systems

Engine noise characteristics depend primarily on the basic engine
design or configuration, size, and speed. Figure 2.24 depicts an
A-weighted diesel engine noise spectrum at different engine speeds
(Grover and Perry, 1979). Ninety percent of the perceived noise is
being radiated between 500 and 5,000 Hz at all engine speeds, with
maximum perceived noise being radiated between 800 and 2,000 Hz. These
two ranges correspond to the frequencies of maximum hearing sensitivity
and to the frequencies of maximum importance to speech, respectively.
A 95.5 dB(SPL) curve is also shown to allow comparison with the
unweighted sound pressure levels. The SPL curve demonstrates that
although the perceived sound level below 500 Hz is not significant in
itself, it still has a great deal of energy. This energy contributes
to the perceived noise level indirectly by exciting structural
resonances that lie in the 500- to 4,000-Hz regions. During engine
warm-up and at idle, it is primarily the components of the noise
spectrum between 800 and 4,000 Hz that are perceived as "diesel knock"
(Oetting and Papez, 1979).

Figure 2.25 compares the overall noise levels of several types of
engines (Priede, 1979). Naturally aspirated engines, when running at
their rated speeds, emit nearly the same noise levels. At a given
engine speed, however, the spark ignition engine noise level is lower
than that of the high-speed indirect injection engine; both noise
levels are lower than that of the larger direct injection truck engine.
Applying a turbocharger to the direct injection engine reduces the
noise level. In general, the direct injection diesel engine generates

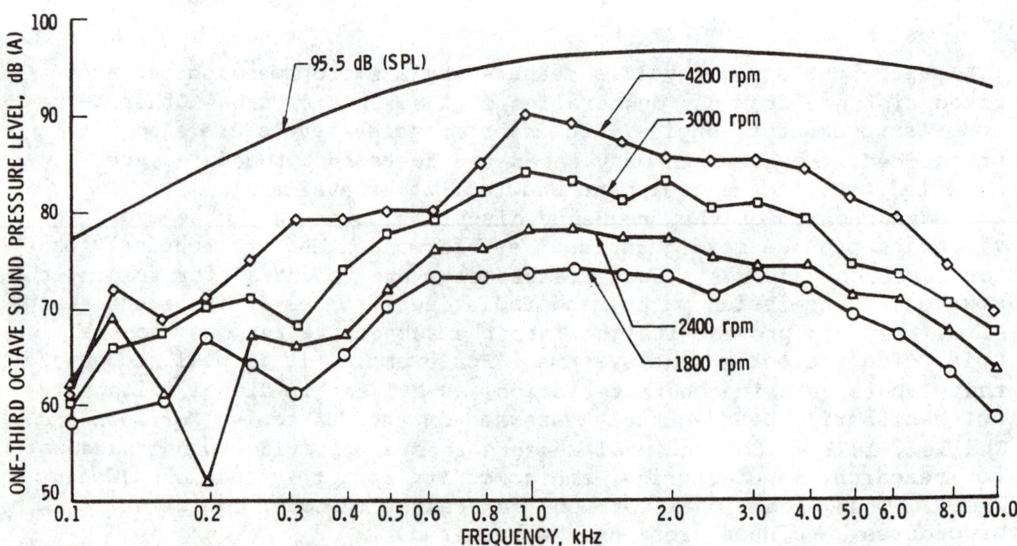

FIGURE 2.24 Diesel engine noise spectrum at various engine speeds (1
meter, unshielded). SOURCE: Grover and Perry, 1978.

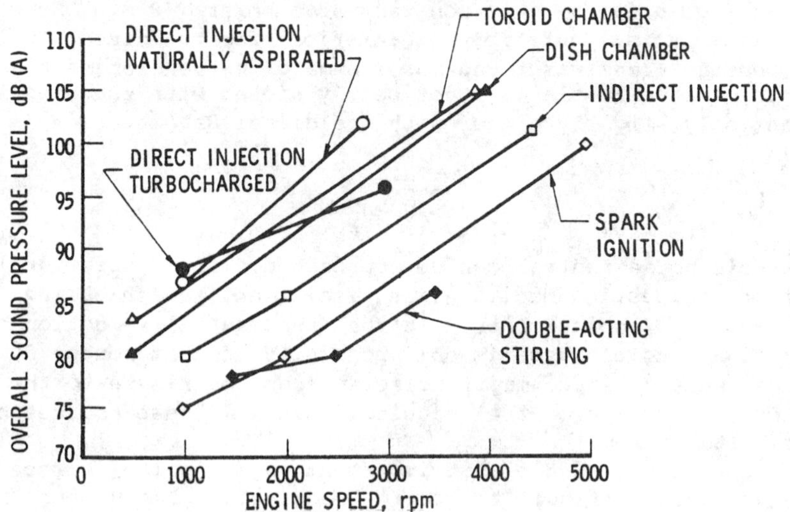

FIGURE 2.25 Comparison of engine noise versus engine speed for several engine types.
SOURCE: Priede, 1979.

more noise than the indirect injection type. The influence of combustion chamber design on noise is also evident. The low-swirl, shallow "dish" chamber is quieter than the toroidal chamber. The effects of engine size are illustrated in Figure 2.26.

Figure 2.27 compares the noise spectrum of a spark ignition engine with that of a comparable naturally aspirated indirect injection diesel, both running at their respective rated speeds (Priede, 1975). As Figure 2.25 indicated, at maximum rated speed the overall noise levels of the two engines were approximately the same: The spark ignition engine measured 101.5 dBA, and the diesel 103 dBA. However, there are noticeable differences in the noise spectrum. For the diesel engine, 70 percent of the noise lies in the range of 500 to 4,000 Hz, with an additional 30 percent contributed by the 125-Hz peak. The spark ignition engine shows just the opposite effect, with 76 percent of its noise in the 160- to 315-Hz peak and only 22 percent in the 500- to 4,000-Hz region. Even though the overall noise levels are similar, the diesel will probably be subjectively judged more annoying because of the speech-masking effect of its noise spectrum.

Overall vehicular noise levels are determined by the basic engine noise level and the degree of attenuation provided by the engine compartment and vehicle. Typically, a 15- to 20-dBA attenuation is achieved. A summary of representative sound levels from diesel and gasoline cars under different driving conditions is shown in Table 2.18.

The SAE drive-by exterior ratings show the diesel Cutlass to be 5 dBA louder than the gasoline-powered Cutlass; the Volkswagen Rabbit gave the same dBA reading for both engines. For both makes, interior measurements were slightly higher with the diesel version during the acceleration portion of the SAE test.

The exterior drive-by at a constant 48.3 km/hr (30 mph) showed
slightly higher noise (exterior and interior) for the diesel Cutlass,
but the opposite trend was found for the gasoline version of the
Rabbit. Idle noise levels were noticeably higher with the diesel
Cutlass and only slightly higher with the diesel Rabbit.

Control Approaches

Engine/vehicle noise control can be approached through structural
modification, combustion modification, isolation, and shielding. An
important aspect of noise control is that substantial reduction by
treatment of a single source is only possibley if that source dominates.
In practice, many separate major noise sources contribute to the overall
noise. Thus, effective control requires that all these sources be
treated simultaneously.

Combustion is one of the most important engine noise sources and is
certainly the most difficult to treat effectively. Because the noise
is inherently part of, and caused by, the complex chamber conditions
during combustion, it can only be reduced by modifying the combustion
process itself. In particular, control of the rate of heat release and

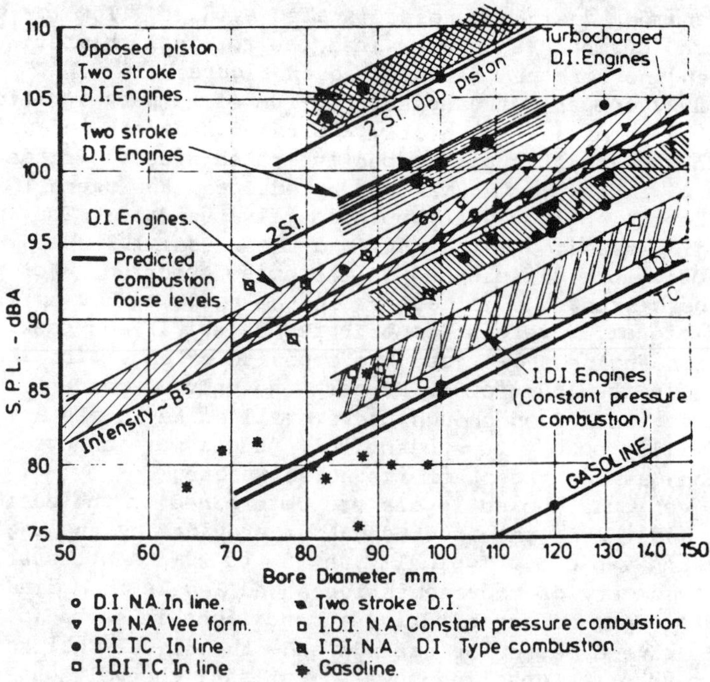

FIGURE 2.26 Relation between overall noise and
bore size at 2000 rev/min, full load.
SOURCE: Priede, 1979.

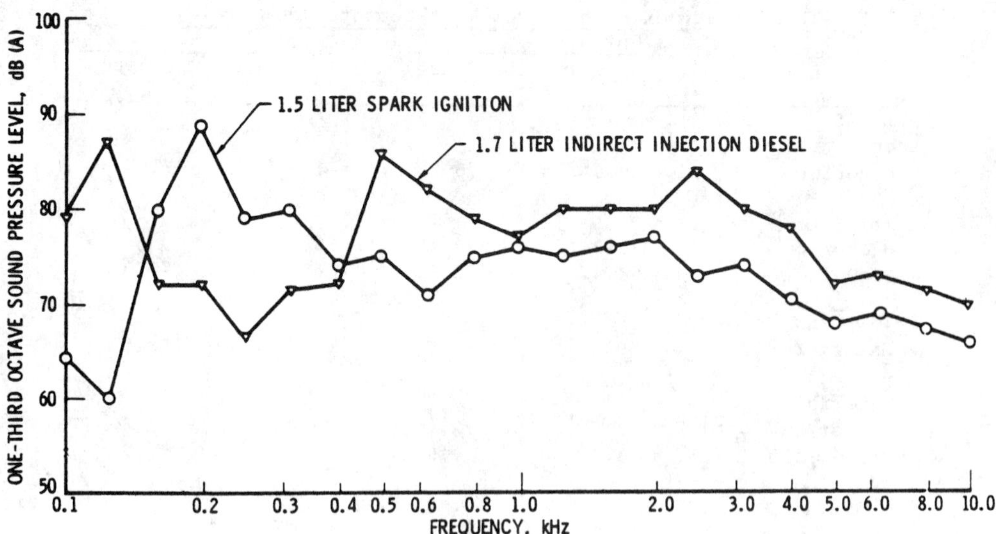

FIGURE 2.27 Comparison of spark ignition and diesel engine noise spectrums (full load, rated speed, 1 meter, unshielded). SOURCE: Priede, 1979.

the peak cylinder pressure is required. Unfortunately, such modifications often adversely affect fuel economy and emissions.

The discussion of combustion modification effects on engine noise, including these trade-offs and interrelations, is incorporated in Chapter 3. But a summary of combustion noise reduction approaches and trade-offs is provided in Table 2.19. Noise has always been a basic design factor in diesel engine development work. Thus, it is not surprising that today's engines are reasonably quiet.

The application of noise reduction techniques to mechanical noise is also limited, as in the case of combustion noise. In this case the limiting factors are not the combustion process but rather lubrication requirements, engine wear, thermal expansion allowances, clearances, and parts inertia. The fundamental cause of mechanical noise is the existence of abrupt, transient forces during engine operation. Most techniques for mechanical noise reduction seek not the elimination or reduction of the sources of the transient forces themselves but rather the reduction of their ability to generate and transmit noise in the mechanical parts. Table 2.20 summarizes the options and their impacts. It is noted that reducing piston slap by decreasing clearances requires that pistons with controlled thermal expansion rates be used. Furthermore, application of these techniques is limited by economic and durability constraints.

Structural response noise treatments group naturally into three major categories: stiffening, isolation, and shielding. Stiffening a structure generally involves adding integral ribbed crossmembers, adding corrugations, or selectively thickening certain areas. The

TABLE 2.18 Summary of Sound Level Measurements (dBA Scale)

Conditions	Cutlass		Rabbit	
	Gasoline	Diesel	Gasoline	Diesel
SAE J986a				
Accel. drive-by				
Exterior	68.8[a]	73.8[a]	71.0	71.5
Interior	80.0			
Blower on[b]	73.2[a]	74.2[a]	78.2	
Blower off	68.8[a]	70.5[a]	76.5	79.5
48.3-km/hr drive-by				
Exterior	58.8	61.2	60.5	58.5
Interior				
Blower on[b]	71.5	72.2	73.5	71.8
Blower off	60.5	64.0	70.5	68.0
Engine idle				
Exterior[c]	64.5	70.0	65.0[d]	67.0
			72.5[e]	
Interior				
Blower On[b]	71.5	71.0	69.5	69.5
Blower off	48.5	51.5	58.0	62.5

[a] Acceleration in first gear.
[b] Windows up, fresh air blower on high.
[c] At 3.05 m.
[d] Electric, radiator, cooling fan off.
[e] Electric, radiator, cooling fan on.

SOURCE: Springer, 1979.

increased material usually results in increased engine weight. The
overall reduction after extensive stiffening is 1 to 4 dBA.

Isolation requires that the radiating component be physically
separated from the engine structure by an absorbent intervening
structure. For example, the oil pan may be cut close to its point of
attachment to the block and remounted with a resilient silicon
elastomer. Isolation is restricted to those components that do not
bear heavy loads. Reductions of 2 to 4 dBA are typical of isolation
treatments.

Shielding an engine component involves placing a thin, light-weight
cover over the component. A resilient, absorbent layer is placed
between component and cover, often fiberglass, synthetic rubber, or
silicon elastomer. Antivibration mountings hold the cover in place.
Reductions of 5 to 10 dBA are typical for shields, depending on type
and location.

TABLE 2.19 Combustion Noise Reduction

Controlling Factor	Treatment	Engine Rating Point	Overall Noise Reduction, dBA	Impact of Treatment		
				BSFC	NO_x	Smoke
Rate of pressure rise	Matched turbocharging	Peak torque	3–4	Moderate increase	Minor increase	Significant decrease
		Full-Load rated speed	10–12	Minor decrease	Moderate decrease	Significant decrease
Injection rate	Increase injection rate and reduce ignition delay	Full-load rated speed	3	Negligible effect	Significant decrease	Minor decrease
Ignition delay	Fumigation	Peak torque	5	Negligible effect	Negligible effect	Moderate decrease
		Full-load rated speed	Negligible effect	Significant increase	Negligible effect	Moderate decrease
Fuel composition	Increase cetane number	Peak torque	4	Negligible effect	Minor increase	Unknown
		Full load	3			
				Negligible effect	Moderate increase	Unknown
Rate of pressure rise and ignition delay	Exhaust gas recirculation	Idle	4–5	Minor increase	Moderate decrease	Negligible effect
Chamber turbulence	Low-turbulence chamber	2/3 load, 2,500 rpm	5–6	Significant increase	Minor decrease	Minor increase

SOURCE: Roessler et al., 1980.

Table 2.21 summarizes potential structural response noise treatments. Economic constraints provide the primary limitation on the application of these techniques. Major engine redesign and engine encapsulation are also possibilities for noise control. As is indicated by Figure 2.24, engine redesign entails increasing the number of cylinders and reducing the bore and stroke. Also, redesign allows better incorporation of structural damping and isolation approaches.

Encapsulation can involve reduction of the size of engine compartment openings, application of sound absorbent linings, or use of a separate engine enclosure. One prototype engine noise enclosure yielded a 19.5-dBA noise reduction (Thien, 1979). Encapsulation has several disadvantages, including reduced engine cooling, increased weight, and reduced engine accessibility. Solutions to these problems are being actively pursued. Also, as with the other approaches, application of encapsulation is limited by economic constraints.

In one recent study, diesel passenger car noise reduction measures and estimated fuel economy, weight, and cost trade-offs were examined (Morrison et al., 1980). The results are summarized in Tables 2.22 and 2.23. The majority of noise reduction measures considered gave noise reductions between 1 and 2 dBA, with redesign and encapsulation somewhat higher at 6 to 8 dBA. No simple correlation of noise reduction with vehicle economy, weight, and cost was found.

TABLE 2.20 Mechanical Noise Reduction

Noise Source	Treatment	Noise Reduction, dBA	Impact of Treatment
Pistons	Offset gudgeon pins	2	Possible ring wear, reduced piston lubrication; chamber redesign may be necessary
	Polymer piston inserts	4-5	Possible reduction in long-term piston thermal stability and mechanical integrity
	Reduced clearances	2-3	Pistons with controlled thermal expansion necessary
	Modified piston lubrication	2-3	Potential reduction in oil consumption, unless upper scraper ring fails
	Articulated pistons	1-2	Better chamber sealing
Timing gears	New gear designs	2	Increased gear wear
	Chain/belt drive	8	Potential weight reduction
Fuel injection equipment	Reduced backlash	2	Increased gear wear
	Modified mounting	2-3	Negligible impact
	Change needle/holder masses	2-3	Negligible impact

SOURCE: Roessler et al., 1980.

In summary, a variety of noise control techniques are available, and diesel noise reductions are certainly technically possible. Given sufficient lead time, noise reductions of 3 to 5 dBA should be achievable through system and component improvements, without significant increases in cost or weight. Comparable improvements can be made to spark ignition engines as well. Further reductions may be possible but will require major engine redesign or engine encapsulation. The latter will involve substantial cost and weight penalties.

TABLE 2.21 Structural Response Noise Treatment

Noise Source	Noise Treatment				Impacts of Treatment
	Stiffen/Damp 1-4 dBA	Isolate 2-4 dBA	Shield 5-10 dBA	Other	
Cylinder head	X		X		Major redesign; weight penalty
Piston-connecting rod assembly				Use light-weight pistons	Pistons may not withstand forces in turbocharged engine
Crankshaft and main bearings	X		X	Torsional damping; increase number of bearings	Major redesign; weight penalty
Block and crankcase	X		X		Major redesign; weight penalty
Oil pan	X	X	X		Weight penalty
Cover panels	X	X	X		Weight penalty
Hydraulics				Reduce clearances, increase viscosity	Thermally controlled components necessary

SOURCE: Roessler et al., 1980.

TABLE 2.22 Diesel Passenger Car Noise Reduction Measures and Estimated Fuel Economy Weight and Cost Trade-Offs

Noise Reduction Measure	Noise Reduction, dBA	Vehicle Weight Change, %	Fuel Economy Change, mile/U.S. gal.	Cost Increase, %	Comments
Reduced engine speed by 10%					
(a) By changing final drive ratio	1 to 2	0	+1	0	Performance loss
(b) By changing intermediate ratio	1 to 2	0	<+1	0	Some performance loss
(c) Increase final drive ratio and increase engine swept vol.	0	+1 to +2	<+1	+1	Performance restored
(d) Increase final drive ratio and turbocharge	1 to 2	+2 to +2	+1	+7	To restore performance
Engine size/configuration change					
(a) Reduce bore:stroke ratio	1 to 2	<+1	−1	0	Same swept vol. and speed
(b) Reduce bore and stroke; increase no. of cyls. from 4 to 6	1 to 2	+2	−1	+4	
(c) Reduce swept vol. and turbocharge to restore power	1 to 2	0 to −1	−1 to +2	+5	Swept vol. from 2μ to 5μ
Combuston process changes					
(a) Retard injection by 5°	1	0	−1 to 0	0	
(b) Increase compression ratio by 1	<1	0	−1	<+1	Production tolerance difficulties
(c) Turbocharge (power increase by 35%)	(increase by 1 to 2)	+2 to +3	−2	μ+7	Performance increase
(d) fumigation	<1	<+1	−1	<+1	
(e) EGR (10% at full load)	0 to 1	<+1	0	<+1	
(f) Pilot injection	0 to 1	<+1	−1	<+1	

SOURCE: Morrison et al., 1980.

TABLE 2.23 Passenger Car Noise Reduction Measures and Estimated Fuel Economy, Weight, and Cost Trade-Offs (Common to Both Diesel and Gasoline Engines)

Noise Reduction Measure	Noise Reduction, dbA	Vehicle Weight Change, %	Fuel Economy Change, mile/U.S. gal.	Cost Increase, %
Shielding/structural changes				
(a) Damping/mass loading components	0 to 2	<+1	0	<+1
(b) Isolation of engine components	0 to 2	<+1	0	<+1
(c) Comprehensive shielding	1 to 3	+1 to +2	μ-1/2	+1 to +2
(d) Enclosure (tunnel type)	4 to 6	+1 to +3	μ-1/2	+2 to +3
(e) Major engine structure redesign	4 to 8	0	0	μ+5
Gas Flow Noise Reduction				
(a) Improved exhaust muffler	1 to 2	+1 to +2	-1 to -2	+1 to +2
(b) Improved intake muffler	0 to 1	μ-1	μ-1	μ+1
Other				
(a) Redesigned piston (close clearance)	1 to 2	0	0	0 to 1

SOURCE: Morrison et al., 1980.

Environmental Impact

EPA has determined that two noise exposure limits are needed to protect the long-term health and welfare of the public (EIC, 1978). They set a 70 dBA maximum level for 24-hour exposure (L_{eq}) for prevention of hearing loss, and a night-weighted 55-dBA level for outdoor 24-hour residential noise exposure (L_{dn}) for interference and annoyance. EPA further determined that 40 to 70 million people are currently subject to L_{dn} noise levels of 60 dBA or more to urban traffic. Attempts to quantify the associated annoyance, as well as potential control strategies, are currently underway (Cermak, 1979a,b; Cermak and Cornillon, 1976; Cermack and von Buseck, 1978; Cermak et al., 1979; Marshall, 1980).

The major classes of urban vehicles are medium- and heavy-duty trucks, light-duty vehicles, buses, and motorcycles. In terms of both specific noise emission levels of individual vehicles and the overall vehicle noise contributions to total noise, much of the traffic noise was found to be due to medium- and heavy-duty trucks, and they were identified as major noise sources. Automobiles and light-duty trucks were found to be significant contributors as well.

On the basis of these factors and the fact that current diesel vehicles are noisier than gasoline vehicles by 3 to 5 dBA, widespread introduction of diesel passenger cars may have a significant effect on urban traffic noise annoyance. Also, the problems are likely to be more severe than a simple dBA weighting might indicate, due to the special spectral distribution of the diesel noise. If required, diesels are capable of meeting the same dBA noise level as gasoline engines (at a cost), even though this overall level may not adequately represent the true effect of increased diesel noise.

Summary and Conclusions

In general, noise measurements are straightforward and well defined. However, impulsive noise characteristics during diesel idling and the increased masking of speech in the 800- to 1,900-Hz frequency range with diesel engines are not adequately assessed by the dBA scale. Hence, both the sound pressure level and the frequency distribution have to be considered in characterizing diesel vehicle noise. Additional work on measurement methods, equipment, procedures, and human response is needed.

The three major sources of noise in diesel engines are: combustion noise, mechanical noise, and structural resonance noise. In light-duty indirect injection engines, combustion noise and mechanical noise are the predominant sources. Combustion noise depends on cylinder pressure histories (rate, peak, fluctuations) and is particularly pronounced with long ignition delays (idle and cold start diesel knock). Mechanical noise depends on the abrupt forces acting on the engine parts; these are particularly pronounced in diesels with high compression ratios and high pressure injection systems.

The overall noise levels of current gasoline and diesel engines are nearly the same when they are operated at their rated speeds, with the spark ignition engine noise level being lower (5 dBA) than that of the indirect injection diesel, and this in turn being lower than that of the direct injection diesel (3 to 5 dBA). However, there are also noticeable differences in the composition of the noise spectrum. The diesel noise spectrum contains a 125-Hz peak (diesel knock) and is concentrated (70 percent) in the 500- to 4,000-Hz range. Spark ignition engines provide only a small contribution (22 percent) in the same frequency range. Thus, even at the same overall noise levels, diesels are subjectively judged more annoying because of the speech masking effect of its noise spectrum (in the 800- to 1,900-Hz region).

It has been determined that 40 to 70 million people are presently exposed to urban traffic noise at levels above that set for minimal interference and annoyance. Thus, widespread introduction of diesel passenger cars may have a significant impact on urban traffic noise annoyance, due to both the higher overall noise level and the more intrusive spectral distribution characteristics of diesel noise.

Control technologies for reducing regulated pollutants generally tend to exacerbate noise. Combustion and injection modifications for particulate control (more premixing) result in higher noise. However, fuel modifications for particulate control (tighter specifications limits on cetane number and composition) will reduce noise emissions. Poorer quality fuels will have a negative impact.

In general, diesel combustion noise has already been optimized because noise has always been a basic design constraint for customer appeal and acceptance. Thus, only small additional reductions are expected in current systems. Nonetheless, diesel engines are capable of meeting the same noise levels as gasoline engines (at a cost) through engine and structure redesign, shielding, and encapsulation. However, the annoyance factor still needs to be assessed.

EFFECT OF FUEL PROPERTIES

One of the important determinants of diesel engine performance is fuel composition. Unlike spark ignition engines, diesel engines can burn a variety of fuels (within wide property bounds) without major engine modifications. Although this is an important advantage, it also makes it important to evaluate the effects that different fuel properties have on emissions. To help define the relationships between fuel composition and engine performance, investigators have studied the effect of cetane number, fuel density, sulfur content, percent aromatics, cloud point temperature, and distillation temperatures. The data show that fuel composition affects diesel emissions. It appears, however, that using fuel composition as a control technique will not significantly improve diesel engine emission characteristics. Reasonable control of diesel fuel properties will, however, minimize operational, cold-start, durability, and emission (primarily hydrocarbons and particulates) problems.

Fuels with different properties do not necessarily respond the same way in engines with different combustion systems (direct injection, swirl chamber, or prechamber designs; and naturally aspirated or turbocharged) fuel injection systems. This is probably related to the differences in the actual chemical and physical processes within the engine cycle and to the fact that the fuel properties do not directly correlate with the dynamics of the combustion and emission formation processes. Furthermore, diesel automotive fuel must be satisfactory for heavy-duty trucks and buses as well.

Because of the relationships between engine design and fuel composition, proper matching of fuel and engine can help minimize particulate emissions. Higher quality fuels, with low sulfur levels, low aromatic content, increased volatility, and cetane numbers greater than 48, have generally resulted in lower particulate emission levels in most engines. These results also suggest that fuel composition, particularly the cetane number, should be controlled during emission testing procedures to eliminate possible sources of variation. The works of Smaby and co-workers (1979) and Roessler and co-workers (1980) contain extensive references and discussions of the literature on fuel effects on emissions.

Tests using two diesel automobiles--a Mercedes 240D and a Volkswagen Rabbit--with five different fuels (Table 2.24) during eight different driving cycles showed that differences between vehicles and operating schedules were greater than those between fuels (Hare and Baines, 1979). Tables 2.25 and 2.26 illustrate these differences in the particulate and the organic soluble fraction of the particulate, and Tables 2.27 and 2.28 provide a summary of the hydrocarbon, carbon monoxide, and NO_x emission data obtained when these five fuels were used. As the data show, most of the gaseous emissions were low, with the exception of the comparatively high hydrocarbon values for the Volkswagen Rabbit when operated on the minimum quality No. 2 fuel. This is probably related to the low cetane index (41.8) of this fuel.

Another component of the particulate emissions from diesel engines is the sulfate fraction that comes from the sulfur in the fuel. The primary oxide in the exhaust is sulfur dioxide. Approximately 2 to 4 percent of the fuel sulfur appears as sulfate in the exhaust (Khatri et al., 1978; Hare and Bradow, 1979). Table 2.29 shows data on sulfate levels in particulate emissions for a Mercedes 240D and Volkswagen Rabbit diesel for various cycles and fuels. In general, the sulfate component in the particulate ranged from a fraction of a percent to more than 5 percent for the Mercedes and from under 1 percent to about 4 percent for the Volkswagen (Hare and Baines, 1979). Figure 2.28 shows sulfate emission data from a number of diesel vehicles tested during the FTP cycle when fuels with different sulfur contents (Roessler et al., 1980) were used. Uncontrolled diesels using fuel with less than 0.25 percent sulfur generally have sulfate emissions similar to those of catalyst-equipped spark ignition vehicles, even though the gasoline sulfur content is only approximately 0.03 percent in unleaded fuel.

Frisch and co-workers (1979) have studied the effect of fuel properties on particulate emissions. Three fuels were blended to obtain approximately the same cetane number, but with varying sulfur content,

TABLE 2.24 Summary of Test Fuel and "National Average" Fuel Properties

Fuel Code and/or Description	Gravity, °API[a]	Cetane, Index[b]	Sulfur wt. %[c]	Distillation, °C[d]			Aromatics, %	Nitrogen, %
				IBP	50%	EP		
No. 2 fuel, averaged properties, 1976 survey (40)	35.7	48.8	0.253	190	261	333	e	e
No. 1 fuel, averaged properties, 1976 survey (40)	42.2	49.5	0.081	176	220	274	e	e
EM-238-F, 2D emissions	36.0	48.6	0.35	192	257	349	29.8	0.005
EM-239-F, "Nat'l. Avg." No. 2	36.1	48.7	0.23	186	257	337	21.6	0.005
EM-240-F, No. 1 or Jet A	44.1	47.4	0.04	162	201	268	13.0	0.006
EM-241-F, "Minimum Quality" No. 2	32.8	41.8	0.26	182	258	327	34.6	0.024
EM-242-F, "Premium" No. 2	38.7	53.0	0.26	183	254	327	12.4	0.008

[a] ASTM D287.
[b] ASTM D976.
[c] ASTM D129 or D1266-70.
[d] ASTM D86 thermal distillation.
[e] No data.

SOURCE: Hare and Baines, 1979.

TABLE 2.25 Particulate Mass Emissions for Two Diesel Vehicles

Grams Particulate per Kilometer by Operating Cycle or Mode

Fuel	Cold FTP	Hot FTP	(Calculated) 1975 FTP	CFDS	FET	NYCC	Steady States Idle[a]	50 km/hr	80 km/8 hr
Mercedes 240D									
EM-238-F	0.335	0.324	0.329	0.261	0.212	0.680	2.99	0.150	0.196
EM-239-F	0.319	0.311	0.314	0.226	0.192	0.565	3.16	0.142	0.165
EM-240-F	0.251	0.223	0.235	0.166	0.140	0.317	1.50	0.114	0.136
EM-241-F	0.408	0.358	0.380	0.257	0.258	0.808	4.00	0.150	0.231
EM-242-F	0.299	0.286	0.292	0.203	0.181	0.563	2.71	0.131	0.195
VW Rabbit Diesel									
EM-238-F	0.252	0.204	0.225	0.206	0.173	0.363	1.93	0.090	0.167
EM-239-F	0.250	0.194	0.218	0.194	0.143	0.384	2.12	0.068	0.148
EM-240-F	0.209	0.152	0.177	0.149	0.138	0.295	0.742	0.047	0.103
EM-241-F	0.565	0.231	0.375	0.222	0.174	0.450	2.84	0.197	0.189
EM-242-F	0.221	0.174	0.194	0.156	0.175	0.402	2.10	0.052	0.164

[a] Grams per hour instead of grams per kilometer.

SOURCE: Hare and Baines, 1979.

TABLE 2.26 Organic Soluble Content by Weight Percent and Major Elements of Particulate

Weight Percent Organic Solubles in Particulate by Operating Schedule[a]

Fuel	Cold FTP	Hot FTP	CFDS	FET	NYCC	Idle	50 km/hr	85 km/hr	Mean Percentage
				Mercedes 240D					
EM-238-F	10.7	11.5	7.7	6.4	11.4	9.3	10.2	7.7	9.4
EM-239-F	9.7	9.8	7.9	7.6	7.9	6.0	9.0	8.6	8.3
EM-240-F	12.2	11.9	10.4	11.7	25.7	20.1	12.7	9.8	14.3
EM-241-F	8.3	7.0	5.5	4.7	3.1	4.1	4.7	6.0	5.4
EM-242-F	9.5	8.5	7.0	2.3	3.9	6.5	7.3	4.9	6.2
Mean percentage	10.1	9.7	7.7	6.5	10.4	9.2	8.8	7.4	8.7
				VW Rabbit Diesel					
EM-238-F	12.2	9.8	14.6	9.1	14.4	10.0	14.9	14.0	12.4
EM-239-F	11.6	14.4	13.2	14.3	27.5	18.4	22.7	12.7	16.8
EM-240-F	13.2	16.4	15.2	15.3	18.7	19.6	15.7	13.4	15.9
EM-241-F	11.8	18.4	15.4	15.4	18.0	16.3	15.4	21.6	16.5
EM-242-F	13.5	13.7	14.3	12.9	11.6	11.4	16.3	7.4	11.4
Mean percentage	12.5	14.5	14.5	13.4	18.0	13.1	17.0	13.8	14.6

TABLE 2.26 (continued)

| Fuel | Weight Percent Element(s) in Organic Solubles | | | | | |
	Carbon	Hydrogen	Nitrogen	Sulfur	Oxygen	CHNSO
				Mercedes 240D		
EM-238-F	82.8	12.4	0.10	0.40	4.2	99.9
EM-239-F	83.5	12.2	0.08	0.36	3.8	100.0
EM-240-F	83.2	12.4	0.10	0.39	3.7	99.8
EM-241-F	83.9	12.2	0.08	0.36	3.4	100.0
EM-242-F	83.7	12.4	0.13	0.41	3.3	99.9
Mean values	83.4	12.3	0.10	0.38	3.7	99.9
				VW Rabbit Diesel		
EM-238-F	83.9	12.7	0.12	0.37	2.9	100.0
EM-239-F	84.2	12.1	0.08	0.41	3.2	100.0
EM-240-F	83.7	12.8	0.21	0.35	2.9	99.9
EM-241-F	84.2	12.4	0.16	0.43	2.7	99.9
EM-242-F	83.8	12.6	0.11	0.38	3.0	99.9
Mean values	84.0	12.5	0.14	0.39	2.9	99.9

[a] Averages used where possible.

SOURCE: Hare and Baines, 1979.

TABLE 2.27 Regulated Gaseous Emission Data for a Mercedes 240D Operated on Five Diesel Fuels

| Fuel | Item | Emissions (g/km) and Fuel usage (1/100 km) by Driving Schedule | | | | | | | | |
| | | 3-Bag FTP Number | | | | | | Steady States | | |
		1	2	3	CFDS	FET	NYCC	Idle[a]	50 km/hr	85 km/hr
Em-238-F,	HC	0.13	No	0.11	0.09	0.06	0.27	2.22	0.08	0.08
2D emissions	CO	0.57	Data	0.57	0.39	0.35	1.11	6.63	0.27	0.36
	CO_2	225.00		228.00	188.00	172.00	354.00	1630.00	124.00	172.00
	NO_x	0.79		0.77	0.84	0.68	1.17	5.88	0.47	0.84
	Fuel	8.42		8.54	7.03	6.43	13.2	0.616	4.64	6.44
EM-239-F,	HC	0.14	0.26	0.16	0.08	0.06	0.27	2.10	0.06	0.06
"Nat'l Avg."	CO	0.65	0.64	0.63	0.45	0.40	1.31	6.18	0.27	0.41
No. 2	CO_2	239.00	232.00	220.00	194.00	175.00	382.00	1530.00	132.00	175.00
	NO_x	0.79	0.82	0.75	0.72	0.73	1.27	5.70	0.50	0.77
	Fuel	8.91	8.68	8.22	7.24	6.53	14.3	0.578	4.95	6.56
EM-240-F,	HC	0.04	0.12	0.10	0.06	0.04	0.15	1.38	0.04	0.05
"Jet A"	CO	0.58	0.56	0.56	0.45	0.41	1.32	6.12	0.25	0.40
No. 1	CO_2	230.00	223.00	230.00	201.00	188.00	401.00	1820.00	119.00	181.00
	NO_x	0.73	0.74	0.73	0.69	0.69	1.34	6.48	0.41	0.70
	Fuel	8.59	8.33	8.60	7.52	7.01	15.00	0.685	4.43	6.76
EM-241, F,	HC	0.19	0.22	0.20	0.08	0.05	0.33	2.97	0.06	0.06
Minimum	CO	0.71	0.72	0.71	0.48	0.40	1.55	7.59	0.30	0.40
Quality"	CO_2	257.00	253.00	241.00	210.00	188.00	410.00	1740.00	131.00	184.00
No. 2	NO_x	0.88	0.88	0.87	0.83	0.80	1.40	7.47	0.49	0.70
	Fuel	9.63	9.49	9.03	7.86	7.03	15.4	0.655	4.91	6.86
EM-242-F,	HC	0.11	0.13	0.12	0.07	0.05	0.20	1.38	0.06	0.05
"Premium"	CO	0.60	0.72	0.71	0.45	0.41	1.25	5.94	0.25	0.40
No. 2	CO_2	230.00	271.00	253.00	183.00	173.00	349.00	1510.00	124.00	159.00
	NO_x	0.77	0.93	0.86	0.71	0.69	1.24	5.61	0.45	0.68
	Fuel	8.60	10.1	9.45	6.85	6.48	13.1	0.567	4.64	5.96

[a] Emissions in grams per hour instead of g/km; fuel in 1/hr.

SOURCE: Hare and Baines, 1979.

TABLE 2.28 Regulated Gaseous Emission Data for a VW Rabbit Diesel Operated on Five Diesel Fuels

Fuel		Emissions (g/km) and Fuel usage (1/100 km) by Driving Schedule						Steady States		
		3-Bag FTP Number								
	Item	1	2	3	CFDS	FET	NYCC	Idle[a]	50 km/hr	85 km/hr
Em-238-F,	HC	0.15	0.15	0.18	0.08	0.11	0.55	7.38	0.10	0.08
2D emissions	CO	0.48	0.50	0.48	0.36	0.32	1.08	14.4	0.27	0.33
	CO_2	151.00	149.00	158.00	127.00	116.00	227.00	1110.00	92.00	119.00
	NO_x	0.61	0.59	0.56	0.56	0.53	0.81	4.65	0.31	0.55
	Fuel	5.66	5.57	5.91	4.74	4.35	8.54	0.432	3.44	4.45
EM-239-F,	HC	0.19	0.24	0.19	0.12	0.11	0.39	6.48	0.07	0.10
"Nat'l Ave."	CO	0.54	0.50	0.49	0.42	0.33	1.16	12.5	0.23	0.36
No. 2	CO_2	151.00	147.00	151.00	131.00	121.00	228.00	1090.00	95.00	117.00
	NO_x	0.62	0.67	0.65	0.50	0.57	0.98	4.89	0.34	0.57
	Fuel	5.66	5.51	5.63	4.92	4.52	8.58	0.421	3.55	4.38
EM-240, F,	HC	0.14	0.18	0.20	0.12	0.15	0.33	3.60	0.05	0.13
"Jet A"	CO	0.55	0.55	0.56	0.42	0.46	1.18	10.6	0.29	0.46
No. 1	CO_2	150.00	151.00	156.00	130.00	121.00	224.00	1190.00	91.00	119.00
	NO_x	0.57	0.58	0.57	0.51	0.48	0.79	5.73	0.31	0.84
	Fuel	5.63	5.65	5.85	4.87	4.55	8.44	0.456	3.40	4.47
EM-241-F,	HC	0.67	0.67	0.79	0.20	0.15	1.35	17.5	1.06	0.02
"Minimum"	CO	0.76	0.80	0.87	0.43	0.34	1.90	28.0	0.92	0.33
Quality"	CO_2	157.00	167.00	163.00	134.00	123.00	241.00	1310.00	94.00	120.00
No. 2	NO_x	0.59	0.58	0.57	0.57	0.52	0.89	5.52	0.30	0.53
	Fuel	5.97	6.33	6.22	5.02	4.61	9.23	0.527	3.68	4.51
EM-242-F,	HC	0.19	0.25	0.18	0.10	0.11	0.37	6.39	0.07	0.08
"Premium"	CO	0.47	0.59	0.49	0.38	0.39	0.96	11.9	0.19	0.35
No. 2	CO_2	157.00	166.00	147.00	136.00	126.00	232.00	1120.00	95.00	121.00
	NO_x	0.60	0.68	0.60	0.57	0.58	0.89	4.77	0.38	0.60
	Fuel	5.90	6.20	5.58	5.09	4.71	8.70	0.432	3.54	4.54

[a] Emission in grams per hour instead of g/km; fuel in 1/hr.

SOURCE: Hare and Baines, 1979.

TABLE 2.29 Summary of Sulfate Data for Two Diesel Vehicles

SO_4 in wt % of Particulate by Operating Schedule

Fuel	Cold FTP	Hot FTP	CFDS	FET	NYCC	Idle	50 km/hr	85 km/hr
					Mercedes 240D			
EM-238-F	2.60	2.41	3.83	6.13	1.91	2.61	2.27	4.90
EM-239-F	2.60	2.22	4.25	4.32	1.20	1.36	1.41	4.73
EM-240-F	1.12	0.72	1.45	1.71	0.47	0.47	0.45	1.10
EM-241-F	2.35	2.60	4.67	4.65	0.92	1.55	1.87	3.51
EM-242-F	2.47	3.39	5.42	4.09	1.51	2.10	1.91	4.31
					VW Rabbit Diesel			
EM-238-F	3.13	2.50	4.47	4.05	1.65	2.28	1.56	3.89
EM-239-F	2.48	2.27	3.66	3.08	1.12	3.30	1.32	3.38
EM-240-F	1.72	0.92	1.54	0.72	1.42	1.62	0.68	1.36
EM-241-F	1.95	2.21	3.60	2.87	1.69	3.87	0.56[a]	2.59
EM-242-F	2.35	1.78	4.29	4.29	1.22	2.57	2.31	2.80

[a]Estimated from incomplete data.

SOURCE: Hare and Baines, 1979.

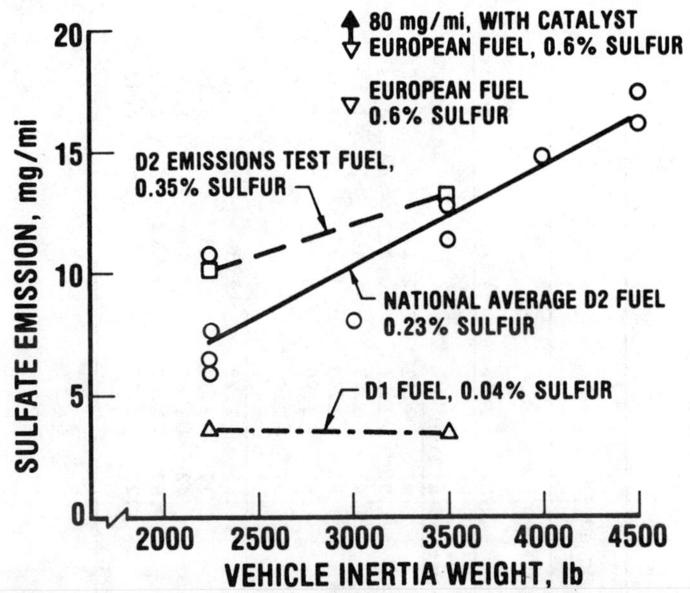

FIGURE 2.28 Sulfate emission of diesel vehicles operated over the FTP. SOURCE: Roessler et al., 1980.

90 percent distillation temperature, and aromatic content. These fuels were used to run a Caterpillar 3208 direct injection engine, at steady state, for two modes of the EPA 13-mode heavy-duty cycle. The properties of the fuels are shown in Table 2.30. Figure 2.29 shows the fractional breakdown of the particulate emissions at mode 3 of the cycle. The data show an increase in both the solid and the soluble organic fractions in going from the No. 1 fuel to the No. 3 fuel, although the percentage breakdown of the seven fractions of the soluble organic fraction for the three fuels stayed approximately constant. It would be possible to infer from these data that the aromatic content, the 90 percent distillation temperature (or the percent mass above 500°F), and the sulfur content affected the particulate emissions, but insufficient fuel blends were run to draw this conclusion.

General Motors has recently completed a study using fuels, supplied by the petroleum companies, of widely differing properties (Burley and Rosebrock, 1979). These fuels were evaluated by dynamometer tests using a 1978-1979, 5.7-liter V-8 production diesel engine operated at five steady state speeds; poppet-type injectors were substituted for the production two-hole pencil-type injectors. The five operating conditions used are representative of the FTP. Appropriate weighting factors were applied to these engine operating conditions to calculate an approximate particulate emission index representative of the FTP cycle. Because injection timing requirements vary from fuel to fuel, each fuel, at each load, was tested at three injection timings: standard, 5 degrees advance, and 5 degrees retard. The data were then analyzed by using a multiple linear stepwise regression method to learn

TABLE 2.30 Fuel Properties of Test Fuels

	Fuel 1	Fuel 2	Fuel 3
Gravity, °API	46.0	38.5	30.3
Viscosity, CST, at 100°F	1.4	2.47	6.8
Cetane No. (D-976)	52.0	53.3	53.5
Pour point, °F	−50	0	+50
Cloud point, °F	−42	+4	+50
Carbon, wt %	85.3	86.4	85.6
Hydrogen, wt %	14.3	13.5	12.5
Oxygen, wt %	0.01	0.022	0.21
Nitrogen, wt %	0.0015	0.0051	0.14
Sulfur, wt %	0.11	0.27	0.44
Distillation (D-86), %			
50	353	414	508
50	396	491	614
90	466	590	700
95	492	620	728
EP	518	655	743
Hydrocarbon type by			
mass spectrometry[a], wt %			
Aromatics	15	21	27
Paraffins	51	40	27
Napthenes	32	37	42
Sulfur compounds	2	2	4
Hydrocarbon type by FIAM[b]			
Aromatics	15	24	[c]
Paraffins and napthenes	85	76	[c]
Olefins	0	0	[c]

[a]The 22-component hydrocarbon type analysis is based on a method that utilizes high-resolution mass spectrometry to distinguish between saturated and aromatic species that have the same nominal molecular weights but different accurate masses (e.g., eicosane, $C_{20}H_{42}$, MW = 282.3286; and undecylnaphthalene, $C_{21}H_{30}$, MW = 282.2347), thus eliminating the need for physical separation of the species in order to measure their relative concentrations.

[b]The Fluorescent Indicator Adsorption Method (FIAM) is the ASTM D 1319 test procedure. This is a common method for measuring aromatics; however, it is designed primarily for gasoline and not for diesel fuel. The method specifies an upper limit of 600°F on end point of fuels tested.

[c]No separation was possible in this method.

SOURCE: Frisch et al., 1979.

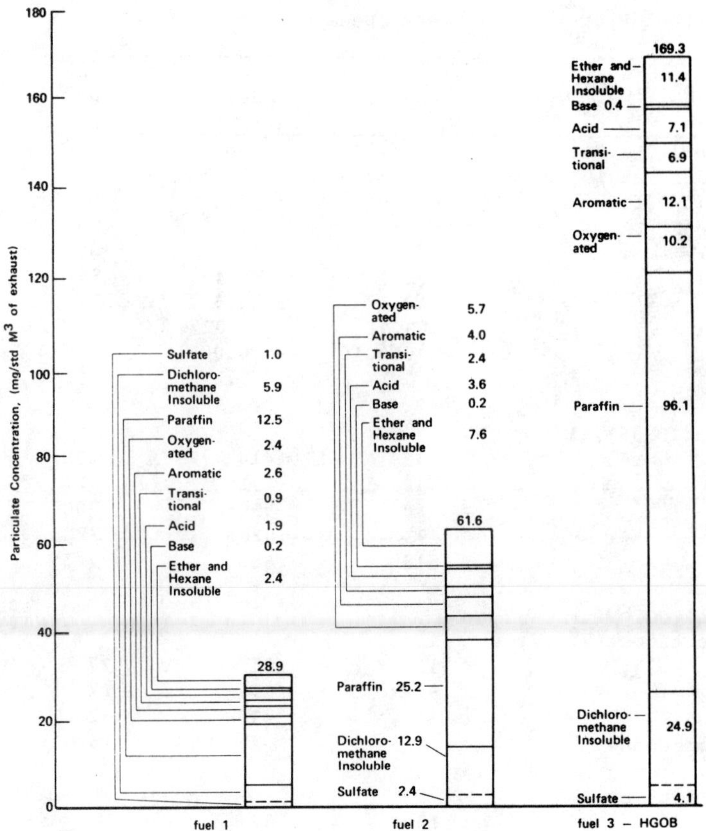

FIGURE 2.29 Fractional breakdown (by mass concentration of soluble organic fraction) for mode 3 at 8:1 volume dilution ratio.
SOURCE: Frisch et al., 1979.

which fuel characteristics related to the various emissions. The regression analysis showed that aromatic content and 90 percent boiling point were the fuel characteristics that affected particulate emissions. Figure 2.30 shows a graph of the resulting regression equation. In determining how well the equation can predict the particulate level of a fuel, a comparison was made between predicted particulates and measured particulates for each of the test fuels. The result was that 73 percent of the measured data were found to be within +10 percent of the predicted value. The boxes in this figure indicate that when comparing typical No. 1 and No. 2 (see Table 2.31) diesel fuels, particulate emissions from a diesel using No. 1 fuel would be expected to be about 25 percent lower.

Another important observation from this study was that when an ignition or cetane improver was blended with certain light petroleum products, thereby raising the cetane number to 46, there was a

TABLE 2.31 Diesel Fuel Specifications

Property	Low-Particulate Fuel	Federal Register 85873-10, 11-15-72	
		D-1	D-2
Cetane number	44	48–54	42–50
Distillation range			
IBP, °F	359	330–390	340–400
10% point, °F	372	370–430	400–460
50% point, °F	376	410–480	470–540
90% point, °F	386	460–520	550–610
EP, °F	435	500–560	580–660
Gravity, °API	48.5	40–44	33–37
Specific gravity	0.804	0.825	0.840
Total sulfur, %	0/.01	0.05–0.20	0.2–0.5
Hydrocarbon composition			
Aromatics, %	0–6	8–15	27 (min)
Paraffins, napthenes, olefins	Remainder	Remainder	Remainder
Flash point, °F Min.	140	120	130
Viscosity, CST (40° C)	1.5	1.6–2.0	2.0–3.2
Pour point, °F			−40
Calc. energy, MJ/kg	46.02		
Calc. energy, Btu/lb	19,805		

154

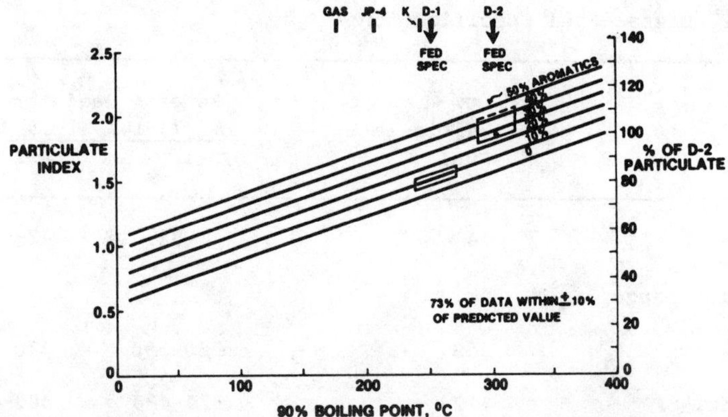

FIGURE 2.30 Graph of equation relating particulate index to fuel aromatic content and fuel 90 percent boiling point. SOURCE: Burley and Rosebrock, 1979.

significant increase in particulates independent of injection timing (Burley and Rosebrock, 1979).

As a result of this study, General Motors conducted another study using one of the best low-particulate fuels available to power several 1979 Oldsmobiles (see Table 2.31) (GM, 1980). Test data showed reductions of over 40 percent in particulates, with little change in other emissions or fuel economy (Dimick, 1980).

Figures 2.31 to 2.35 show the effect of cetane number, aromatics, and 90 percent distillation temperature on the smoke and gaseous emissions based on data taken by Toyota using the Japanese 3-mode and 6-mode cycles for loads and speeds, as shown in Figure 2.36.

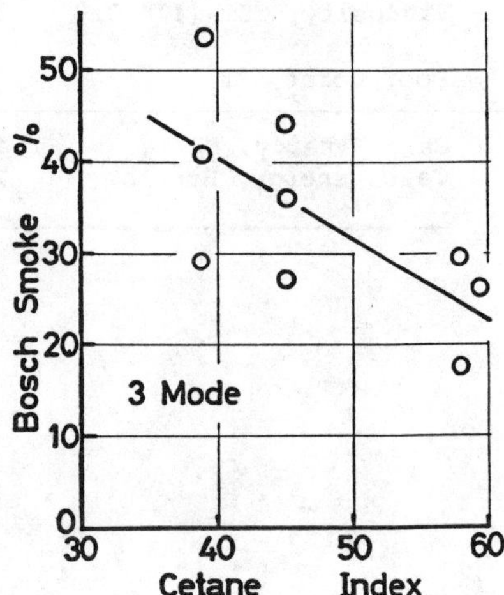

FIGURE 2.31 Effects of fuel properties on smoke. SOURCE: Toyota, 1980.

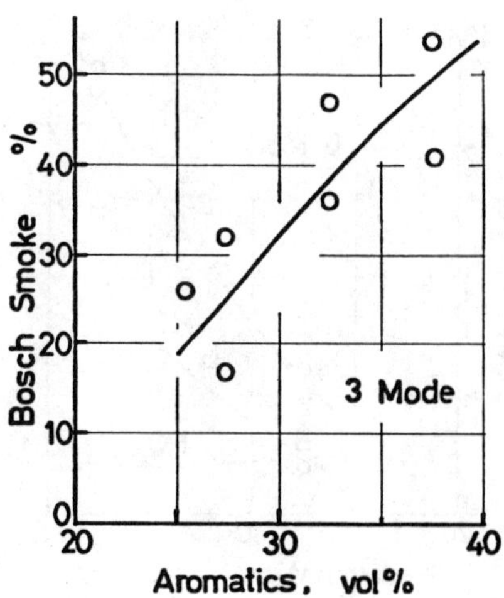

FIGURE 2.32 Effects of fuel properties
on smoke. SOURCE: Toyota, 1980.

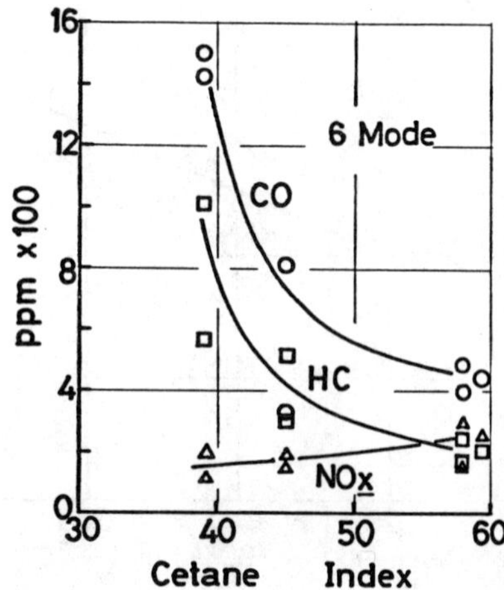

FIGURE 2.33 Effects of fuel properties
on emissions. SOURCE: Toyota, 1980.

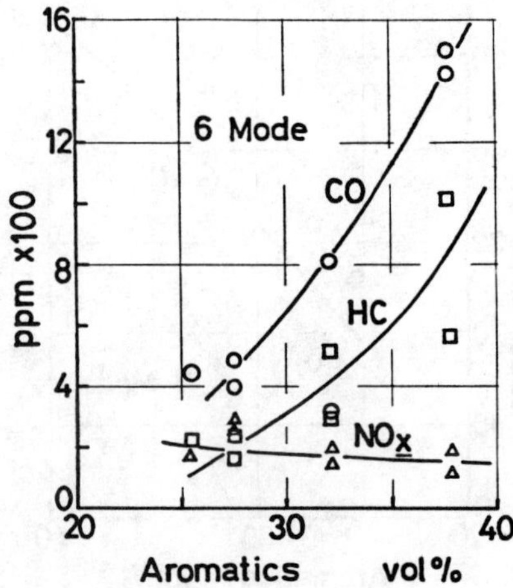

FIGURE 2.34 Effects of fuel properties
on emissions. SOURCE: Toyota, 1980.

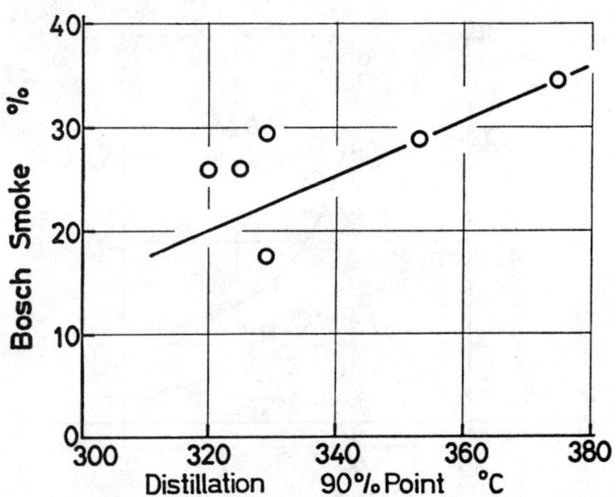

FIGURE 2.35 Effects of 90 percent
distillation temperature on
particulate.
SOURCE: Toyota, 1980.

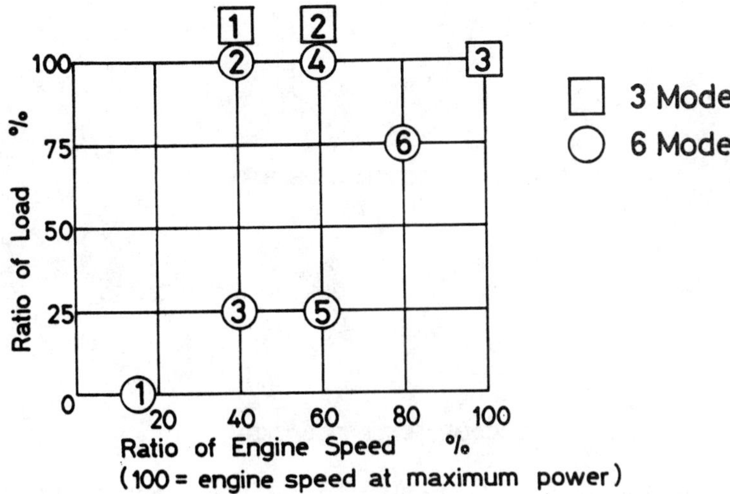

FIGURE 2.36 Required engine conditions for 3-mode
and 6-mode cycles emission test modes.
SOURCE: Toyota, 1980.

Clearly, the petroleum industry, in conjunction with the automotive
industry, needs to do more studies on the feasibility of producing a
low-particulate diesel automotive fuel. Such a study should look at
the trade-off between cost and fuel availability as a function of fuel
quality. The data would tend to indicate the general property limits
listed in the following tabulation:

Property	Limits
Cetane number	>48
Aromatics	<20 percent
90% distillation point	<316°C (600°F)
Sulfur	<0.25 percent by mass
Cloud point temperature	<15°F in summer
	<5°F in winter
	<0°F in cold climates

As diesels become more common in passenger cars and light- and
medium-duty trucks, fuel blending for cold temperature operation needs
to be done at the refinery to insure satisfactory operation of diesels
in individual consumer operation.

3

EMISSION CONTROL TECHNOLOGY

INTRODUCTION AND BACKGROUND

With few exceptions, through the 1981 model year, automotive manufacturers had little difficulty producing light-duty diesels that meet the emission standards.* Standards for the 1982 model year and thereafter, however, require substantial reductions in hydrocarbons, oxides of nitrogen (NO_x) and particulates, especially for larger vehicles. Compliance will necessitate the addition of engine emission controls. Basic approaches to diesel engine emission control fall into three major categories:

• engine modifications, including combustion chamber configuration and design, fuel injection timing and characteristics, blowing and turbocharging, forms of supercharging, and exhaust gas recirculation (EGR);

• exhaust aftertreatment, including traps, trap oxidizers, and catalysts; and

• fuel modifications, including control of fuel properties, fuel additives, and nonconventional fuels and fueling configurations such as emulsions, alcohol blends, fumigation, and synthetic fuels.

As is illustrated in Figure 3.1, each control measure or modification interacts with most engine systems and other control hardware.

Control and/or accommodation of emission control measures can range from simple passive or mechanical adjustments (e.g., for fuel changes), to open-loop control (for modulated EGR and programmed injection scheduling), to complete closed-loop electronic control (for continuous optimization of all systems). To obtain emission reductions for the various pollutants, several control measures must be used simultaneously. Furthermore, the control systems must be packaged in an integrated system, even though their actions are sometimes antagonistic.

----

*Table 1.5 provides a schedule of exhaust emission standards applicable to gasoline- and diesel-powered light-duty vehicles.

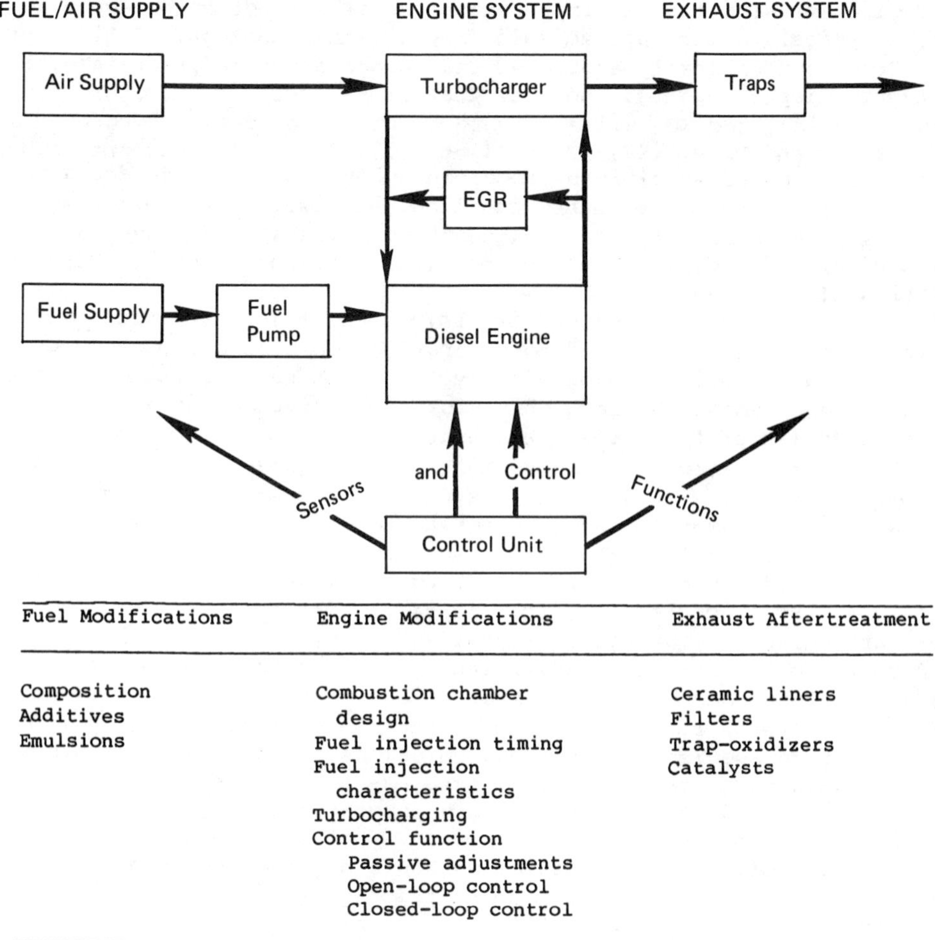

Fuel Modifications	Engine Modifications	Exhaust Aftertreatment

Composition
Additives
Emulsions
	Combustion chamber
	  design
	Fuel injection timing
	Fuel injection
	  characteristics
	Turbocharging
	Control function
	  Passive adjustments
	  Open-loop control
	  Closed-loop control
		Ceramic liners
		Filters
		Trap-oxidizers
		Catalysts

FIGURE 3.1 Morphology of diesel emissions control techniques.

When modest levels of emission control are required, the system functions, and the effects of system interactions and trade-offs can be accommodated relatively simply. As the emission standards become more stringent, however, control systems become more complex, system interactions and trade-offs become more critical, and more precise overall system control becomes essential. Indeed, a complete systems approach to vehicle operation and emissions control is, therefore, required.

Engine Operation and Emission Trade-offs

The effects of vehicle and drivetrain parameters on emissions and control strategies can be illustrated by analyzing engine emission maps. This approach also provides insight for the various emission control trade-offs.

Engine emission maps indicate emissions as a function of speed and load; a typical particulate emission map is shown in Figure 3.2. Test data, from both naturally aspirated and turbocharged indirect injection light-duty diesel engines, indicate that brake-specific hydrocarbon, carbon monoxide, and NO_x emissions are functions of engine speed and load (French and Pike, 1979; Bassoli et al., 1979). At constant higher power levels, brake-specific hydrocarbon and carbon monoxide emissions can either increase or decrease with load, at lower power settings, they always decrease with load. Although particulate emissions are minimized at midload, there are no consistent trends at other operating conditions (GM, 1979a).

For a given vehicle-engine-drivetrain combination, vehicle power requirements can be determined for operation, on a second-by-second basis, according to a given driving cycle. This can then be translated to engine requirements by accounting for the operating characteristics and efficiencies of the tires, rear axle, transmission, torque converter, and accessories. An example of a torque-speed-time matrix is provided in Figure 3.3. When this matrix is combined with emission maps, a map of the modal contributions to total emissions results. Typical maps illustrating the percentage contribution of NO_x and hydrocarbons, as a function of load (torque) and speed, are shown in Figure 3.4. Small corrections for transient operation and for cold-start emissions must also be applied.

A number of points are illustrated by these maps. The emission maps show islands and peaks at different engine operating conditions. Hence, modifying or restricting engine vehicle operations for control of one emission species may increase emissions of others. Furthermore, control measures for the various species may have to be directed toward, and applied to, different engine operational regimes. It is possible that even though a short time is spent at a particular operating condition, that condition can be a major contributor to total emissions if it corresponds with an emission peak. A large amount of time with low emission output will also contribute significantly to the total emission output.

Changes in axle ratio, transmission, etc., will modify the torque-speed-time matrix. Figure 3.5 illustrates the effect of axle ratios on engine speed-load envelopes. A decrease in axle ratio causes the engine to operate at the same power level, but at lower engine speeds and higher brake-mean-effective-pressure (BMEP) levels. This decrease also results in different emission characteristics.

In general, lower axle ratios reduce emissions, improve fuel economy, and degrade acceleration performance. For instance, in tests using a 3,000-pound inertia weight vehicle with a 2.5-liter engine, a change from a 3.2:1 to a 2.8:1 axle ratio reduced hydrocarbon, carbon monoxide, and NO_x emissions by 6, 12, and 6 percent, respectively (Bassoli et al., 1979). In addition, fuel economy for the vehicle improved by 8 percent during the combined FTP cycles, and acceleration time from zero to 60 mph increased by 2 percent.

Tests with another 3,000-pound vehicle with a 2.1-liter engine showed even greater reductions (GM, 1979b). When the axle ratio was reduced from 4.15:1 to 3.34:1, hydrocarbon, carbon monoxide, and NO_x

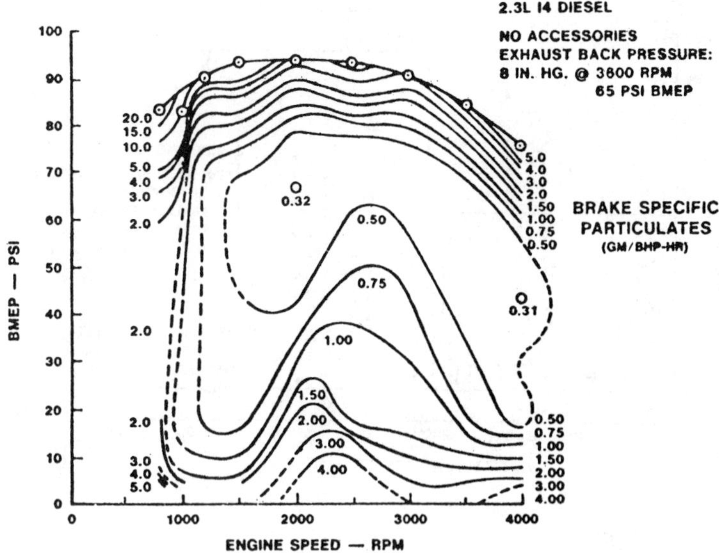

FIGURE 3.2 Brake-specific particulate island map
for a 2.3-liter L4 swirl chamber diesel.
SOURCE: Wade, 1980.

emissions were reduced by 26, 14, and 17 percent, respectively, and
fuel economy improved by 14 percent. However, the zero to 60 mph
acceleration time increased 22 percent.

The effects of other vehicle engine parameters can be analyzed
through a similar process. For example, decreasing the engine
displacement/inertia weight ratio causes the engine to operate at
higher BMEP levels. Thus, the effect should generally be the same as
decreasing axle ratio.

The emission, vehicle design, and vehicle operation trade-offs
described above emphasize the need for a total systems approach to
solving the problems of producing a vehicle that simultaneously
minimizes emissions and maximizes fuel economy, and one that is also
drivable, maintainable, and capable of pulling heavy loads such as
trailers, and operating in cold and hot temperatures.

Design Targets

Low-mileage, research-prototype vehicles are used in most emission
tests. In addition, the data sometimes come from a single vehicle or
test and, thus, often do not accurately reflect the emission levels of
production vehicles. Prototype-to-certification slippage, car-to-car
variability, test-to-test variability, and in-use deterioration factors
(the factors used to adjust 4,000-mile emission levels to 50,000 miles
because standards are based on this mileage) can all affect emissions.
The factors must be considered carefully in interpreting advanced

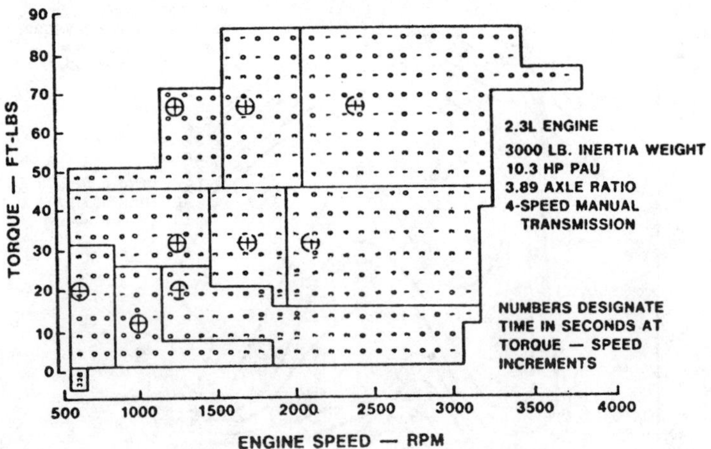

FIGURE 3.3 A torque-speed-time matrix with
mapping point regions superimposed.
SOURCE: Wade, 1980.

control technology data relative to the technological feasibility of
certain emission levels. Hence, it is necessary to set emission design
targets for the research prototypes significantly below the standards
that production and certification vehicles must ultimately meet.

The established safety margin for production slippage, and testing
and vehicle variables, typically ranges from 10 to 20 percent,
depending on manufacturer and emission species. For hydrocarbon
emissions, variability in the test procedure and in fuel sensitivity
can increase this figure to 30 percent or more. If the recommended
modifications in the hydrocarbon test procedure were implemented and if
there were less variation in diesel fuel properties, the hydrocarbon
safety factor would be more consistent with those for other emissions.

The actual in-use deterioriation factor is less certain. Based on
the durability and stability of diesel engines, it is expected that
with no control devices, in-use deterioration factors will be small.
Data from 1978 and 1979 certification test field studies (see Figures
2.9 and 2.10, and Table 2.7) support this premise. As emission control
devices are added, however, in-use deterioration factors become less
certain. For example, when EGR was added to a diesel to control NO_x
emissions, particulate deterioration factors reportedly reached 1.4.

A production safety margin of 20 percent is adequate for vehicles
that require only minor emission controls. (This figure is 30 percent
for hydrocarbons if the procedure is not modified as recommended.) For
more complex and more severe control approaches, a production safety
margin of 30 percent should be used. The production safety margin can
be adjusted downward as the control technology is developed and
engine/vehicle operating experience is acquired.

The remainder of Chapter 3 focuses on diesel emission control
technologies. Each of the basic control concepts is described and

evaluated according to its effects on engine/vehicle performance and exhaust emissions. The following effects and factors are considered, as appropriate, in the analysis:

* effects on particulate emission levels;
* effects on NO_x, hydrocarbon, carbon monoxide, and unregulated emissions;
* effects on engine brake-specific fuel consumption and vehicle fuel economy;
* effects on engine performance and durability;
* need for active control, the degree of sensitivity;
* complexity;
* degree of maintenance required;
* relative cost;
* ease of integration within the engine; and
* status of technology.

The discussions apply to indirect injection light-duty diesels; direct injection diesels are discussed in the section on direct injection engines.

ENGINE MODIFICATIONS

In attempts to improve the performance of light-duty diesels, manufacturers are investigating a number of engine design modifications. The

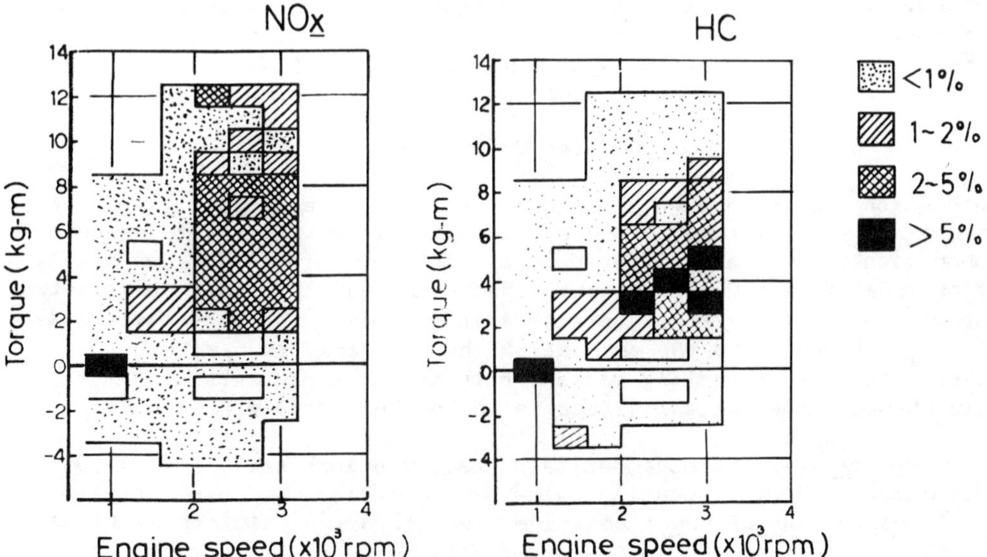

FIGURE 3.4 Modal contribution to mass emissions over the Federal Test Procedure (FTP) cycle. SOURCE: Toyota, 1980.

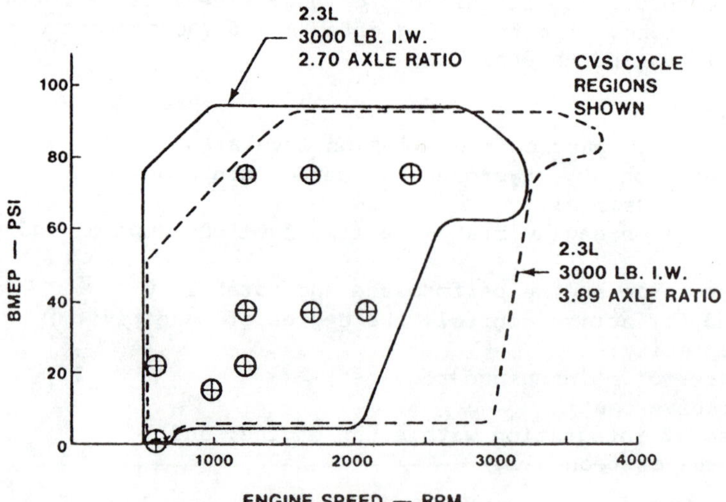

FIGURE 3.5 Envelopes of FTP
brake-mean-effective-pressure-engine speed regions
for 2.3-liter, 3,000-lb inertia-weight vehicles
with 2.70 and 3.89 axle ratios.
SOURCE: Wade, 1980.

basis of these modifications are changes in the engine's combustion
system--improvements in combustion chamber design and the fuel injection
system. In addition, engine add-ons, such as turbochargers and exhaust
gas recirculation (EGR) systems, are also being considered as means for
improving the overall performance of the light-duty diesel.

Combustion Chamber Design

Combustion chamber design and engine performance are closely related.
In engines with divided chambers, fuel-air mixing and the subsequent
combustion process are affected by swirl rate, turbulence, and the flow
of residual fuel and combustion products through the passageways between
the prechamber and the oxygen-rich main chamber. These conditions are
in turn influenced by the combustion chamber geometry. Within the
prechamber, the connecting passage between the two chambers, and the
main chamber, the following factors influence engine performance:

* __prechamber__: prechamber/main chamber volume ratio, swirl rate,
turbulence, injector location, glow plug location, and crevice areas;
* __connecting passage__: passage size (throat area), number of
holes, shape (cross section and approach contours), direction of flow
(prechamber and main chamber), and location;
* __main chamber__: vertical or angled valves, head side geometry
(cutouts); piston top geometry (cutouts), top ring position.

In addition, bore, stroke, compression ratio, manifolding, etc., also affect performance. Table 3.1 shows the combustion chamber design and fuel injection factors that influence emissions. (Note that injection timing and combustion chamber geometry affect emission levels more than any other variables.)

Compression Ratio

Typically, compression ratios greater than 20:1 are used in light-duty diesel engines. These high compression ratios improve performance and cold starting, and reduce emission levels during cold-temperature operation. Figure 3.6 illustrates the effects that compression ratios can have on emissions for two injection timings (note the high level of hydrocarbons at the 7°C timing).

Prechamber Design

Figure 3.7 illustrates a swirl combustion chamber and indicates the various angles and radii that were examined in one study (Isuzu, 1980).

TABLE 3.1 The Effects of Combustion Chamber Design and Fuel Injection Variables on Emissions From an Indirect Injection Diesel Engine

| | Gaseous Emissions | | | Particulates[a] | |
| | | | | Unburned | |
	HC	NO_x	CO	Hydrocarbons	Smoke
Injection timing	+++	+++	+	+++	++
Injection rate	++	++	+	++	++
Spray cone angle	+	+	+	+	+
Secondary injection	+++	+	+	+++	+
Combustion chamber geometry	+++	+++	+	+	+++
Injector location	++	++	+	+	+

[a] Particulates are divided into two categories because hydrocarbons contribute to the soluble organic fraction and smoke is an indicator of the solid particulate fraction.

+++ Strongly dependent.
++ Moderately dependent.
+ Slightly Dependent.

SOURCE: Toyota, 1980.

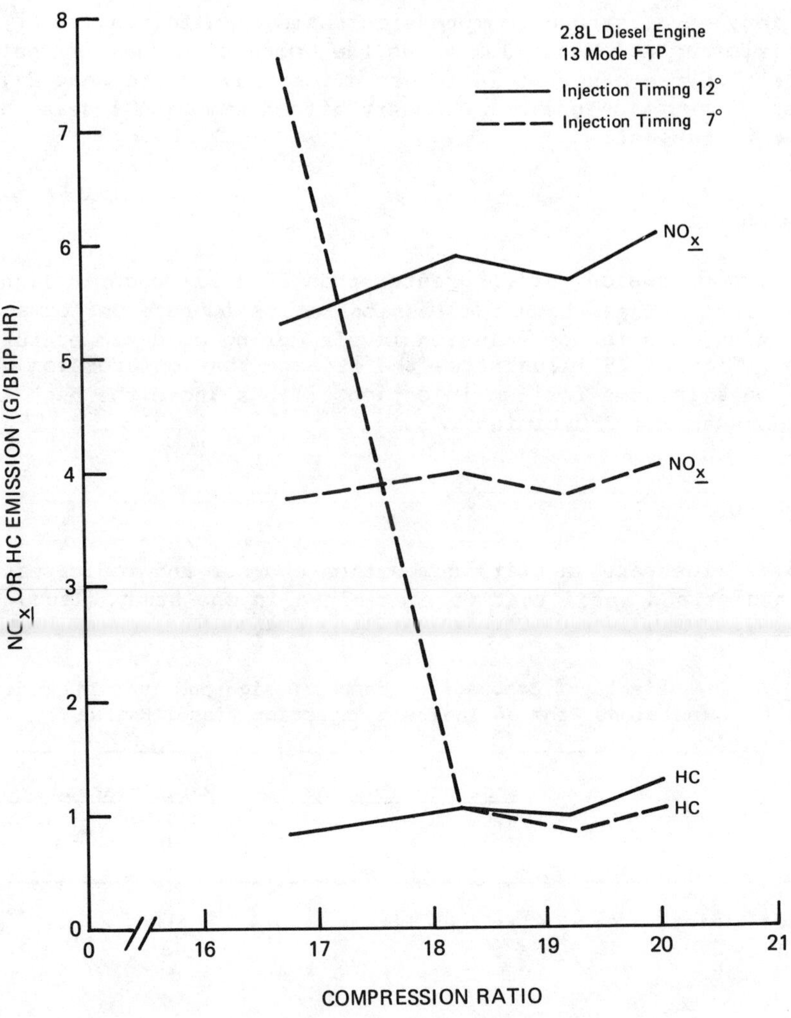

Specification

Type of Combustion Chamber	Nozzle Angle deg	Throat Angleß, deg	Spray Center line Eccentricity r, mm	Spray Radius r Sphere Radius r
1(base)	39.0	35.0	2.8	0.19
2	37.5	35 0	5.0	0.34
3	39.0	35.0	5.0	0.34
4	37.5	27.5	5.0	0.34
5	39.0	27.5	5.0	0.34
6	37.5	27.5	6.4	0.43

FIGURE 3.6 Effect of compression ratio on NO_x and hydrocarbon emissions. SOURCE: Isuzu, 1980.

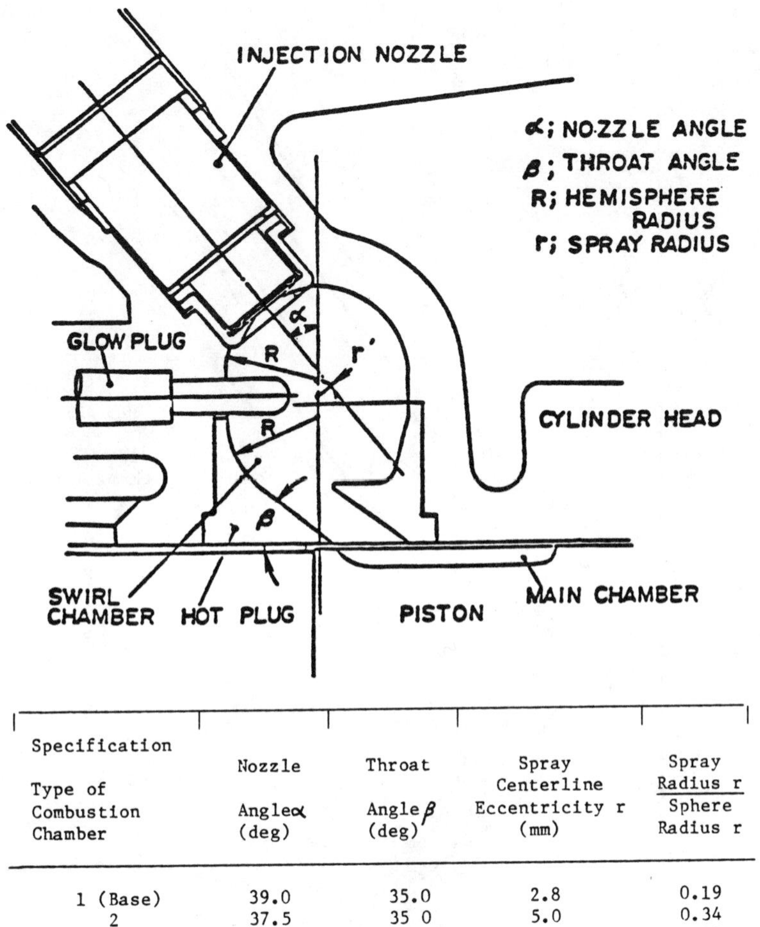

Specification Type of Combustion Chamber	Nozzle Angle α (deg)	Throat Angle β (deg)	Spray Centerline Eccentricity r (mm)	Spray Radius r / Sphere Radius r
1 (Base)	39.0	35.0	2.8	0.19
2	37.5	35 0	5.0	0.34
3	39.0	35.0	5.0	0.34
4	37.5	27.5	5.0	0.34
5	39.0	27.5	5.0	0.34
6	37.5	27.5	6.4	0.43

FIGURE 3.7 Combustion chamber configuration for 2.4-liter diesel engine. SOURCE: Isuzu, 1980.

Figure 3.8 provides emission data, as a function of injection timing, for the six designs tested. The data indicate the general emission trade-offs and the sensitivity of emissions to chamber design.

In addition to swirl chamber configuration, the throat is also an important chamber design variable (see Figure 3.7). Figure 3.9 shows idle noise data on two sides of an engine for two throat-to-total-bore area ratios. The data illustrate the need to optimize for both engine noise and emissions. Figure 3.10 shows emission data illustrating the effect of a +20 percent change in throat area (Toyota, 1980). These data also show that swirl ratio increases with a decrease in throat

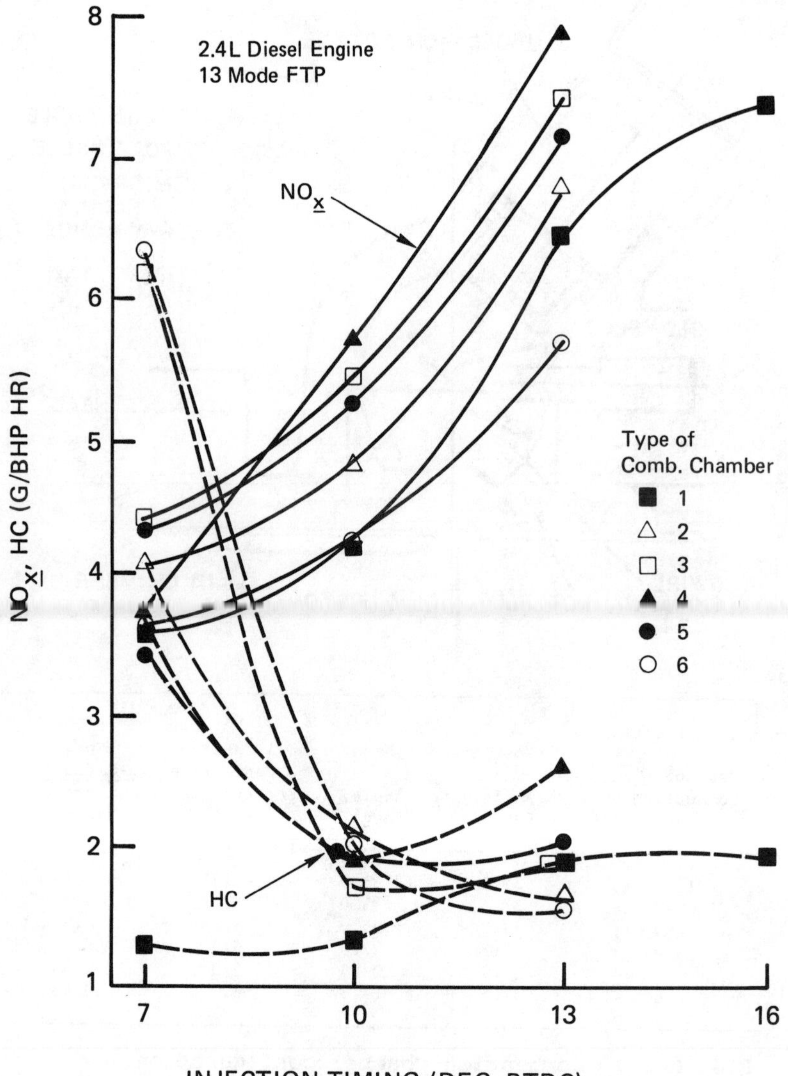

FIGURE 3.8 Effect of combustion chamber configuration on NO$_x$ and hydrocarbon emissions. SOURCE: Isuzu, 1980.

area. Furthermore, they provide an example of the trade-offs between NO$_x$, hydrocarbon, and particulate emission levels. A smaller throat area (-10 percent from production-base) reduces particulates and hydrocarbons but increases NO$_x$ emissions; in addition, it gives poorer startability (Nissan, 1980).

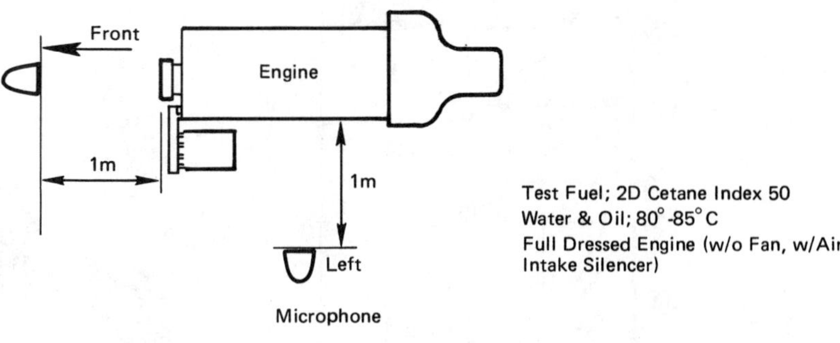

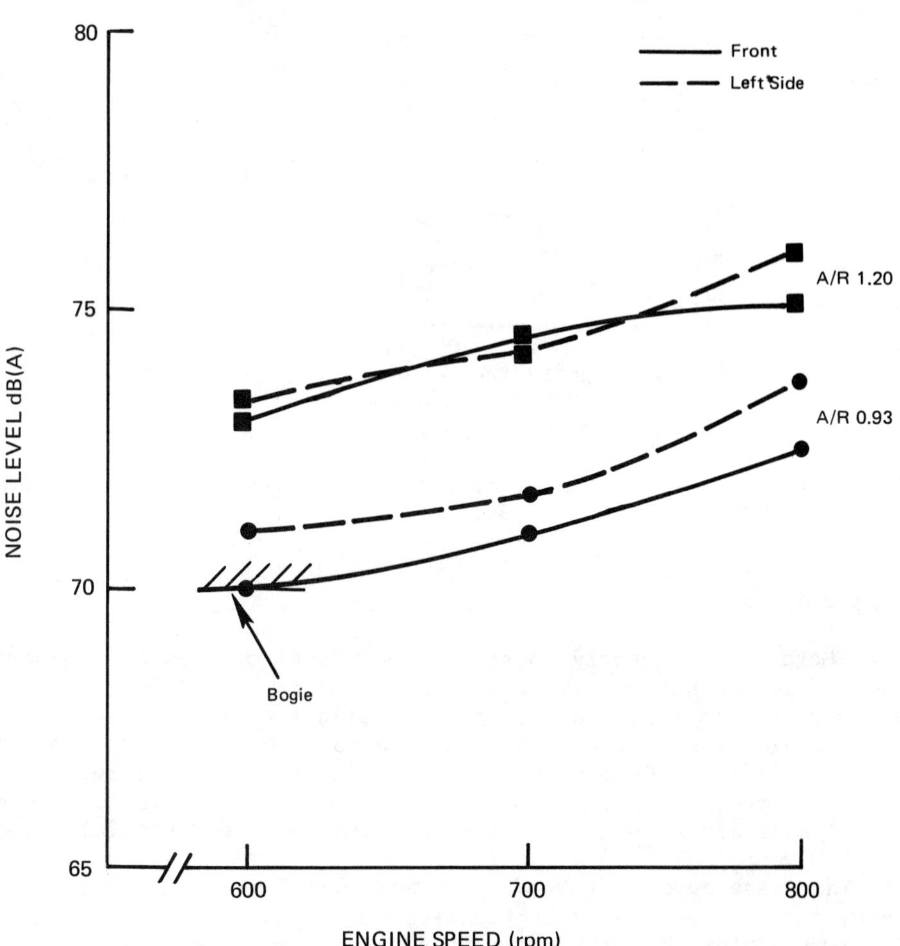

FIGURE 3.9   Idle noise test results.   SOURCE: Isuzu, 1980.

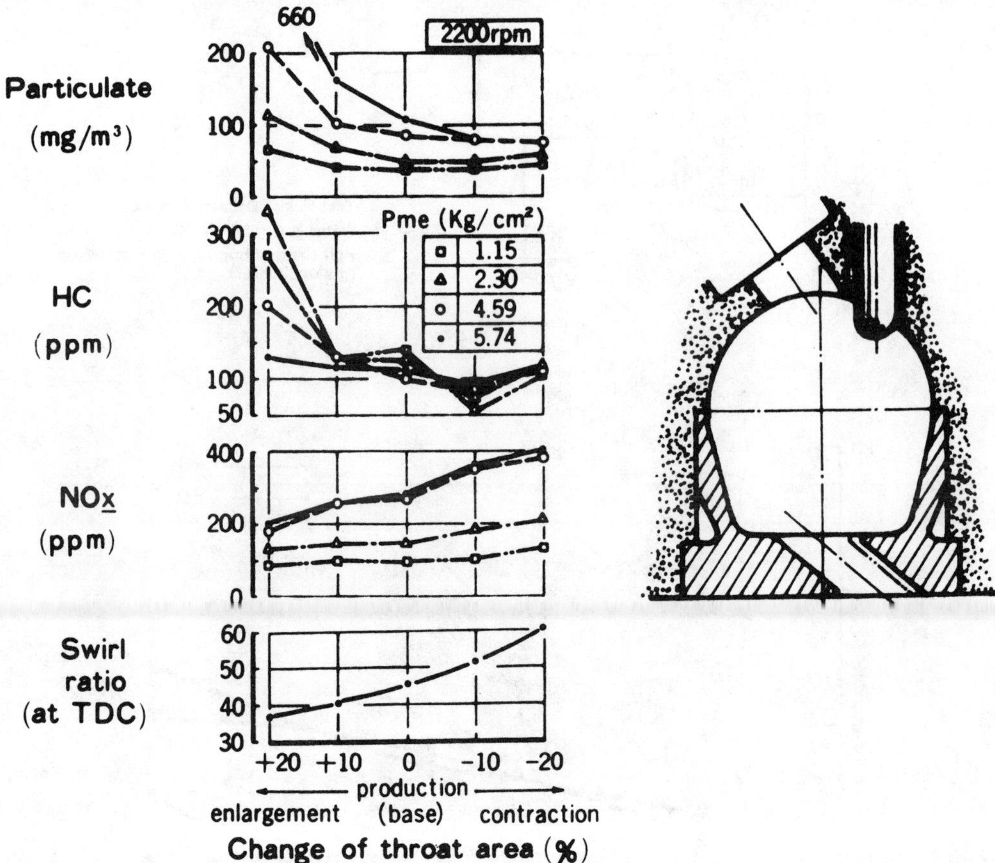

FIGURE 3.10 Combustion chamber modification--effect of throat area on emissions. SOURCE: Toyota, 1980.

Glow Plug Geometry

General Motors has recently looked at the effect of glow plug geometry on diesel particulate emissions (GM, 1980). Previously General Motors research showed that removal of the glow plug from the swirl chamber of an Opel 2.1-liter diesel resulted in a 50 to 60 percent reduction in particulates over a 10-point engine operating cycle. In a recent experimental program, which sought to show whether the particulate was a thermal or a mixing effect, General Motors concluded the following:

* The presence of a glow plug changes the fuel-air mixing process, thereby increasing particulate emissions.
* A glow plug that minimizes flow disturbances is needed to reduce particulates.
* The reduction of flow disturbances by the glow plug will increase NO_x emissions.

There is also an important interaction between the fuel injection system and the combustion chamber geometry.

Fuel Injection System

The fuel injection system is the heart of a diesel engine. It has important design and performance interactions with combustion chamber geometry. Its performance and control capability effectively determine engine performance. The following injection design variables will ultimately determine engine performance: spray location, spray character, injection timing, injection duration and rate, and maximum fuel quality. Injection timing and duration (including rate of injection) are the main variables affecting engine performance (Wessel and Joachim, 1979).

Research has shown that fuel-air mixing, because it controls burn time (which is almost 50°C), is of primary importance in diesel combustion (Greeves et al., 1980). Experiments show that fuel economy improves with shorter burn times and that faster fuel-air mixing reduces particulate formation (Greeves et al., 1980). However, the accompanying higher rates of mixing and combustion can result in higher rates of NO_x formation and cylinder pressure rise. Both of these effects can be minimized by retarding injection timing.

The two principal sources of hydrocarbon emissions in diesel engines are late release of fuel from the nozzle sac volume and incomplete combustion in the over-lean regions in the premixed fuel-air mixture (Greeves et al., 1980). The emission of hydrocarbons, therefore, can be minimized by injector nozzle design and by providing a high end compression temperature (achieved by high compression ratio and turbo-charging) to minimize ignition delay and premixing. A high cetane number fuel (greater than 48) reduces the amount of premixing and, therefore, hydrocarbon emissions. Matching the initial rate of fuel injection to the ignition delay of the fuel also helps.

The Effects of System Variables In addition to these parameters, the injection process needs to be stable and repeatable, with a sharp end-of-injection cutoff and no secondary injections. Closing the injection nozzle needle valve causes a pressure wave to return to the injection pump. Part of this pressure wave is reflected at the pump back to the injection nozzle. If the reflected wave is strong enough, it will reopen the spring-loaded needle valve, causing a secondary injection. Secondary injections cause performance deterioration, particularly regarding hydrocarbon and smoke emissions. Correct design of the following injection system components can eliminate secondary injections: fuel injection tubing length and diameter, nozzle-opening pressure, needle valve and spring mass, needle valve lift, and pump discharge valve characteristics (Wessel and Joachim, 1979).

Figure 3.11 shows five different rates of injection that were run in a single-cylinder engine; Figure 3.12 shows the emission and performance data for constant injection timing as a function of the mean injection rate (MIR) for one load. Reduced rates (B) from the production system (A) appear to give lower NO_x emissions, with minor trade-offs among other parameters. Note that since the start of injection was essentially fixed, the reduced injection rate includes a timing retard.

Figure 3.13 shows the effect of injection timing on fuel consumption and emissions for an engine speed of 2,000 rpm and two loads. Further retarded timings result in reduced NO_x emissions but higher hydrocarbons. These data clearly show the importance of setting the injection timing accurately and precisely to minimize hydrocarbons, NO_x, and particulates. Present production timing tolerances are typically + 1°C. There are a number of factors that contribute to timing variations, and a systems approach to all these variables must be used to reduce the overall tolerances. These factors are:

* speed timing advance (mechanical) variations;
* initial timing setting;
* load sensor for timing;
* nozzle opening pressure;
* diameter of injection lines;
* retraction volume; and
* cam/follower and other pump component variability.

Development work is under way to reduce timing variations and thus optimize emission control better.

Figure 3.14 provides data on hydrocarbon, NO_x, and particulate emissions, as a function of timing, for a number of vehicles. The sensitivity of NO_x and hydrocarbons to timing is apparent. Particulate emissions show less variation, because the optimum timing region shown in this figure is concave and fairly flat. Figure 3.15 shows that fuel economy is fairly insensitive to timing variations. These data include effects of cylinder to cylinder timing variations-- variations which represent another source of emission changes because each cylinder is not timed at the optimum point.

The basic pump-line-nozzle system with distributor pumps, modified for improved timing control as a function of load and speed, and for improved injection rates to match injection characteristics to combustion chamber designs, will continue to be used in most light-duty vehicles through 1984. Other control functions used on current light-duty engines are: automatic cold-start timing advance, temperature-controlled idle stop (for fuel quantity change), and temperature-controlled starting fuel quantity. Turbocharged engines also use aneroid devices (which sense engine compressor boost pressure during acceleration) to limit maximum fuel delivery and thereby prevent overfueling until the boost pressure reaches a preset level.

<u>Electronic Fuel Injection Systems</u> Electronically controlled injection pumps, which will lower hydrocarbon, NO_x, and particulate emission levels, should become available by 1985. With recent developments in electronics, particularly microprocessor technology, the electronic control of fuel injection equipment becomes a much more

173

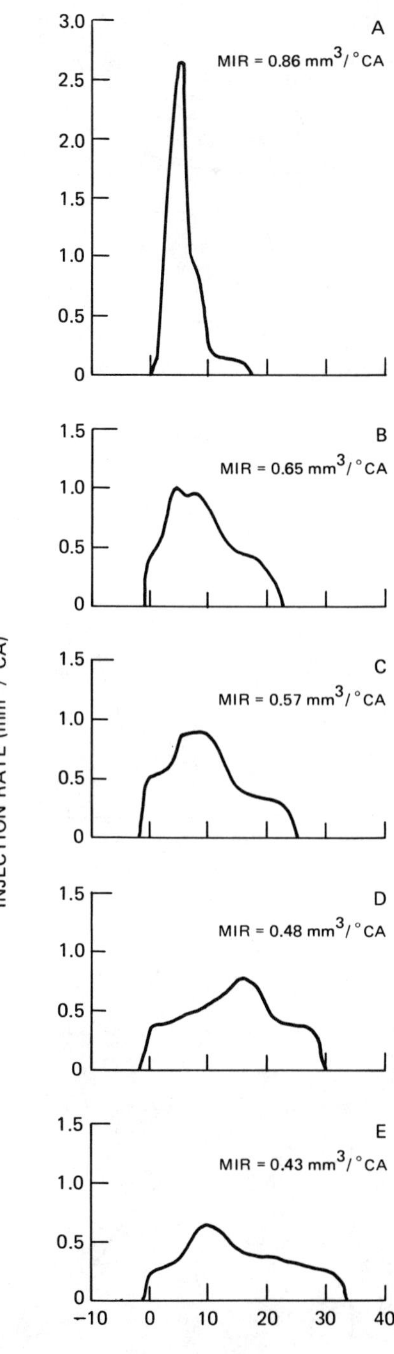

FIGURE 3.11 Effect of injection rate.
SOURCE: Nissan, 1980.

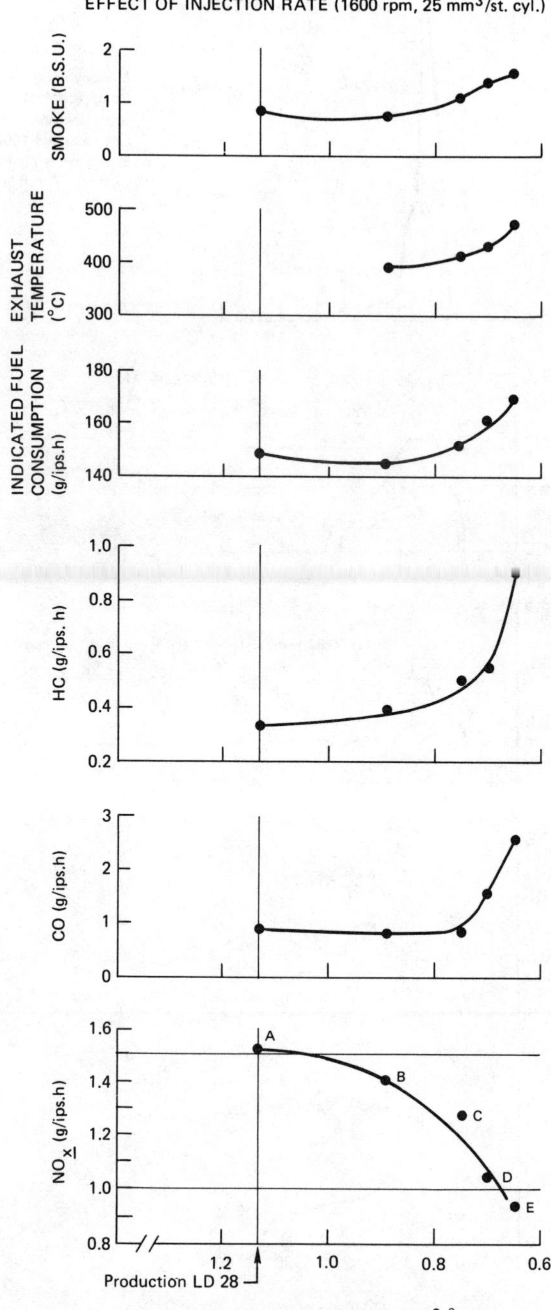

FIGURE 3.12 Effect of injection rate.
SOURCE: Nissan, 1980.

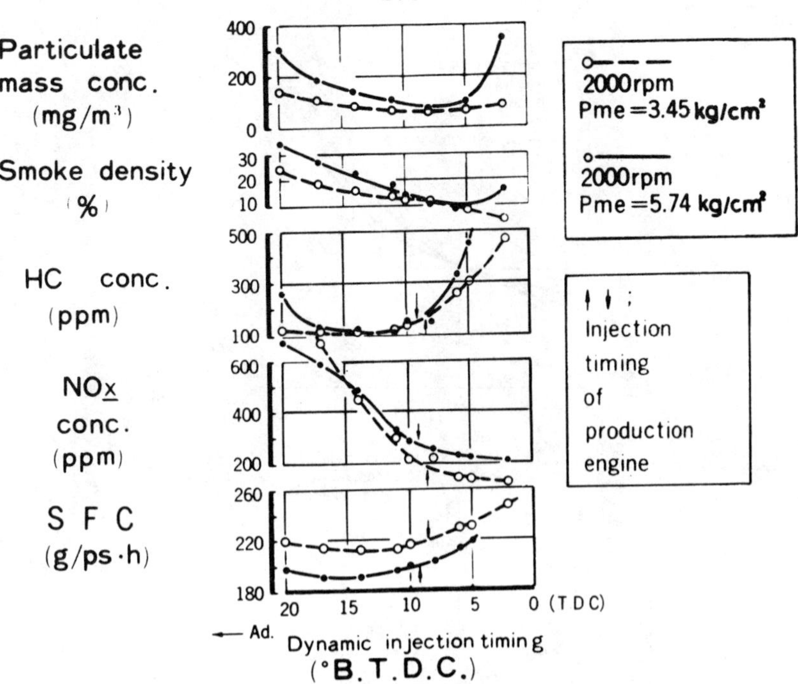

FIGURE 3.13 Effect of injection timing on **emissions**.
SOURCE: Toyota, 1980.

attractive proposition. These controls are compatible with the all-
altitude requirement that currently will require development of a
separate analog control system applicable only to the 1984 model year
(GM, 1980). In addition to the technical advantages of electronically
controlled systems, they will require a smaller capital investment than
the advanced technology needed for fully controlled mechanical systems.

Figure 3.16 shows a schematic arrangement of a system being
developed to provide electronic control of a distributor pump (Gliken
et al., 1979). With this system, electronic sensors pick up signals
representing fuel rate and advance position; these signals are
transferred to an electronic control box. The preprogrammed control
box determines optimum fuel delivery rate and timing setting from
information reported by the sensors and driving demand. Signals are
then sent from the control box to actuators that set fuel delivery and
timing for the optimum conditions. A block diagram of the control
system is shown in Figure 3.17. The maximum fuel delivery curve as a
function of speed, and the timing map as a function of speed and load
(or derivatives), are stored in the memory of the microprocessor
(Gliken et al., 1979).

The electronic technology for such a system is available, and a
number of prototype cars with electronic fuel injection control systems
are running, but further development is needed in the sensor and
actuator area.

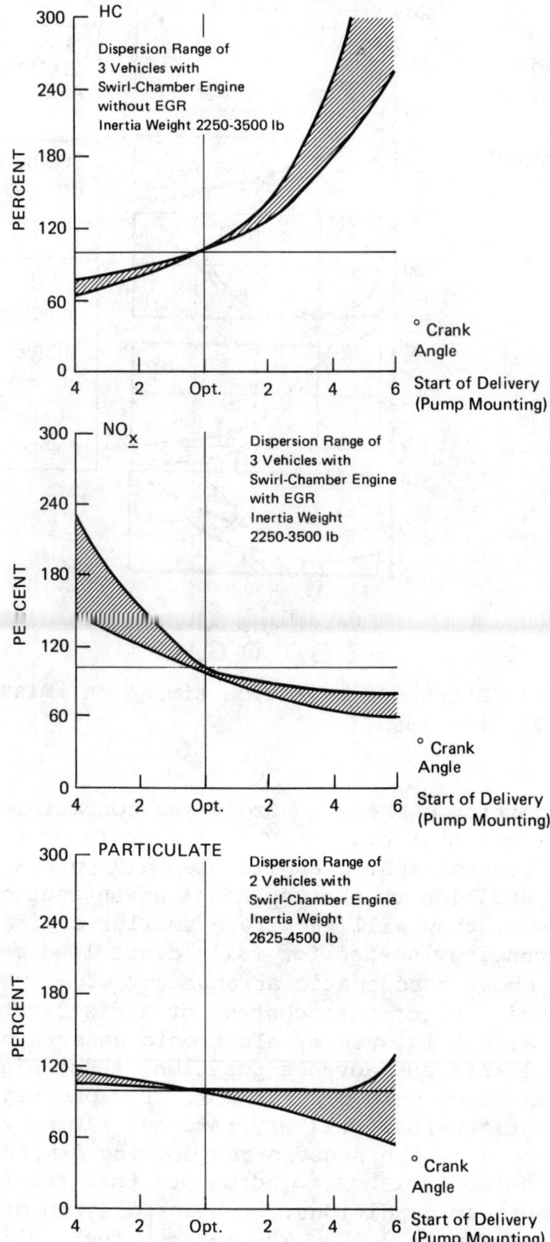

FIGURE 3.14 Dispersion range of emissions, as a function of injection timing, for three vehicles with swirl-chamber engines without exhaust gas recirculation. SOURCE: Robert Bosch, 1980.

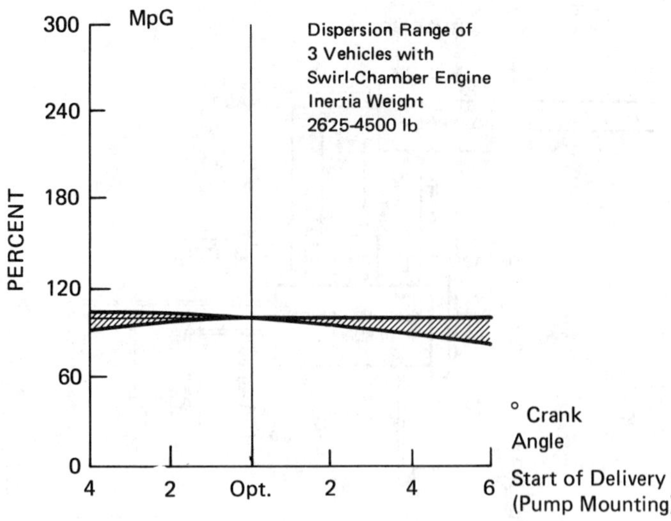

FIGURE 3.15 Dispersion range of fuel economy, as a function of injection timing, for a swirl chamber engine. SOURCE: Robert Bosch, 1980.

<u>The Injector Nozzle</u> The other important part of the injection system is the injector nozzle. A recent development in this area is the CAV microinjector described in the work of Howes (1980). The conventional pintle nozzle used for indirect injection engines prior to 1980 cannot be easily reduced below a certain size. Today's diesel passenger cars, because of limited space in engines in the 0.4- to 0.7-liter/cylinder capacity, need nozzles with extremely small diameters. To obtain a substantially smaller injector with a satisfactory spray pattern for swirl chamber engines, Lucas CAV developed a design based on a poppet valve. In a conventional injector, the needle valve moves inward within a precision-fitted body. This leads to a costly design that also usually requires a back-leak pipe. Reversing the movement of the valve and thereby allowing it to move outward with the fuel, like a check valve, reduces cost and system complexity and overall bulk. Figure 3.18 compares the CAV microinjector to a conventional pintle nozzle injector. Some improvement in cylinder head configuration and weight can be gained, as shown in Figure 3.19. Figure 3.20 shows the developed performance of this injector at 1,600 rpm as compared with the reference fuel injection system (1978-1979 production) for the General Motors 5.7-liter engine fitted with a multiorifice inward opening nozzle. These data show significant hydrocarbon and smoke improvements. Development tests showed that the glow plug and the injector had to be reversed, presumably because too much fuel was being put on the wall by the air swirl (Howes, 1980).

The microinjector developments described above illustrate the magnitudes of improvements (10 to 20 percent) and trade-offs that can occur through continued development. Other manufacturers are doing

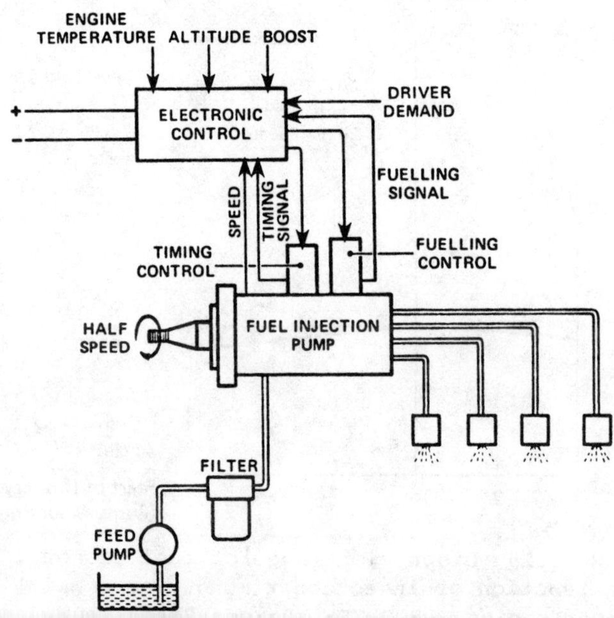

FIGURE 3.16 Electronic fuel injection
system. SOURCE: Gliken et al., 1979.

development tests with this and other injectors. Redesign and development tests of combustion chamber geometry to optimize the performance
of different injectors can take several years.

The other important nozzle variable for hydrocarbon control in
direct injection engines is the sac volume. Multiorifice injectors
with inward-opening needles generally have four or more small (0.006-
to 0.012-inch-diameter) orifices. Figure 3.21 shows heavy-duty engine
hydrocarbon data plotted as a function of sac volume for three engines
operated over the EPA 13-mode steady state test cycle. Sac volumes of
0.6 mm^3 are feasible, whereas volumes of 0.3 mm^3 or lower have not
been proven durable. These data clearly show that hydrocarbon reductions can be obtained when sac volume is decreased. General Motors has
evaluated various sac volumes in its 5.7-liter swirl chamber engine.
These results are shown in Table 3.2. Some reductions in hydrocarbons,
carbon monoxide, and NO_x emissions, as well as improvements in fuel
economy, were achieved, particularly with the 0.2 mm^3 sac volume (GM,
1979c).

By 1985, it would appear that electronic fuel injection, in
conjunction with EGR (for NO_x control), can be used to reduce the
NO_x and particulate emissions to levels approximately 30 percent
below those achieved for 1982. Approximately half of this particulate
improvement would come from combustion chamber, rate of injection, and
nozzle design; the other half would come from the electronically
controlled fuel system, which would provide better speed-load and
start-up timing control. This conclusion is based on general develop-

ment trends of the past and a general assessment of the additional
capability provided by electronic control. Because the FTP is a
transient cycle, possible emission improvements can only be determined
by experimental research and development tests.

Turbocharging

Early interest in turbocharging the light-duty diesel to control
particulate emissions arose because of successful experience with
turbocharging heavy-duty diesels. Through the use of a radial inflow
turbine, turbochargers convert thermal energy in engine exhaust gas
into rotational mechanical energy. With turbocharging, an exhaust-
driven turbine is coupled to a compressor in the intake system. The
compressor, by converting the rotational mechanical energy from the
turbine, increases the pressure and density in the intake manifold
supplying air to the engine. Because the thermodynamic cycle of the
engine begins at a higher initial pressure, the turbocharged engine can
achieve the same brake mean effective pressure at a lower fuel/air
ratio than the naturally aspirated engine. Figure 3.22 illustrates a
complete turbocharger system. Note should be made of the waste gate
control valve, which bypasses exhaust gases under high-speed and -load
conditions directly to the exhaust manifold, thereby limiting intake
boost pressure. As is shown in Figure 3.23, carbon monoxide, hydrocarbon, and particulate emission levels all decrease with a decreasing
fuel/air ratio (increasing air/fuel ratio) up to a point where further
increases in fueling rate may result in poor combustion and increases

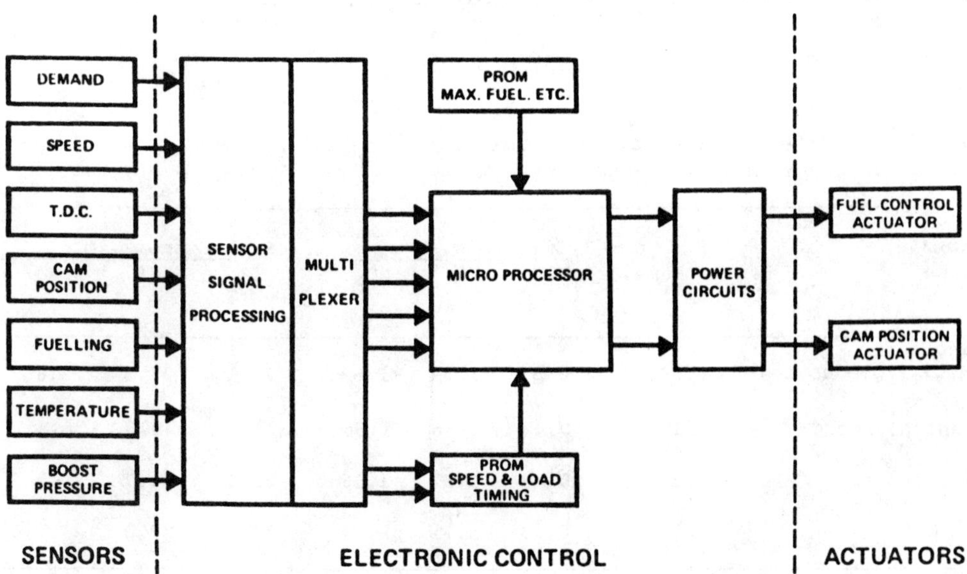

FIGURE 3.17 Block diagram of electronic control. SOURCE: Gliken *et al.*, 1979.

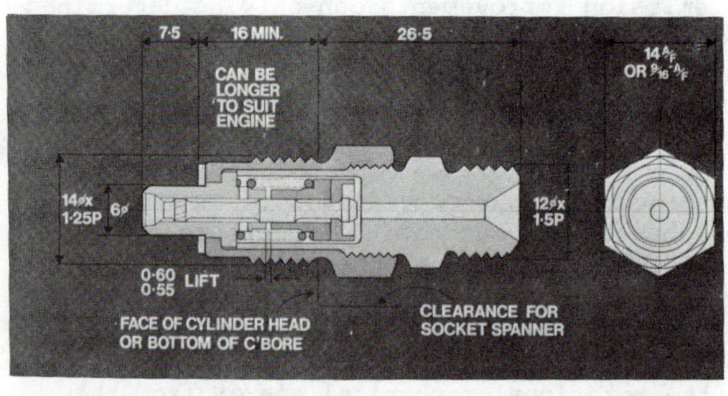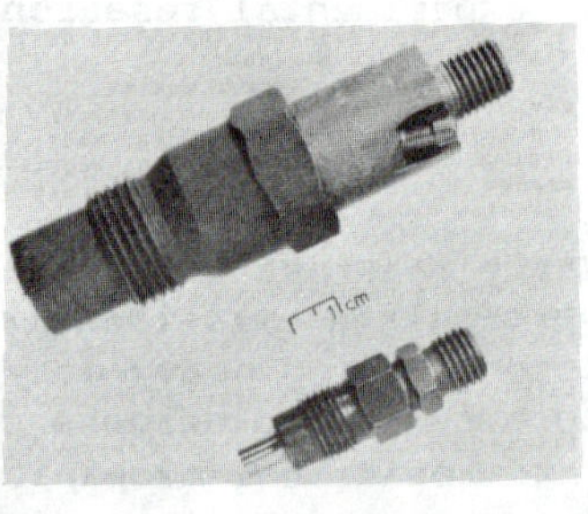

FIGURE 3.18 Microinjector. SOURCE: Howes, 1980.

in emissions and specific fuel consumption. The greatest advantage of turbocharging is the resulting increase in specific power output. Turbocharging makes it possible to obtain two or more power levels from the same basic engine--a marketing and manufacturing advantage.

Effects on Emissions, Fuel Economy, and Performance

As is shown in Table 3.3, attempts at turbocharging light-duty diesels, without other vehicle or driveline modifications, generally have not

TABLE 3.2 Effect of Sac Volume on Emissions and Fuel Economy for a Vehicle With a 5.7-Liter Diesel Engine

Nozzle Type	Sac Volume, mm^3	Emissions, g/mi			Fuel Economy, mpg		
		HC	CO	NO_x	FTP	FET	Comb.
Short sac (1979 production)	0.59	0.53	1.89	1.69	21.1	27.5	23.6
Conical sac	0.35	0.48	1.66	1.65	23.6	30.8	26.4
Tapered sac	0.20	0.43	1.52	1.56	23.3	31.7	26.5
VCO[a]	0.10	0.43	1.71	1.79	22.1	29.0	24.7

[a] Valve covers orifice.

SOURCE: Roessler et al., 1980.

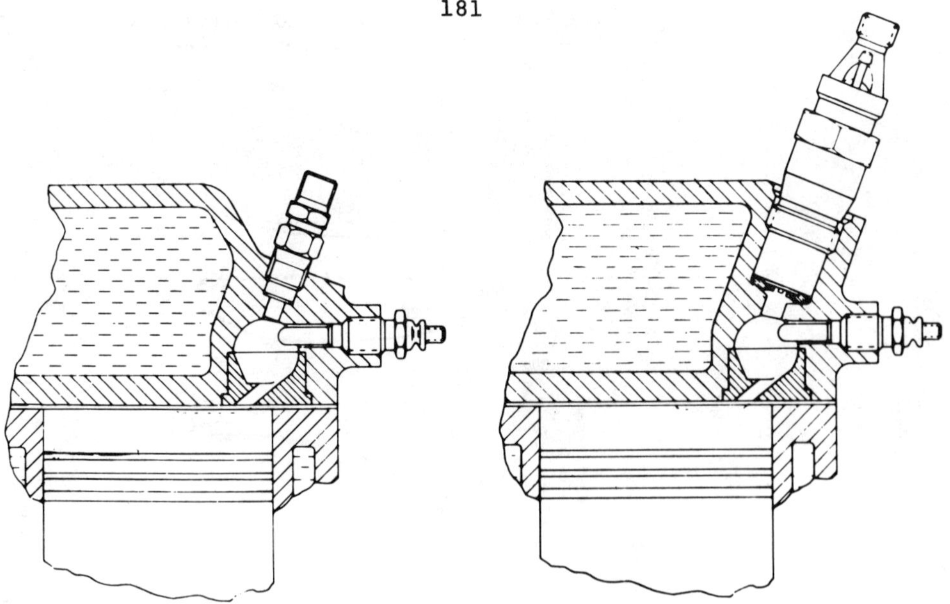

FIGURE 3.19 Cylinder head configuration of a microinjector and a conventional injector. SOURCE: Howes, 1980.

resulted in particulate reductions. Particulate emission levels from turbocharged engines show no consistent pattern; NO_x levels were higher more often than lower; and fuel economy was somewhat degraded. Hydrocarbon emissions, however, were reduced by up to 30 percent. The performance (acceleration) of the turbocharged vehicles ranged from a 1 percent improvement to a 3.6 percent degradation.

The lack of significant acceleration improvements in turbocharged diesels is easily explained. Figure 3.24 shows that for a typical turbocharger-engine match the turbocharger does not supply appreciable boost at low engine speeds, and Figure 3.25 shows the low turbocharger efficiency associated with low engine speeds. During the zero- to 60-mph-acceleration test, the turbocharger has time to get through the transient delay and supply adequate boost to offset the inefficiency exhibited at low engine speeds (Figure 3.26). It should be noted that this is a maximum fueling rate test and not a driving cycle test. The engines of Table 3.3 had the same fuel delivery system as the naturally aspirated engines. As a result, the turbocharged engines could not achieve their peak fuel/air ratio and peak power potential. As is shown in Table 3.4, when the naturally aspirated fuel system was replaced with a system optimized for turbocharged operation, performance improved significantly.

The data of Table 3.4 indicate that the maximum deliverable fueling rate does not affect either emissions or fuel economy. These data would seem to indicate that during the Federal Test Procedure (FTP) cycle either there are no full-throttle acceleration regimes for these vehicles or such transients do not contribute significantly to emissions

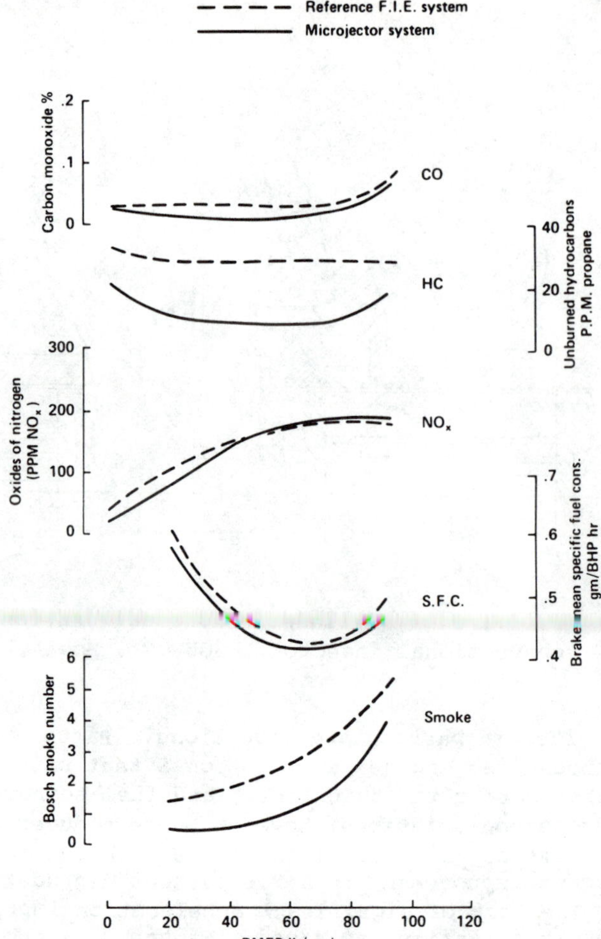

FIGURE 3.20 Developed performance of micro-injector. SOURCE: Howes, 1980.

or fuel consumption. Previously it has been assumed that turbocharging an engine with a naturally aspirated fuel delivery system would lower particulate emissions. The data in Table 3.4 indicate that this is not generally the case. However, an underpowered naturally aspirated vehicle might emit lower levels of particulates when it is turbocharged because of leaner operation during the full load acceleration periods of the FTP.

The low turbocharger efficiency during the FTP cycle illustrated in Figure 3.25 has been cited as the reason for the lack of emission control with turbocharging. Previous investigators (French and Pike, 1979) suggested that the turbocharger acted as a throttling device during much of the FTP cycle. If these investigators were correct, turbochargers would cause a decrease in the thermal and volumetric efficiencies of the engine and an increase in the fuel/air ratio needed to achieve a specific load. The decreased engine efficiency would result in a higher level of fuel consumption, and the increased

fuel/air ratio would result in increased NO_x and particulate
emissions. This reasoning does not explain the decrease in hydrocarbon
emissions with turbocharging. However, investigators at AiResearch
(Roessler <u>et al</u>., 1980) report that turbochargers do not throttle the
intake, even at low engine speeds. As is shown in Figure 3.27, this
finding was supported by earlier research of Wiedemann and Hofbauer
(1978). This figure does not, however, show the back pressure from the
exhaust turbine, which restricts air flow through the engine. Hence,
the turbocharger effect on emissions cannot be explained by intake
pressure measurements alone. During all but high fuel/air ratio
operation the back pressure can be higher than the boost pressure.

When the back pressure is higher than the boost pressure, the
residual fraction is higher and the volumetric efficiency is lower than
they would be in a naturally aspirated engine. This results in what
might be envisioned as "internal" recirculation of exhaust gas. A
slight amount of EGR generally decreases hydrocarbon and NO_x
emissions, but it has little effect on carbon monoxide, particulates,
and fuel consumption. However, the intake air temperature is raised by
residual backflow during the valve overlap period and by compression
with the slight turbocharger boost. As a result, NO_x emissions

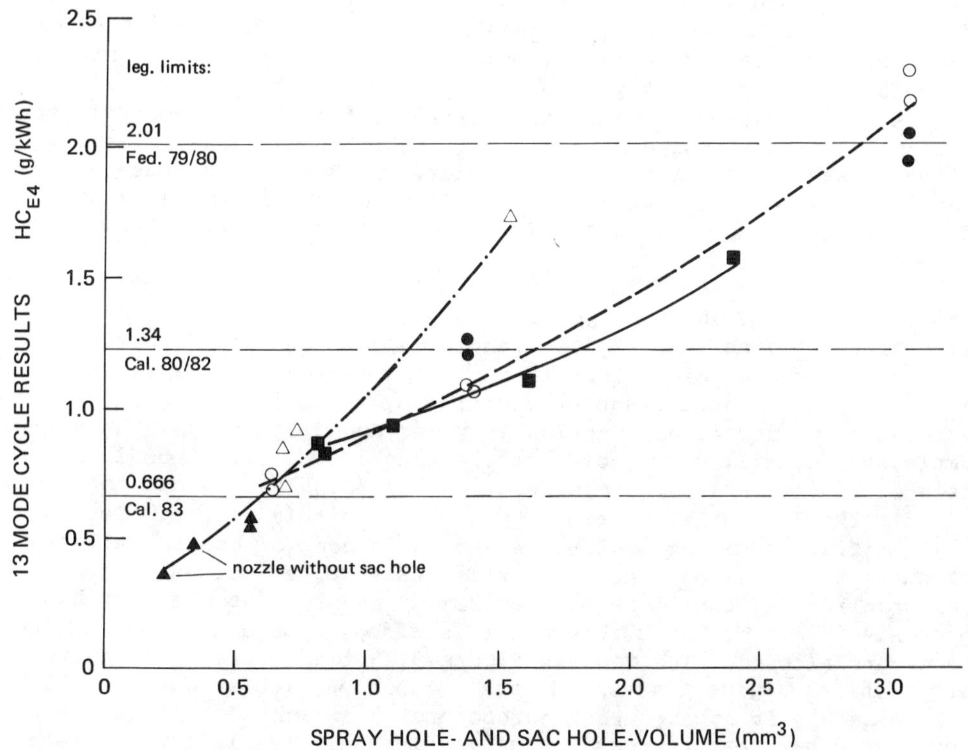

FIGURE 3.21 Heavy-duty hydrocarbon data as a function of sac volume.
SOURCE: Robert Bosch, 1980.

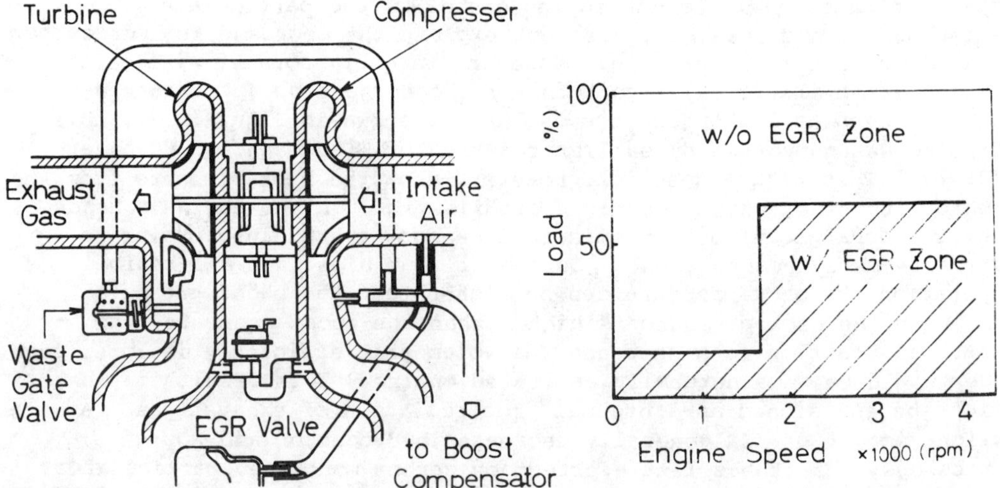

FIGURE 3.22 The turbocharger system. SOURCE: Toyota, 1980.

increase slightly, and carbon monoxide and hydrocarbon emissions decrease. A slight decrease in the air/fuel ratio compensates for the slight decrease in volumetric efficiency, and fuel economy decreases and emission levels increase. The general trend for this regime appears to be lower NO_x, carbon monoxide, and hydrocarbons, as well as higher particulates and fuel consumption.

When boost is greater than back pressure, the thermal and volumetric efficiencies of the engine, the air/fuel ratio, the intake pressure, and the intake temperature are all higher. No EGR occurs in this condition. Again, these various factors may have opposing effects. The general trend appears to be higher NO_x but lower particulates and fuel consumption.

This emission picture is further complicated by two additional factors. To obtain the same power output (with the same gear ratio), turbocharged engines require lower engine speeds and higher brake mean effective pressures than naturally aspirated engines. As a result, the time available for heat transfer from the cylinder is increased, but the rate may be decreased, particularly per pound of charge. As an example, the overall effect of turbocharging on NO_x emissions is presented in Figure 3.28. In effect, the turbocharger pushes the engine to the left up a constant power curve in this figure. Depending on the initial operating conditions and the amount of boost available, turbocharging can either increase or decrease NO_x emissions. The final complicating factor is that emissions and fuel economy are highly engine specific. When a turbocharger is added, both emissions and fuel economy are altered. The maps in Figure 3.29 illustrate how a turbocharger shifts engine load-speed maps for one particular engine.

In general, it appears that turbocharging a light-duty diesel with no changes other than increasing the maximum deliverable fueling rate will result in better acceleration, lower hydrocarbon emissions and fuel economy, and higher NO_x emissions. Whether particulate emissions

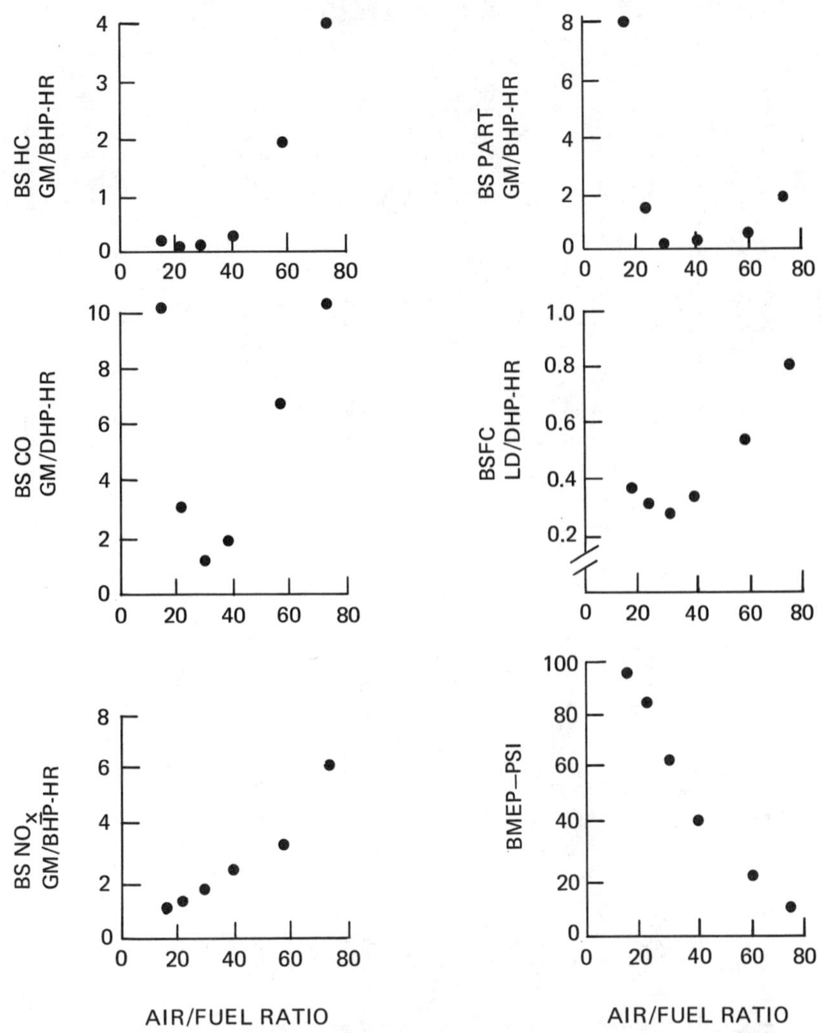

FIGURE 3.23 Effect of air/fuel ratio on emissions with a four-cylinder, 2.3-liter indirect injection diesel at 1,500 rpm. SOURCE: Adapted from the data of Wade (1980).

increase or decrease is dependent on the particular engine. Because of the complex relationships among boost pressure, back pressure, and related parameters, it is possible that a turbocharger that decreases particulate emissions during the FTP may increase them during another cycle, such as the low speed New York City cycle. However, it seems apparent that increased levels of particulate emissions are primarily due to engine operating conditions where back pressure exceeds boost pressure and that increased NO_x levels are primarily due to higher turbocharger boost operation.

TABLE 3.3 Effects of Turbocharging a Light-Duty Diesel With No Other Engine or Vehicle Modification

| Vehicle | Inertia Weight, lbs | Engine, liters | Turbo- Charged | Emissions, g/mi | | | | Comb. Fuel Econ., mpg | Accel. 0-60 mph, sec |
				HC	CO	NO_x	Part.		
Oldsmobile Toronado	4,500	5.7	No	0.039	1.2	1.55	0.26	24.3	15.7[a]
			Yes	0.32	1.2	1.76	0.28	22.2	16.2[a]
Oldsmobile Delta 88	4,500	4.3	No	0.26	0.90	1.68	0.32	24.0	24.9[a]
			Yes	0.31	0.84	1.60	0.30	24.9	–
Mercedes	4,000	3.0	No	0.29	1.1	1.20	0.27	27.0	26.0[a]
			Yes	0.20	1.0	1.48	0.30	26.2	26.0[a]
Mercedes	4,000	3.0	No	0.18	0.9	1.66	0.80	26.6	21.2[b]
			Yes	0.14	0.9	1.75	0.50	24.6	22.4[b]
Mercedes	4,000	3.0	No	–	–	1.88	0.65	–	–
			Yes	–	–	1.87	0.63	–	–
Lancia/Opel	3,000	2.1	No	0.49	1.3	0.95	0.23	35.7	19.9[a]
			Yes	0.47	1.3	1.14	0.24	33.6	19.7[a]
Opel	3,000	2.1	No	–	–	1.26	0.31	–	–
			Yes	–	–	1.36	0.33	–	–
Volkswagen	2,225	1.5	No	–	–	1.20	0.15	–	–
			Yes	–	–	0.90	0.26	–	–

[a]GM, 1979b.
[b]French and Pike, 1979.

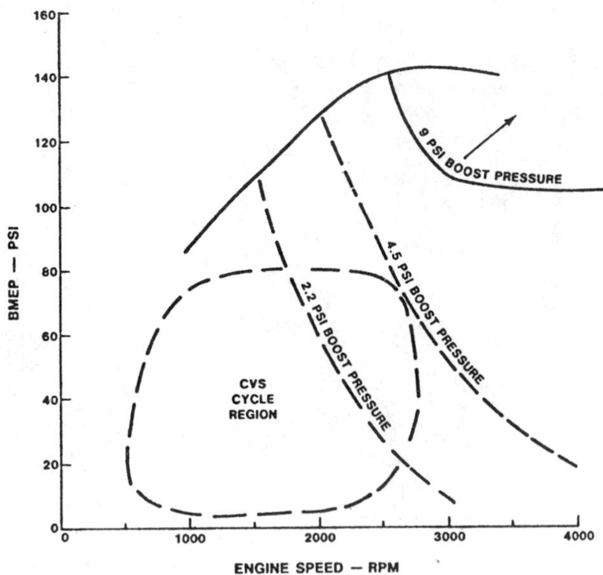

FIGURE 3.24   CVS driving cycle region.
SOURCE: Wade, 1980.

Turbocharging and Lowering the Rear Axle Ratio

The difficulties resulting from poor turbocharger efficiency during the
FTP driving schedule can be partially overcome by using lower rear axle
ratios and thereby causing the engine to operate at lower average speeds
and higher fuel/air ratios. As is shown by the emission and fuel
economy island maps of Figures 3.2 and 3.29, lowering engine speeds and
increasing brake mean effective pressure (for constant power) generally
decrease hydrocarbon, carbon monoxide, NO_x, and particulate emissions
and improve fuel economy. In addition, lowering the rear axle ratio
effectively shifts the FTP cycle region on the brake-mean-effective-
pressure engine speed map.

The effect of the rear axle ratio is illustrated in Figure 3.30 and
Table 3.5. When the rear axle ratio is decreased, hydrocarbon, carbon
monoxide, NO_x, and particulate emissions decrease (although particulate emissions may increase again if the ratio becomes too low) and
both fuel economy and maximum vehicle speed improve. The acceleration
of the vehicle suffers, but it is still better than the naturally
aspirated version. For this reason the rear axle ratio on a naturally
aspirated vehicle cannot be lowered to achieve emission control--the
acceleration would become too poor for acceptable drivability.

Comparing two naturally aspirated vehicles with rear axle ratios of
3.2 and 2.8, and a turbocharged vehicle with a rear axle ratio of 2.8,
demonstrated that lowering the axle ratio was the most important factor
in hydrocarbon reduction, although the turbocharger contributed to the
carbon monoxide decrease. The turbocharged vehicle exhibited superior

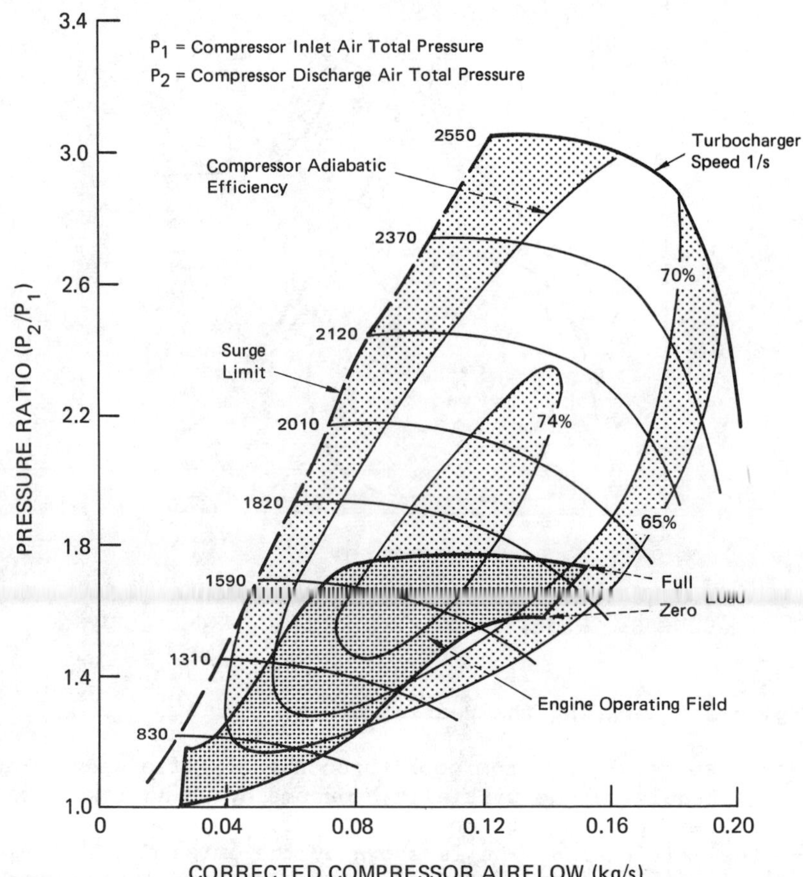

FIGURE 3.25 Turbocharger compressor characteristics and engine operating regime. SOURCE: Oblander _et al._, 1978.

acceleration and top speed. In addition, for the turbocharged engine, NO_x emissions were higher, and fuel economy was worse, than for the 2.8 ratio naturally aspirated engine and better than for the 3.2 ratio naturally aspirated engine. Unfortunately, it is difficult to quantify the relative importance of the turbocharger and axle ratio in explaining the decrease in particulate emissions for the 2.5-liter engine, but the Opel data indicate that the axle ratio was more significant.

Comparing the naturally aspirated engine with the low-rear-axle-ratio turbocharged systems in Table 3.5 shows that performance (acceleration times), fuel economy, top speed, and hydrocarbon and carbon monoxide emissions all improved when the engine was turbocharged and the rear axle ratio lowered. NO_x and particulate emissions were comparable or lower for the turbocharged version. Pushing the rear axle ratio even lower, to the point of comparable performance, should show an improvement in all other parameters, including particulates.

However, this may not be the case for all engines, because particulate maps are engine-specific and lowering the rear axle ratio too far may move the FTP cycle region into a high-particulate region on the map.

Improving Turbocharger Efficiency

Lowering the rear axle ratio may overcome some of the problems associated with poor turbocharger efficiency during the FTP cycle but will not directly solve this difficulty. An alternate approach would be to turbocharge an engine with a smaller displacement and thereby achieve the same performance as is achieved with a larger naturally aspirated engine. For the same rear axle ratio, the smaller turbo-

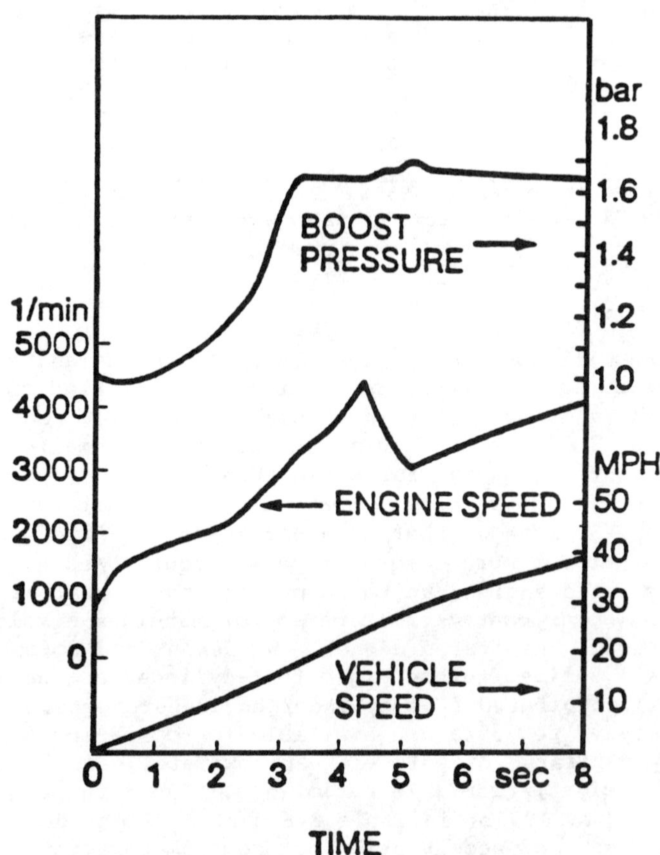

FIGURE 3.26 Transient behavior of 1978 Mercedes-Benz 300 SD with an automatic transmission. Full-load acceleration from idle of turbocharged indirect injection diesel. SOURCE: Oblander et al., 1978.

TABLE 3.4 Effect of Fueling Rate on Emissions, Fuel Economy, and Performance of Naturally Aspirated and Turbocharged Engines

Vehicle	Type	Fuel System	Emissions, FTP, g/mi				Fuel Econ., mpg	Accel. 0-60 mph., sec
			HC	CO	NO_x	Part.		
Oldsmobile 88	NA	NA	0.26	0.90	1.68	0.32	24.0	24.9
4.3	TC	NA	0.31	0.84	1.60	0.30	24.9	–
	TC	TC	0.26	0.79	1.66	0.29	24.3	17.0
Mercedes 300 SD	NA	NA	0.29	1.08	1.21	0.27	27.0	26.3
3.0	TC	NA	0.20	0.98	1.48	0.30	26.2	26.0
	TC	TC	0.20	1.00	1.44	0.30	26.9	15.7
Pre X-car Opel[a]	NA	NA	0.49	1.32	0.95	0.23	35.7	19.9
2.1	TC	NA	0.47	1.33	1.14	0.24	33.6	19.7
	TC	TC	0.47	1.36	1.13	0.28	33.3	14.7
Toyota	NA	NA	0.38	1.38	1.65	0.35	27.0	–
2.2	TC	TC	0.27	1.04	2.07	0.31	24.8	[b]

[a] Not well matched (AiResearch, 1980).
[b] Approximately 72 percent of time necessary for natural aspiration.

SOURCE: GM, 1979b; Toyota, 1980.

charged engine will operate at higher average speeds and brake mean effective pressures than the larger naturally aspirated engine. A larger turbocharged engine, in a vehicle with a lower rear axle ratio, will operate at lower speeds. A turbocharger on a smaller engine would be more efficient during the FTP cycle than a turbocharger on a larger engine, with power equivalent to that of a larger naturally aspirated engine. Figure 3.31 shows that a turbocharged 2.3-liter engine would have up to 43 percent more torque at peak torque speed than a 2.3-liter naturally aspirated engine, up to 18 percent more power at rated speed, and torque and power characteristics approximately equivalent to a 3.0-liter naturally aspirated diesel. Wiedemann and Hofbauer (1978) report that a 1.5-liter turbocharged four-cylinder engine and a 2.2-liter naturally aspirated five-cylinder engine are generally equivalent.

Unfortunately, few data are available for comparing turbocharged and naturally aspirated diesels with approximately equal power and equal inertia weight, rear axle ratio, etc. The data presented by General Motors (GM, 1979b) in Table 3.6 show a slight deterioration in NO_x, particulates, and acceleration but some improvement in hydrocarbon and carbon monoxide emissions for a turbocharged and a larger naturally aspirated engine with equivalent fuel economy. Both vehicles are in the 4,500 pound inertia weight class and have three-speed automatic transmissions. Information about the rear axle ratios was not reported. Wiedemann and Hofbauer (1978) found significant improvements in hydrocarbon, carbon monoxide, NO_x, and particulate emissions; fuel

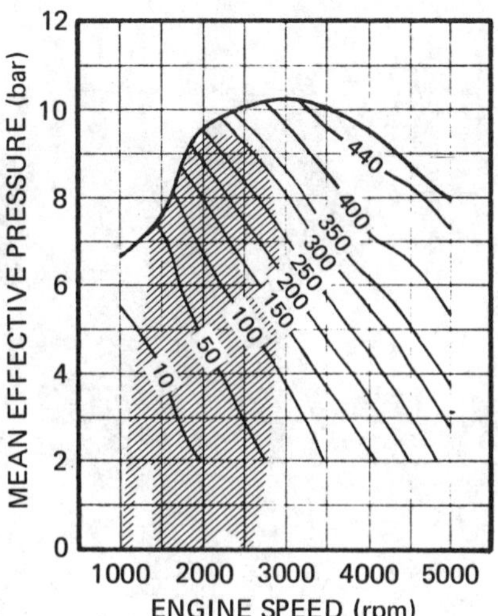

FIGURE 3.27 Boost pressure (torr) map
of 70-hp indirect injection diesel
engine with approximate FTP cycle region
shown. SOURCE: Adapted from Wiedemann
and Hofbauer, 1978.

economy was significantly improved, but performance data were not
reported. Both Volkswagens fall into the 2,250-pound-inertia-weight
category.

Use of a "Comprex" supercharger rather than a traditional turbocharger improves boost at low engine speeds. This device transfers
pressure from the exhaust, through transient waves created by a rotary
valving device at both ends of a pressure exchanger, and thereby
compresses intake air. The data in Table 3.7 illustrate the effects of
the Comprex on emissions, fuel consumption, and performance. Unfortunately, the data are not sufficient to allow a straightforward
analysis of the effects due solely to the Comprex; however, some mixing
of exhaust gas in the intake flow stream, which provides unavoidable
EGR, is known to occur. Data indicate that the Comprex can decrease
NO_x with corresponding increases in fuel consumption and hydrocarbon
and carbon monoxide emissions. Lowering the rear axle ratio to the
point of comparable performance could improve emissions and fuel
economy considerably. Table 3.8 provides recent emissions and fuel
economy data for an Opel Rekord 2300D (3,000-pound-inertia-weight) car
with a Brown Boveri Comprex. Unfortunately, the baseline data for this
vehicle were not available. Because there are no data available for a
comparison of the Mercedes 220D vehicles with and without Comprex, the
220D with the Comprex is compared with the 240D naturally aspirated
engine in Table 3.7. The data indicate that there is no significant
improvement; however, there may be several differences in these engines
or vehicles that may have contributed to these results.

A variable-area nozzle in the turbocharger may improve boost at low
engine speed. However, these devices are still in development stages
and no emission data are available. Known designs are more complex and
expensive than the nozzles of conventional turbochargers.

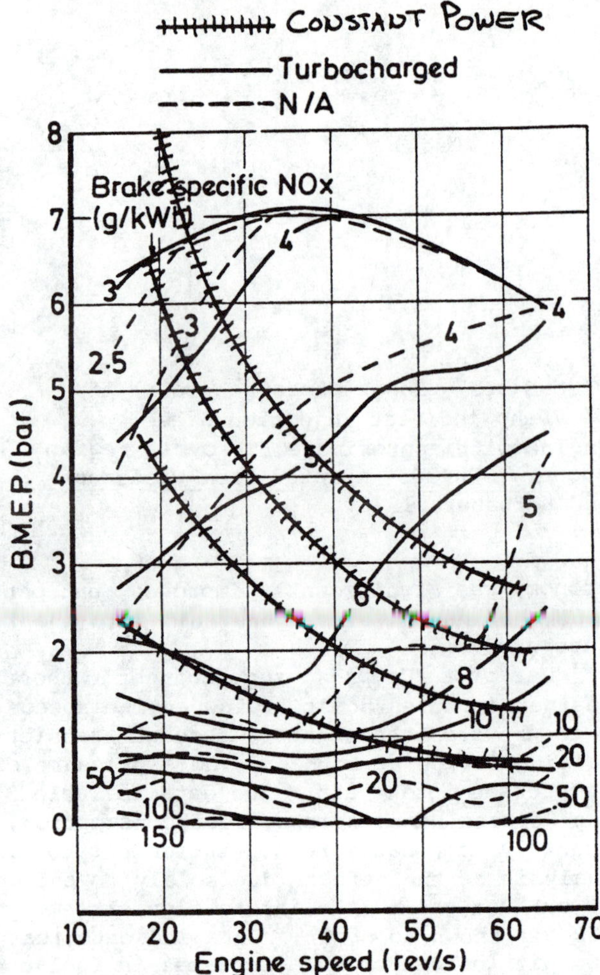

FIGURE 3.28 Constant power curves on NO$_x$ map for naturally aspirated and turbocharged engines: 3.0 liters, prechamber. SOURCE: French and Pike, 1979; power curves courtesy of Aerospace.

Turbocharging in Combination with Other Control Technologies

The available data suggest that use of a turbocharger alone or in conjunction with a lowered rear axle ratio may not lead to emission reductions in all vehicles. However, the use of turbocharging in conjunction with other control technologies (including lowering the rear axle ratio) can lower emissions while maintaining some of the performance improvements of turbocharging. Interactions between turbochargers and EGR are discussed below. The turbocharging of an oxidation catalyst-equipped diesel is discussed in the section on exhaust aftertreatment. Fuel injection scheduling for turbocharged vehicles and production turbocharged diesels is discussed in the following paragraphs.

As was previously noted, retarding static injection timing can decrease NO$_x$ emissions with little effect on carbon monoxide

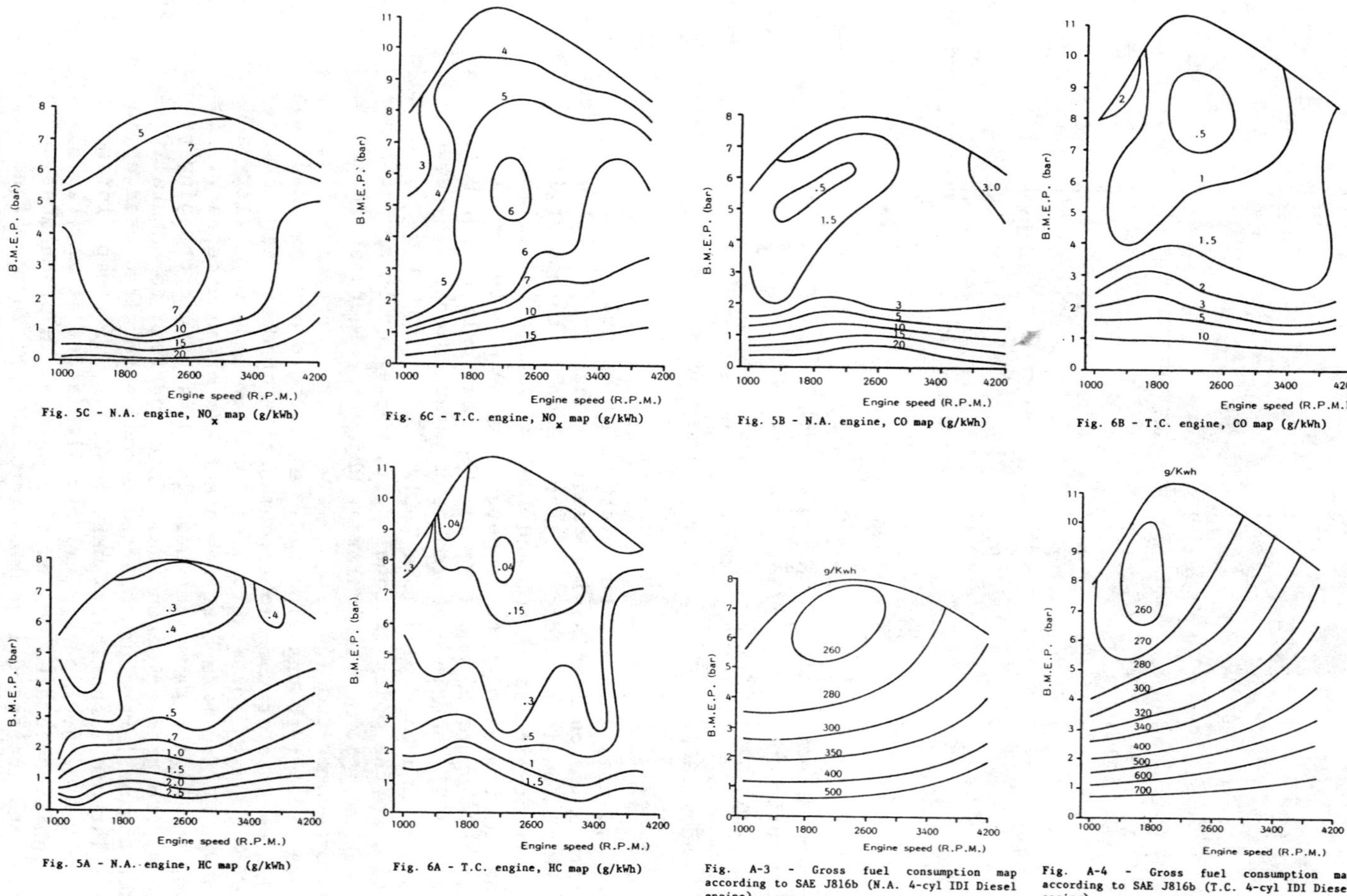

FIGURE 3.29 Effect of turbocharger on NO_x, carbon monoxide, hydrocarbons, and fuel consumption maps. SOURCE: Bassoli et al., 1979.

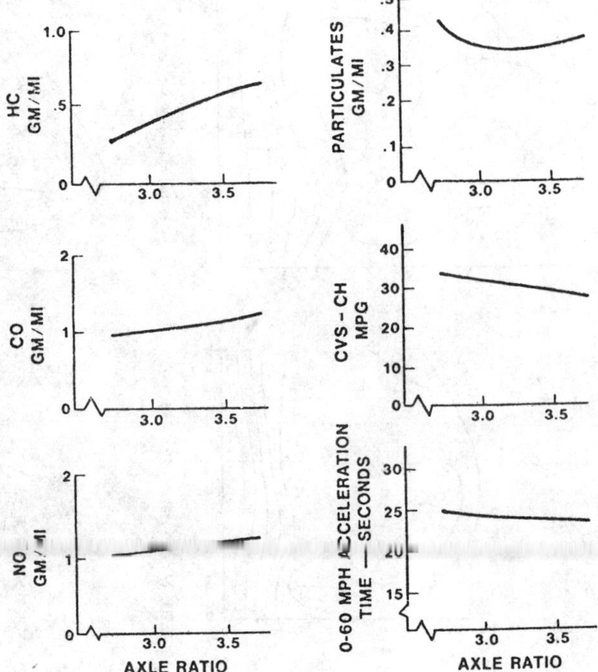

FIGURE 3.30 Effect of
rear axle ratio on
emissions, fuel economy,
and performance.
SOURCE: Wade, 1980.

emissions and fuel economy, but hydrocarbon emissions will increase.
Particulate emission levels will remain unchanged or will even decrease,
unless the timing is retarded too far. Because turbocharging has the
opposite effect—it raises NO_x and lowers hydrocarbon emissions—
retarding the timing of a turbocharged engine may be beneficial. By
employing this approach and lowering the rear axle ratio, Bassoli and
co-workers (1979) obtained improvement in emissions, fuel economy,
acceleration, and top speed (see Table 3.9). Further decreases in rear
axle ratio, to the point of equal performance, would show further
improvement except for particulates, which may increase if the rear
axle ratio becomes too low.

In practice, turbocharging a diesel is more complex than might be
expected. Because of the higher combustion chamber temperatures
generated by the turbocharger, additional cooling capability and
components that are more heat resistant are often necessary. In
addition, the higher pressures associated with turbocharging require
components with greater strengths. In comparison to naturally aspirated
3.0-liter engines, the turbocharged Mercedes required changes in the
piston, wrist pin, crankshaft, main bearings, oil pump, intake and
exhaust valves, injection system, etc. (Oblander et al., 1978). Of
special significance are the required changes in prechamber and nozzle
geometries and increased injection timing. The turbocharged engine is
used in a slightly heavier vehicle than is used for the naturally

TABLE 3.5 Effect of Rear Axle Ratio

Displacement, liters	NA/TC	Rear Axle Ratio	Emissions, g/mi				Fuel Econ., mpg	Accel. 0-60 mph, sec	Top Speed km/hr
			HC	CO	NO$_x$	Part.			
			Pre X-car Opel						
	TC	3.34	0.35	1.17	0.94	0.23	37.8	18.0	*
	TC	3.74	0.37	1.07	1.10	0.25	34.6	17.2	*
	TC	4.15	0.47	1.36	1.13	0.28	33.3	14.7	*
	NA	4.15	0.49	1.32	0.95	0.23	35.7	19.9	*
			Fiat[a]						
3.7	TC	2.50	0.29	1.99	2.57		29.5	14.0	177.5†
	TC	2.70	0.32	2.15	2.80		28.2	13.8	174.0†
	TC	2.90	0.34	2.25	3.00		27.2	13.5	162.0†
			Fiat[a]						
2.5	TC	2.60	0.28	1.33	1.47	0.36	37.7	13.8	170.0††
	NA	3.20	0.31	1.68	1.60	0.55	34.6	17.7	146.0††
	NA	2.80	0.29	1.48	1.50		37.9	18.1	157.0††
	TC	2.80	0.29	1.36	1.59		35.7	13.7	165.0††

[a]Calculated from steady state engine maps--results may be partially due to injection timing.

SOURCE: (*) GM, 1979b; (†) Cornetti, 1979a,b,c; (††) Bassoli et al., 1979.

aspirated engine and incorporates a lower rear axle ratio, the 1980 model incorporates EGR. Table 3.10 compares Mercedes naturally aspirated and turbocharged production vehicles.

Other Effects of Turbocharging

Turbocharging generally decreases noise from diesels. There are few reported data for the effect on unregulated emissions. A turbocharger and its associated components would add from $100 to $400 to the price of the engine, not including the cost of the necessary design changes mentioned in the preceding paragraph.

Exhaust Gas Recirculation

Exhaust gas recirculation is commonly used to reduce NO$_x$ emissions from spark ignition engines. With this control technique, a portion of the exhaust gas is circulated back into the intake system. The exhaust is essentially inert, the charge is diluted without additional oxygen, and the combustion temperature is decreased. The NO$_x$ formation rate

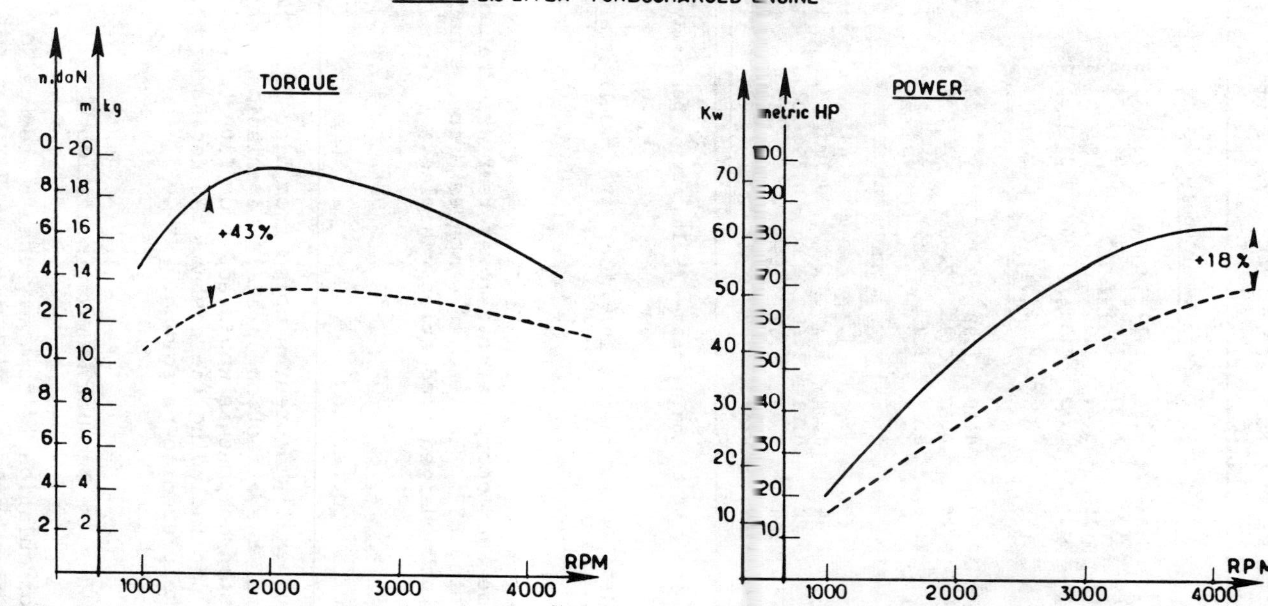

FIGURE 3.31 General equivalence in torque and power characteristics of 2.3-liter turbocharged engine and 3.0-liter naturally aspirated engine. SOURCE: Moulin et al., 1979. (See Wiedemann and Hofbauer (1978) for similar plots.)

TABLE 3.6 Comparison of Turbocharged Diesels With Naturally Aspirated Diesels of Greater Displacement

Vehicle	Displacement, liters	Aspira-tion	Emissions, g/mi				Fuel Econ., mpg	Accel. 0-60 mph, sec	HP
			HC	CO	NO_x	Part.			
Olds Toronado	5.7	NA	0.39	1.21	1.55	0.26	24.3	15.7	125
Olds Delta 88	4.3	NA	0.26	0.90	1.68	0.32	24.0	24.9	–
		TC	0.26	0.79	1.66	0.29	24.3	17.0	–
VW I.5	2.2	NA	0.35	1.70	1.50	0.48	36.0	–	68
VW I.4	1.5	TC	0.11	0.80	0.90	0.25	45.0	–	70
		NA	0.16	1.00	1.20	0.35	41.0	–	50

SOURCE: GM, 1979b; Wiedemann and Hofbauer, 1978.

TABLE 3.7 Effects of Comprex Supercharger on Light-Duty Diesels

Vehicle	Displacement, liters	Aspira-tion	Emissions, g/mi				Fuel Econ., mpg	Accel. 0-60 mph, sec
			HC	CO	NO_x	Part.		
Opel	2.1	NA	0.15	0.79	0.99	–	34.5	23.0
		TC	0.25	0.95	1.24	–	31.5	14.0
		Comprex	1.16	3.70	0.66	–	29.1	12.5
Mercedes	2.2	NA	0.29	0.97	1.27	0.48	25.7	–
Mercedes	2.2	Comprex	0.18	1.30	1.05	0.66	25.9	–

SOURCE: Monaghan, 1979; Springer and Stahman, 1977.

TABLE 3.8   Brown Boveri Comprex Demonstration Car-Opel Rekord 2300D (3,000 lbs)

Test Location	Emissions, g/mi				Fuel Economy, mpg		
	HC	CO	NO_x	Part.	City	Highway	Comp. Fuel Econ.
Ricardo	0.26	1.50	0.64	0.32	30.0	38.8	33.4
EPA	0.42	1.70	1.01	0.25	29.7	41.6	34.0
U.S.-Man.A	0.34	1.56	0.79	0.25	30.9	39.6	34.3
U.S.-Man.B	0.48	1.96	0.86	0.31	27.7	38.0	31.6

SOURCE: EPA, 1980b.

is exponentially dependent on temperature; hence, NO_x emissions decrease with EGR.

In theory, EGR should work equally well for spark ignition and diesel engines. In fact, although EGR can significantly reduce NO_x emissions, there are some difficulties associated with EGR in diesels. The disadvantages of using EGR in diesels are the trade-offs between NO_x, hydrocarbon, and particulate emissions, and maintenance and durability problems. The advantages are that EGR is one of two practical methods for controlling NO_x emissions (the other is to retard fuel injection timing) and that it decreases noise and odor during cold starts.

Types of EGR

Exhaust gas recirculation systems can be either constant or modulated. Constant systems provide relatively steady EGR for all engine operating conditions. Modulated systems proportion the EGR rate as a function of engine operating conditions. Modulated EGR systems range from simple on/off systems (with relatively constant EGR for low to moderate loads and no EGR for high loads) to sophisticated closed-loop electronically controlled systems that proportion the EGR rate to optimize NO_x, hydrocarbon, and particulate control, and fuel economy. Figure 3.32 provides diagrams of a simple two-rate-plus-off system and a fully electronic system modulated with load and speed signals.

Effects of EGR

Figure 3.33 shows the effects of EGR on emission levels, fuel consumption, and other engine parameters for baseline, advanced, and retarded injection timings at one engine operating condition. For baseline injection timing, increasing EGR significantly reduces NO_x but increases particulates. With 40 percent or less EGR, hydrocarbon

TABLE 3.9 Combined Effect of Turbocharging, Timing, Retard, and Rear Axle Ratio on Performance and Emissions of 2.4-Liter Indirect Injection Engine

| Aspiration | S.I.T. | Plunger Diameter | Rear Axle Ratio | Emissions, g/mi | | | | Comb. Fuel Econ., mpg | Accel. 0-60 mph, sec | Top Speed, km/hr |
				HC	CO	NO_x	Part.			
NA	1 btdc	9	3.2	0.31	1.68	1.60	0.55	35.0	17.7	146
TC	5 atdc	10	2.6	0.28	1.33	1.47	0.36	37.7	13.8	170

SOURCE: Bassoli et al., 1979.

TABLE 3.10 Comparison of Production 1978 and 1980 Turbocharged Mercedes With Comparable Naturally Aspirated Light-Duty Diesel Engines

| Vehicle | Rear Axle Ratio | Emissions (g/mi) | | | | Comb. Fuel Econ., mpg | Accel. 0-60 mph, sec |
		HC	CO	NO_x	Part.		
1980 300 SD[a]	3.07	0.265	1.03	1.17	0.49	24.8	–
1980 300D	3.46	0.283	1.13	1.96	–	28.0	–
1978 300 SD	3.07	0.260	1.12	1.78	0.40	23.4	14.0
1978 300D	3.46	0.360	1.77	1.98	0.41	–	21.2

[a] Tentative 1980 300 SD Results (Barth, 1980).

SOURCE: EPA, 1980C; Oblander et al., 1978; Barth, 1980; GM, 1979b.

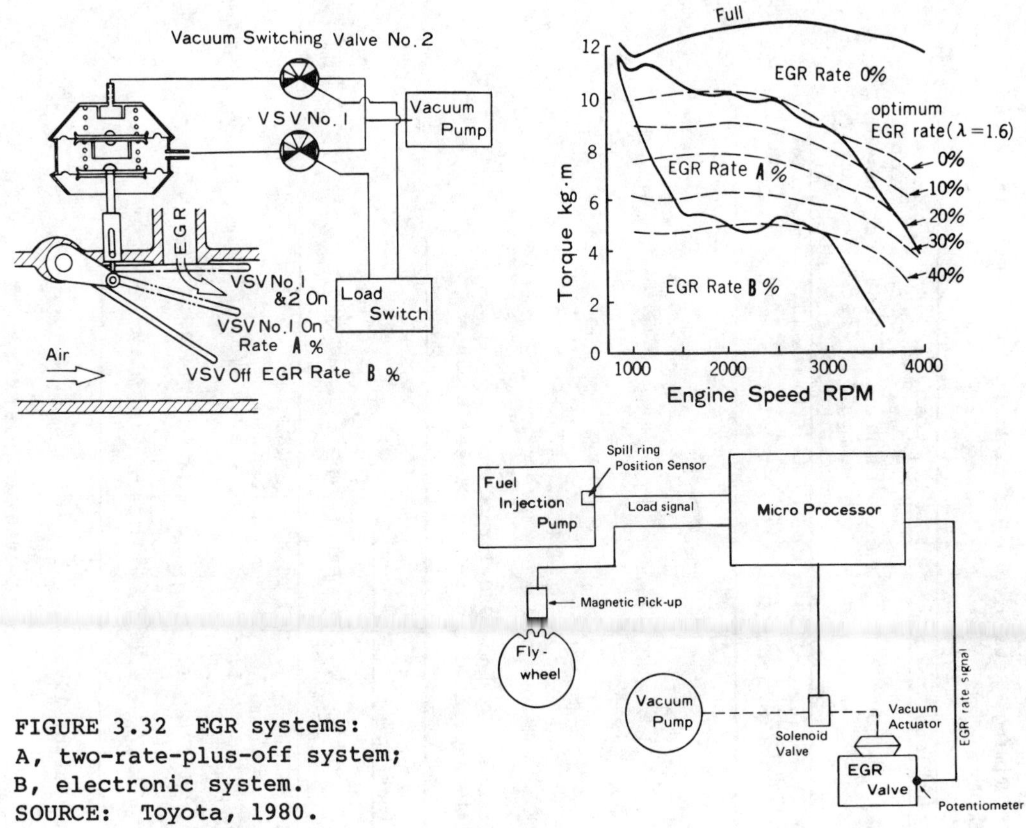

FIGURE 3.32 EGR systems:
A, two-rate-plus-off system;
B, electronic system.
SOURCE: Toyota, 1980.

emission levels are reduced, but when the EGR rate is greater than 40
percent, hydrocarbons drastically increase. Too much EGR also
significantly increases carbon monoxide. EGR slightly deteriorates
fuel economy. Retarding injection timing lowers NO_x even more,
lowers particulates compared to baseline injection, and significantly
increases hydrocarbon emissions. More than 40 percent EGR with retarded
injection results in much higher fuel consumption. Advanced injection
timing increases NO_x and particulates but lowers hydrocarbon emissions
and fuel consumption.

Figure 3.34 further illustrates the NO_x/hydrocarbon and NO_x/
particulate trade-offs for operation during a driving cycle. Figure
3.35 shows the NO_x/hydrocarbon trade-offs with and without EGR for an
injection timing sweep. The figure demonstrates the narrow range of
timing variations that can be accommodated while remaining within the
hydrocarbon/NO_x control region. Figure 3.36 is similar to Figure
3.35 but is for a different engine/vehicle system and shows the effect
of cetane number on emissions with and without EGR. Comparison of
Figures 3.35 and 3.36 illustrates the different effects of EGR and
timing for different engine/vehicle system combinations.

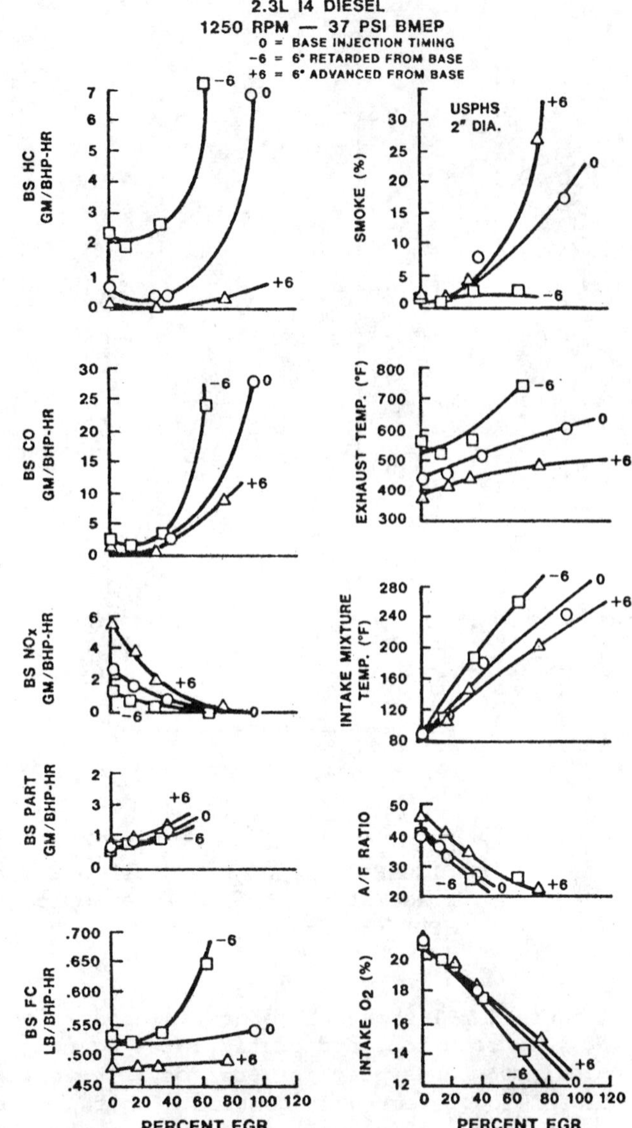

FIGURE 3.33 Effect of EGR rate and injection timing on emissions, fuel economy, and other performance and operating parameters for a 2.3-liter indirect injection engine at 1,250 rpm, 37-psi brake mean effective pressure. SOURCE: Wade, 1980.

Examination of these figures and engine emission maps reveals several important trends. Not only are NO_x, hydrocarbon, and particulate emission levels interdependent, they also are related to injection timing. Timing has opposite effects on particulate and NO_x emissions and hydrocarbon emissions. Retarded timing within certain ranges (see Figure 3.13) decreases NO_x and particulates but increases hydrocarbon emissions. Retarded timing also increases the exhaust temperature. Maximum EGR rates are limited by increased fuel consumption and decreased drivability (Amano et al., 1976). The emission maps

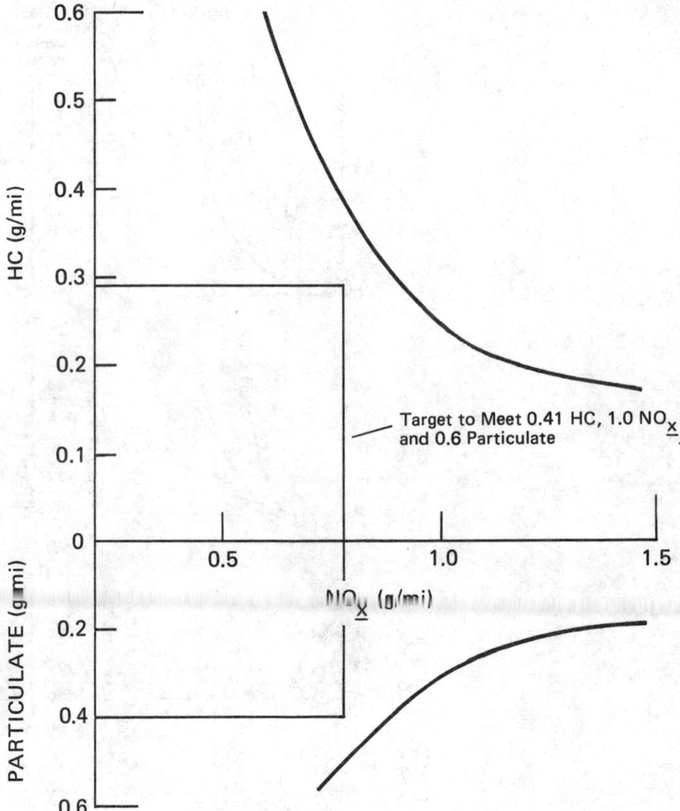

FIGURE 3.34 NO$_x$ hydrocarbon and NO$_x$ particulate
trade-offs compared to example engineering goals.
SOURCE: Nissan, 1980.

(Figure 3.29) show that operating conditions that produce maximum
NO$_x$, hydrocarbon, and particulates do not coincide. Hence, a
modulated EGR system is mandatory. The optimum system would employ
closed-loop electronic control of both EGR rate and injection timing.

Figure 3.37 shows two maps where the EGR rate is optimized as a
function of engine speed and load. Map differences result from the use
of a variety of strategies for emission control, fuel economy, performance, and drivability trade-offs. Simpler systems could be used for
vehicles that emit NO$_x$ at levels approaching the federal standard
without EGR.

Table 3.11 shows the effects of various EGR modulation schemes for
several different vehicles. As is illustrated in the table, reasonable
EGR rates can be used to decrease NO$_x$ without unacceptable increases
in other emissions. Control of NO$_x$ to meet a 1.0-g/mi standard can
only occur, however, with increased levels of hydrocarbon, carbon
monoxide, and particulate emissions for some vehicles.

Few results have been reported for advanced modulated EGR systems. These systems will range from open-loop electronic EGR control to closed-loop electronic control of both EGR and fuel injection scheduling. Electronically controlled EGR systems will be superior to, but more expensive than, the systems discussed above. An alternative to electronic controls would be combinations of turbocharging, modulated EGR, oxidation catalysts, and static injection timing modifications.

EGR in Combination with Other Control Technologies

Light-duty diesels in the 2,000-pound-inertia-weight category, equipped with modulated EGR and/or modulated injection timing should be able to meet the 1.0 NO_x, 3.4 carbon monoxide, 0.41 hydrocarbon, and 0.6 particulate g/mi standards (Wade, 1980). However, more sophisticated controls will be necessary for heavier vehicles to meet these standards.

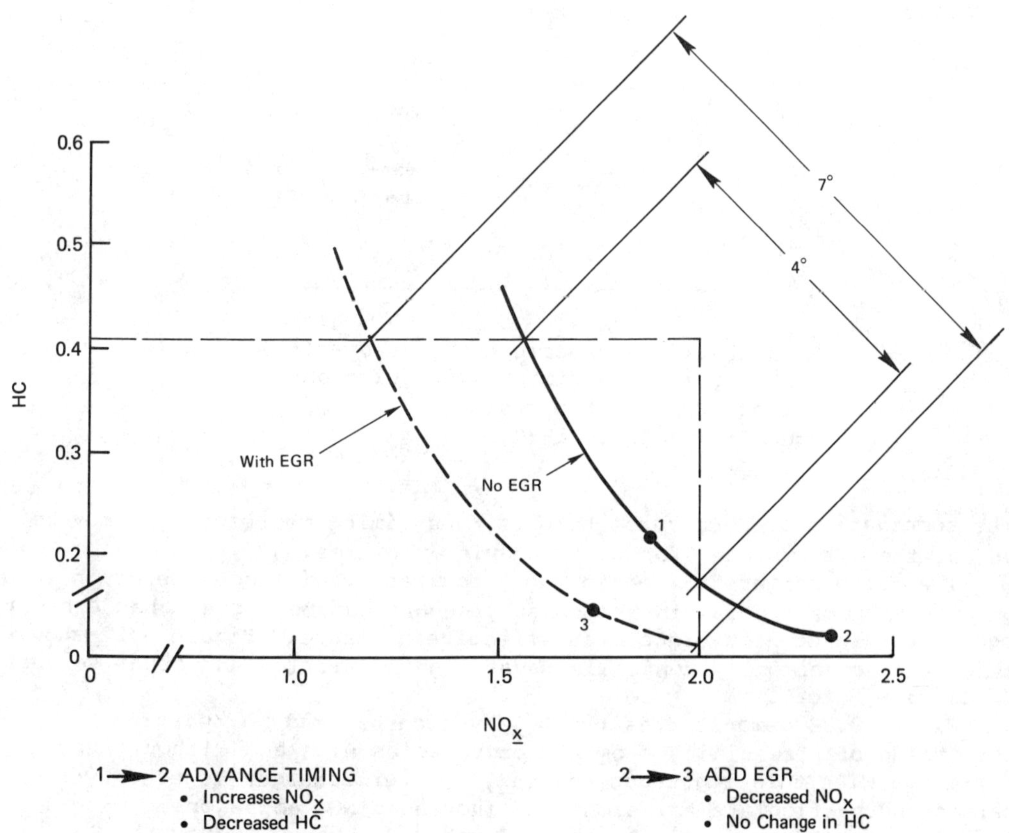

FIGURE 3.35 Hydrocarbon NO_x trade-off: timing sweep with and without EGR. SOURCE: Dimick, 1980.

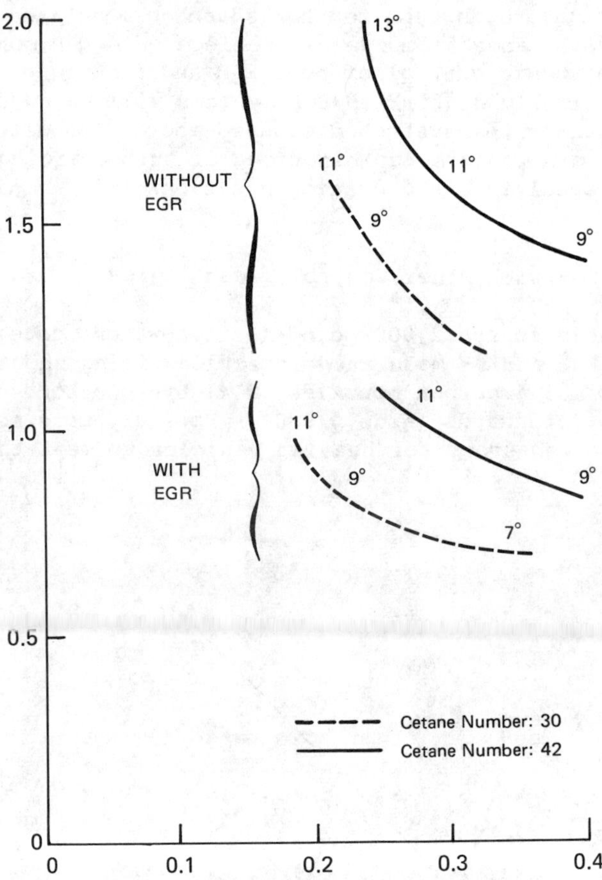

FIGURE 3.36 Hydrocarbon/NO$_x$ trade-off:
static injection timing sweep with and
without EGR for two fuels.
SOURCE: Peugeot, 1979.

The combination of EGR, turbochargers, and timing modifications may be
one of the systems used for these heavier vehicles.

The turbocharger-EGR combination requires special consideration
because at some speeds, intake boost pressure can be higher than exhaust
back pressure at high brake mean effective pressure. Figure 3.22 shows
a fully electronically controlled system using airflow and other sensors
(fuel flow, etc.).

Figure 3.38 demonstrates the NO$_x$/hydrocarbon and NO$_x$/particu-
late trade-offs resulting from EGR on vehicles with and without turbocharging. With this light-duty diesel, the turbocharger lowered hydrocarbon and particulate emissions and increased NO$_x$ emissions.
Applying some EGR to the turbocharged vehicle lowered hydrocarbon
emissions even further but had no significant effect on particulates.
Further increases in the EGR rate rapidly increased hydrocarbon and

particulate emissions. If hydrocarbon and particulate emissions were kept to levels equal or below those exhibited by the naturally aspirated vehicle without EGR, it could be used on the turbocharged vehicle to lower NO_x. This point is further illustrated in Figure 3.39, which demonstrates the dramatic decreases in hydrocarbons and the flattening of the particulate profile possible with a turbocharged EGR system. Improvement in specific fuel consumption is also illustrated. Therefore, turbocharging in conjunction with EGR has advantages over the use of EGR alone, especially for heavier vehicles for which more control is necessary.

The results of a study by French and Pike (1979) of turbocharging with EGR using a Mercedes 300D are presented in Table 3.12. The acceleration results are biased because the naturally aspirated fuel delivery system was retained on the turbocharged vehicle. A turbocharger fuel delivery would not be expected to affect emissions significantly during the driving cycle, but it would improve performance considerably. Turbocharging decreased hydrocarbon and particulate emissions but increased NO_x. The addition of EGR to the turbocharged engine decreased NO_x to very low levels but increased hydrocarbon, carbon monoxide, and particulate emissions. Timing retard decreased NO_x further and improved particulates slightly. However, hydrocarbon and carbon monoxide levels increased significantly. Fuel economy suffered from both EGR and timing retard.

Bassoli and co-workers (1979) conducted an extensive study of emissions control for light-duty diesels. Their results are presented in Table 3.13. Lowering the rear axle ratio from 3.2 to 2.8 decreased emissions, but not to the level required to meet the standards. Timing

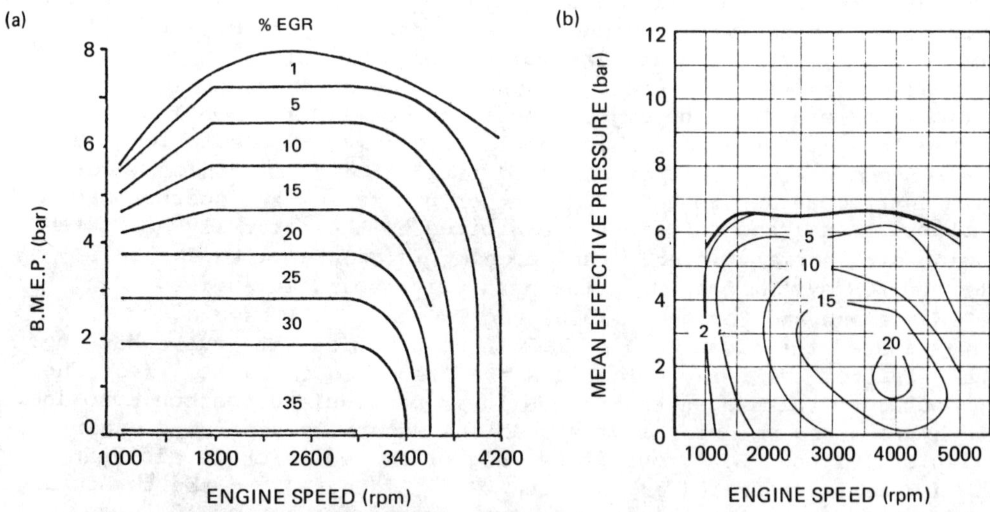

FIGURE 3.37 Example of modulated EGR maps resulting from different trade-off/control strategies for naturally aspirated diesels. SOURCE: (a) Bassoli et al., 1979; (b) Wiedemann and Hofbauer, 1978.

TABLE 3.11 Effect of Modulated EGR System on Hydrocarbon, Carbon Monoxide, and Particulate Emissions, and Fuel Economy

Manufacturer (IW, lbs)	Displ. (1)	Type of System	Emissions, g/mi				Fuel Econ., mpg
			HC	CO	NO_x	Part.	
Toyota*	3.0	None	0.42	1.92	1.32		
(4,000 lbs)		Sys. 1[a]	0.59	2.32	1.08		
		Sys. 2[a]	1.12	3.69	0.92		
		Sys. 3[a]	3.03	6.01	0.85		
Toyota[†]	2.2	None	0.35	1.36	1.63	0.34	25.7
(3,500)		Modul.	0.47	1.92	1.15	0.46	26.4
Olds[††]	5.7	None	0.26	1.14	1.71	0.30	25.2
(4,500)		Modul.	0.30	1.76	0.77	0.65	24.8
Buick[††]	4.3	None	0.28	0.83	1.59	0.22	
(4,000)		Modul.	0.47	1.93	0.74	0.27	
Opel[††]	2.1	None	0.22	0.72	1.38	0.41	-
(2,500)		Modul.	0.22	0.79	0.76	0.44	-

[a]Systems 1, 2, and 3 use progressively higher EGR rates.

SOURCE: (*) Amano et al., 1976; (†) Toyota, 1979; (††) GM, 1979b.

retard was more effective in lowering NO_x, but it increased hydrocarbons by a factor of 3 and particulates by 24 percent. EGR was more effective in lowering NO_x and did not increase hydrocarbons. (The modulated EGR system used was designed to have EGR rates below the high-hydrocarbon sensitivity region of the emission map in Figure 3.29.) Unfortunately, the data are incomplete for particulate emissions. The available results show that turbocharging, reducing the rear axle ratio, and retarding the injection timing can simultaneously decrease particulate emissions and increase fuel economy. NO_x emissions were not reduced to the 1.0 g/mi standard limit. Using EGR in conjunction with a turbocharger reduces NO_x further but raises hydrocarbon and particulate emissions to levels exhibited by the naturally aspirated engine with normal timing. Fuel economy is greater with the turbocharged-EGR system than with the naturally aspirated engine.

Wiedemann and Hofbauer (1978) and Toyota (1980) have also investigated the advantages of EGR in combination with turbocharging. The data from their investigations are presented in Table 3.14. The turbocharged Volkswagen engine has lower particulate, carbon monoxide, and hydrocarbon emission levels than the naturally aspirated engine (with and without EGR), but it has higher NO_x emission levels than the naturally aspirated engine with EGR. Adding EGR to the turbocharged system reduces NO_x to the lowest levels found for any of the cases studied but increases particulates dramatically. The heavier Toyota vehicles consistently emitted less NO_x and fewer particulates when turbocharged. Using less EGR with the turbocharged system should keep

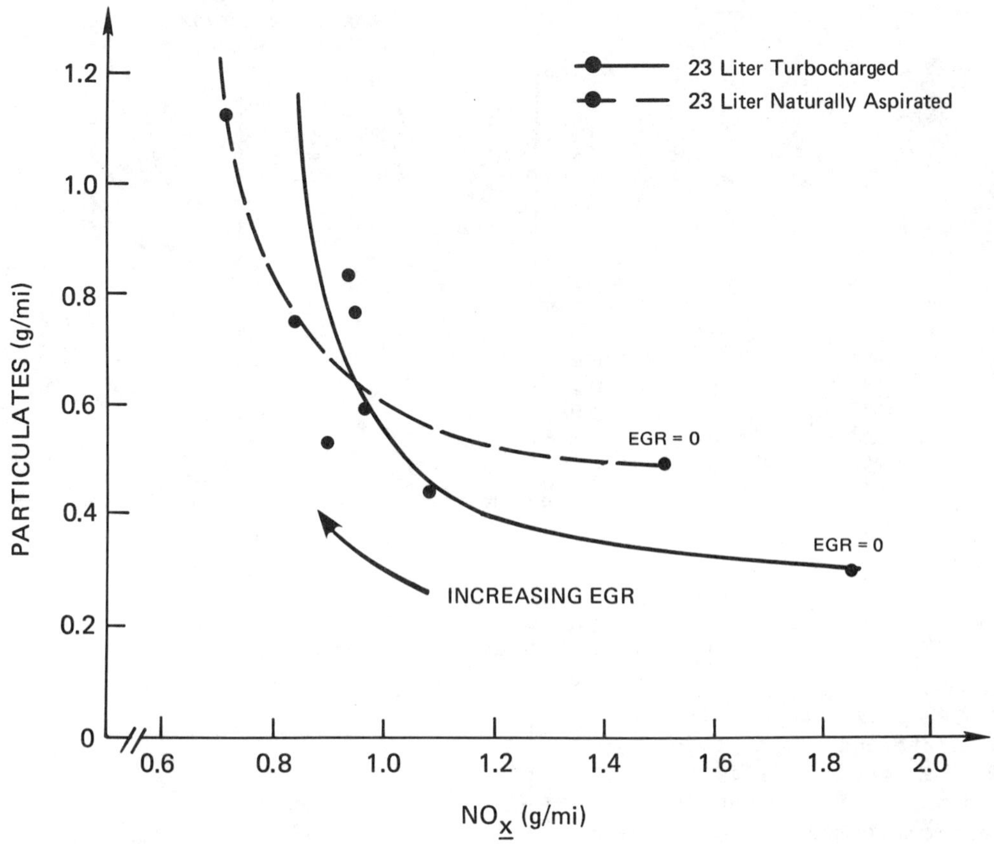

FIGURE 3.38 Emission trade-offs with EGR on turbocharged and naturally aspirated 3,500-lb. inertia weight light-duty diesel. SOURCE: Peugeot, 1979.

NO$_x$ emission levels below the 1.0-g/mi standard while maintaining particulate levels at the 0.6-g/mi standard.

Other Effects of EGR

EGR systems have both advantages and disadvantages with regard to durability, drivability, noise, odor, and unregulated emissions. The most significant disadvantage is a decrease in engine durability. Increased wear of the piston and piston grooves, camshaft and valve train, and cylinder bore have been noted (GM, 1979a; Nissan, 1979, 1980). As is shown in Figure 3.40, this increased wear appears to result from an increased carbon level in the engine oil (Nissan, 1979, 1980; Toyota, 1980). Increased deposits in the intake system have also been observed (Toyota, 1980; French and Pike, 1979).

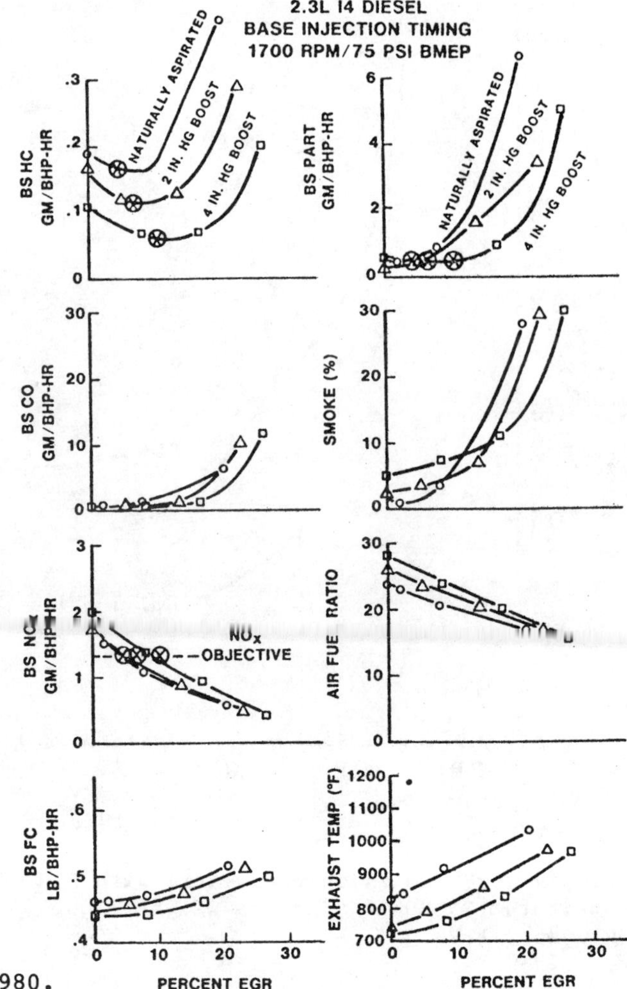

FIGURE 3.39 Effect of EGR rate and boost pressure on emissions, fuel economy, and other engine performance and operating parameters for a 2.3-liter indirect injection diesel at 1,700 rpm, 75-psi brake mean effective pressure. SOURCE: Wade, 1980.

Cadle and co-workers (1979) report that EGR may increase aldehyde emissions; however, Voissard (1979) and Wiedemann and Hofbauer (1978) report no such effect for naturally aspirated light-duty diesels. Voissard reports that turbocharging doubles aldehyde emissions but that a turbocharged engine with EGR emits fewer aldehydes than a naturally aspirated engine. Weidemann and Hofbauer report the opposite results; that is, a turbocharger will decrease aldehydes, and a turbocharger with EGR increases aldehydes up to the levels emitted from a naturally aspirated engine (Table 3.14). The differences in these findings undoubtedly are the result of the type of EGR used and other system modifications employed, such as the rear axle ratio used with the turbocharged version of the light-duty diesel. It has also been noted that a turbocharger may prevent the clogging that is prevalent in

TABLE 3.12 Effects of EGR, Turbocharging, and Timing Retard on Emissions, Fuel Economy, and Performance (Mercedes 300D) With Change in Rear Axle Ratio and With a Naturally Aspirated Fuel Delivery System

Version	Emissions (g/mi)				Comb. Fuel Econ. CVS, mpg	Accel. 0-60 mph, sec
	HC	CO	NO_x	Part.		
NA	0.18	0.94	1.66	0.80	24.30	21.2
TC	0.14	0.94	1.75	0.50	23.80	22.4
TC, EGR 1	0.38	2.38	0.47	-	19.95	-
TC, EGR 1, 4 retard	1.05	2.96	0.38	1.23	18.86	25.3
TC, EGR 1, 7 retard	4.97	6.13	0.32	0.93	17.15	28.6
TC, EGR 2	0.40	2.41	0.53	1.11	19.47	23.5

SOURCE: French and Pike, 1979.

naturally aspirated EGR systems but that EGR may cause carbon buildup on the intake compressor blades.

EGR has other benefits in addition to the reduction in NO_x. Oetting and Papez (1979) reported that EGR reduces cold knock if the exhaust gas temperature is maintained at its highest possible level (through the use of short EGR pipes). Cadle and co-workers (1979) have shown that EGR can reduce the nitrogen dioxide emission rate and the nitrogen dioxide/NO_x ratio. Increased odor and decreased drivability have also been reported as side effects of EGR (Wiedemann and Hofbauer, 1978).

EXHAUST AFTERTREATMENT

Diesel and gasoline engines have not only different emission characteristics but also different exhaust temperature characteristics. These temperature differences affect the performance of many of the aftertreatment systems. The differences in temperature ranges are illustrated by the map of exhaust temperatures (as a function of load and speed) in Figure 3.41. Note that the absence of heat transfer isotherms would be essentially horizontal for the diesel and vertical for the gasoline engine, with very little variation with speed. A detailed exhaust temperature time trace, using the Federal Test Procedure (FTP) cycle for a typical diesel and gasoline vehicle, is provided in Figure 3.42. As shown, the temperature of diesel exhaust typically varies from 200° to 400°C, and at one point to 600°C. This range is due to the diesel's variable air/fuel ratio with mostly part-load operation. Because the gasoline engine is esentially a constant air/fuel ratio

TABLE 3.13 Effects of EGR, Turbocharging, and Timing Retard on a Fiat 2.4-Liter Light-Duty Diesel

Version	Static Injection Timing	Rear Axle Ratio	Emissions (g/mi)				Comb. Fuel Econ., mpg
			HC	CO	NO_x	Part.	
NA	1° btdc	3.2	0.31	1.68	1.60	0.55	34.6
	1° btdc	2.8	0.29	1.48	1.50	–	37.9
	5° atdc	3.2	0.99	1.94	0.94	0.68	–
NA, EGR	1° btdc	3.2	0.31	1.93	0.88	>0.55[a]	–
	1° btdc	2.8	0.30	1.80	0.84	>0.55[a]	–
TC	5° atdc	2.8	0.29	1.36	1.59	–	35.7
	5° atdc	2.6	0.28	1.33	1.47	0.36	37.7
	5° atdc	2.4	0.27	1.28	1.36	–	39.8
TC, EGR	5° atdc	2.6	0.31	2.67	1.19	0.53	35.6

[a] Estimated.

SOURCE: Bassoli et al., 1979.

engine, its exhaust temperature during the cycle ranges between 600° and 700°C.

The basic control systems for diesel exhaust aftertreatment are: the use of reactors or thermal in-stream oxidation; the application of oxidation catalysts, the concept of catching particulate in either an inert trap or a trap that has been catalyzed (this includes the concept of trap-oxidizers--i.e., periodically regenerated traps), and the use of electrostatic precipitators. All of these approaches must be analyzed relative to their control of total particulate mass, the soluble organic fraction, the sulfate fraction of the particulate, and unregulated pollutants such as nitrogen dioxide and relative to their effect on the gaseous pollutants, hydrocarbons, carbon monoxide, and NO_x.

Thermal Reactors--Thermal In-Stream Oxidation

A thermal reactor is nothing more than an enlarged section of the exhaust manifold, which may have interior flow passages and insulation. Its purpose is to provide sufficiently high temperatures and residence times to allow oxidation of carbon monoxide, hydrocarbons, and particulates. This approach to emission control is also referred to as thermal in-stream oxidation.

The preliminary order-of-magnitude kinetic calculations carried out by Murphy and co-workers (1979) show that a temperature of 1727°C would be required to burn a 1-μm particle within 50 ms. At this temperature a 0.1-μm particle would be burned in 5 ms, and a 0.0287-μm particle

TABLE 3.14 Effects of EGR and Turbocharging on Emissions and Fuel Economy of a Light-Duty Diesel

| | Aspira-tion | Regulated Emissions, g/mi | | | | | Unregulated Emissions, mg/mi | | Fuel Econ., mpg |
		EGR	HC	CO	NO_x	Part.	Sul-fates	Alde-hydes	
VW	NA	No	0.16	1.0	1.20	0.35	7.7	34.4	41
(2,250 lbs, IW)		Yes	0.40	2.5	0.47	0.45	6.4	34.0	40
VW	TC	No	0.11	0.8	0.90	0.25	6.5	26.1	45
		Yes	0.20	1.2	0.40	1.48	8.7	33.3	44

SOURCE: Wiedemann and Hofbauer, 1978.

in 1 ms. If the exhaust velocity was 21 m/s, a 50-ms residence time could be obtained with a 1-meter-long system. Order-of-magnitude calculations have shown that the energy needed to heat the exhaust to temperatures ranging from 200° to 1200°C would require half of the amount of fuel used by the engine to produce power (Murphy et al., 1979). This fuel use is obviously unacceptable unless the energy used to raise the bulk exhaust temperature could be recovered. Design of a compact heat exchanger for the temperatures and flow rates involved would be difficult.

Wade (1980) also investigated the potential for reducing particulates in the exhaust gas, through oxidation, by using a manifold reactor. With this method, the exhaust gas from a single-cylinder diesel engine was electrically heated in a 50-foot sample line: particulate emissions were measured at 5-foot intervals. The results of these tests, as compared to a theoretical calculation for the oxidation of carbon, are provided in Figure 3.43 (Wade, 1980). As is shown in the figure, a temperature range well above 1500°F (816°C), and probably above 2000°F (1094°C), would be required to achieve significant oxidation of the particulate emissions within the residence time provided by an exhaust reactor with a volume equal to the engine displacement.

One possible method for attaining these high temperature and reactor volume (residence time) conditions is to add a burner to the exhaust gases. A ceramic gas turbine recuperator could then be used to recover the energy and resupply it to the exhaust gas. This method does not appear to be fuel efficient. Furthermore, the burner and recuperator would be at least as expensive as traps or trap-oxidizer systems.

The other method is to use a catalyst that would reduce hydrocarbons, the soluble organic fraction, solid particles, sulfur dioxide, nitric oxide, etc., depending on the material. One possibility is to add the catalyst to the particles as they are formed, i.e., the catalyst could be a diesel fuel additive (Murphy et al., 1979). Although no exact mechanism is known, several--involving either chemical reaction of the particle with the additive, agglomeration of the combustion and additive particles, or additive particles acting as nuclei for particu-

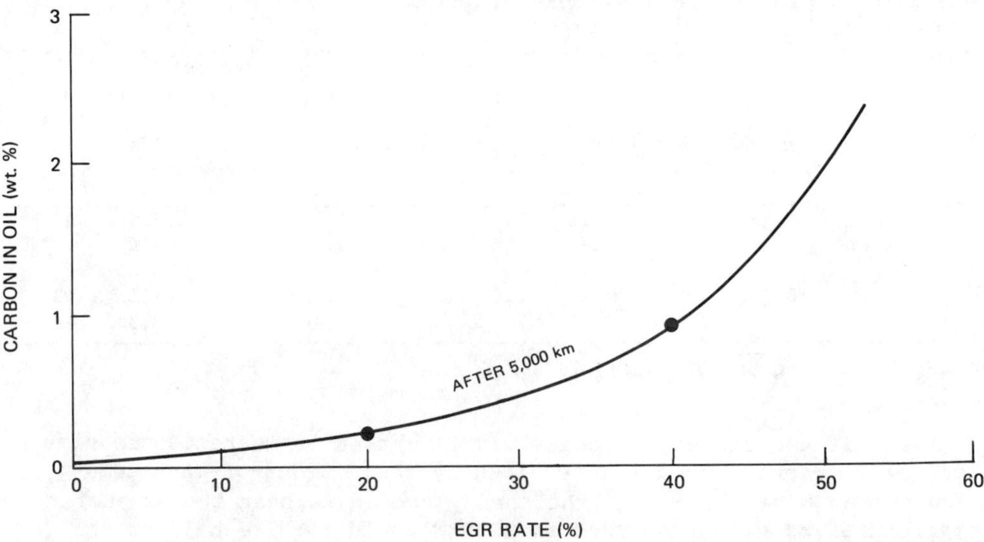

FIGURE 3.40 Effect of EGR rate on carbon contamination of oil in a
axial chamber light-duty diesel operated at 2,400 rpm, 3 kg/mi.
SOURCE: Nissan, 1979.

late formation, or adsorption of a gaseous additive by the particulate--
could be suggested (Murphy et al., 1979).

From a thermal energy conservation standpoint, the thermal reactor
approach seems best suited for oxidizing gaseous hydrocarbons, carbon
dioxide, and any hydrocarbons adsorbed on the particles. It fits the
systems approach to reducing particle emissions, but it will only be
effective in reducing the magnitude of the soluble organic fraction and
in cleaning up hydrocarbons and carbon monoxide that might be increased
by combustion system trade-offs. Supplying energy thermally to oxidize
the carbonaceous particles seems inefficient.

The reactor approach will not reduce the solid particles and will
only partially reduce the soluble organic fraction, unless high
temperatures or a catalyst is used. Oxidation of the soluble organic
fraction and hydrocarbons will be influenced by the load and speed of
the engine: The degree of control is hard to assess because no data are
available. Nitrogen dioxide and sulfate emissions could be increased,
but hydrocarbon, carbon monoxide, and odor emissions should be
decreased. The control system should have minimum effect on engine
brake-specific fuel consumption if no additional fuel is required.
Performance and durability should require little maintenance if
properly developed.

The use of a thermal reactor and catalyst will add to engine
production costs because to be effective at the light loads of the FTP
cycle, installation of a reactor between the exhaust ports and manifold
would be required. The reactor equipment could be integrated into the
engine design, but ceramic and/or steel materials that can withstand

high-temperature operation would be used. Current regulatory standards
apply to total mass: The soluble organic fraction constitutes only 10
to 40 percent of the total particulate from the FTP. To date, the
thermal reactor technology has not been developed because it reduces
only the soluble organic fraction and could actually increase the
sulfate fraction and in turn increase total particulate emissions.
Because the particulate soluble organic fraction is currently of major
importance, a control strategy that focuses only on this function may
be justified and should be examined.

Catalysts

Exhaust oxidation catalysts (pellets and monoliths) have been successfully used to control carbon monoxide and hydrocarbon emissions from
gasoline-powered vehicles. An oxidation catalyst would be even more
advantageous for a diesel, if it worked effectively at the low temperatures illustrated in Figure 3.42. It might also help to reduce the
particulate soluble organic fraction and reduce exhaust odor and
irritancy by promoting the complete oxidation of partially oxygenated
hydrocarbons. Among the disadvantages of a catalyst is that it might
promote the formation of carbonyls and other unregulated pollutants
such as nitrogen dioxide, nitrous oxide, and nitromethane. Unless a
low sulfur fuel (less than 0.15 percent by mass) is used, an oxidation
catalyst will likely promote the formation of excess sulfates. It is
believed that insulated containers and/or catalysts in manifolds will
be necessary if a high level of control of the soluble organic fraction

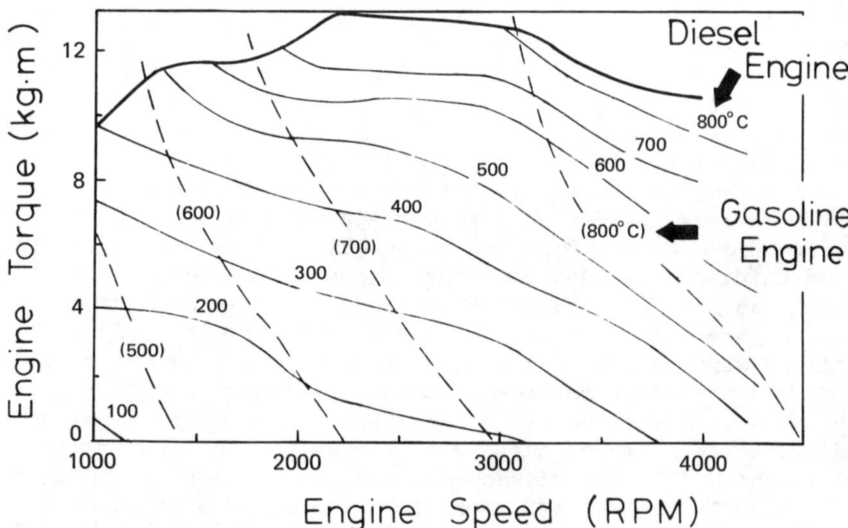

FIGURE 3.41 Characteristics of exhaust gas temperature.
SOURCE: Toyota, 1980.

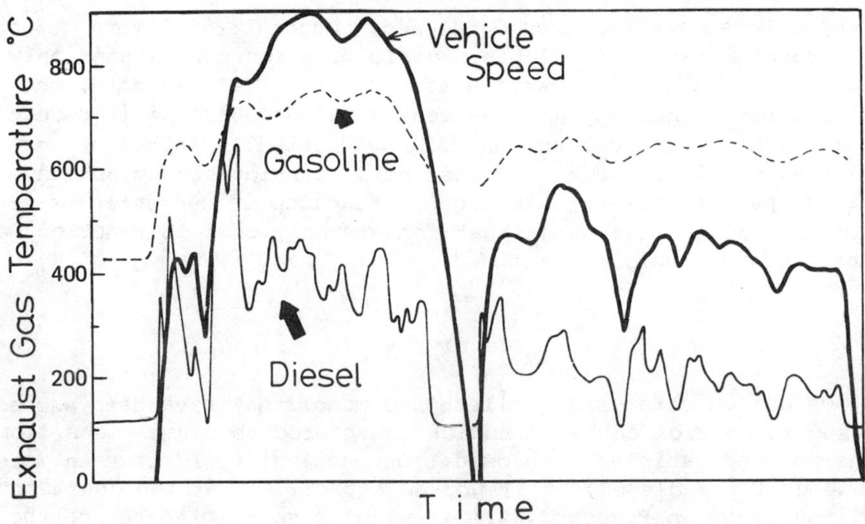

FIGURE 3.42 Exhaust gas temperature during FTP cycle.

is to be obtained. To date, there has been little development of these
types of containers and catalysts because current standards apply only
to total particulate mass.

Catalysts by themselves will probably not be applied to production
vehicles to oxidize the gaseous species. However, a catalyst that
would work only on the solid fraction (which has a lower ignition
temperature) and the soluble organic fraction of the particulate matter
would be attractive. Its application, in conjunction with thermal
reactors, would also be attractive. Use of catalyst materials in
conjunction with traps to work on either gaseous phase hydrocarbons, or
the solid particulate, or both, would also be attractive.

Johnson-Matthey engeers have recently developed a catalyst that,
they believe, reduces the ignition temperature of the solid particulate
matter and oxidizes the adsorbed hydrocarbons. They have also
developed a laboratory technique to measure the "light-off," or
ignition, temperature of diesel particulate matter with various
catalysts. A wire is inserted in the exhaust of an engine to let
particles collect on its surface. It is then inserted in an apparatus
where the wire is heated to various temperatures. Carbon monoxide and
carbon dioxide are measured along with the energy increase of the air.
Figure 3.44 shows data for an uncoated and a coated wire with catalyst
A. The data show that the ignition temperature has been changed from
450° to 350°C. They have also developed catalysts that "light-off" at
200° and 90°C (less feasible because of cost). This catalyst has been
applied to a compressed metal mesh substrate (block form) in a manifold

catalyzed trap and will be discussed further in the section on traps,
catalyzed traps, and trap-oxidizers.

Catalysts Downstream from Exhaust Manifold

Recently, a 1979 Oldsmobile diesel with a 5.7-liter engine was used to
evaluate catalyst effects on emissions. The vehicle was tested with
and without Engelhard (PTX-516) and UOP (UOP-103 and 99) monolithic
catalysts during the FTP and the Fuel Economy Test (FET) cycles
(Bykowski, 1979). No. 2 diesel fuel with a 0.29 percent by mass sulfur
content was used during the tests.
Table 3.15 shows the data with and without the Engelhard catalyst.
The catalysts reduced carbon monoxide emissions by approximately 90
percent and hydrocarbon emissions by approximately 60 percent. NO_x
emissions increased slightly. The application of the dual Engelhard
catalytic converters did not change the fuel economy.
The aldehyde, hydrocarbon, and sulfate data show the extent of
catalyst activity. The reduction in the total amount of aldehydes was
approximately 55 percent. With the exception of methane and ethane,
which remained constant, most of the individual hydrocarbon emissions
decreased. The sulfate emissions increased to a level approximately 5
times greater than the baseline rate.
Table 3.16 shows the data with and without the use of the two UOP
catalysts. The fresh UOP catalyst (UOP-103) reduced carbon monoxide

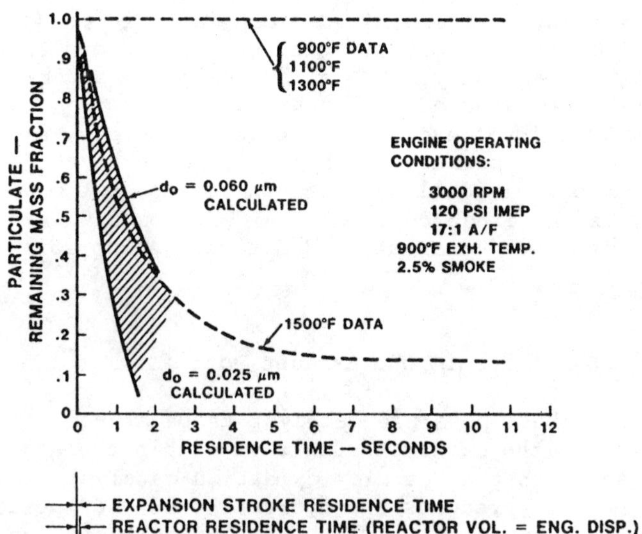

FIGURE 3.43 Measured and calculated.
particulate emission oxidation rates in a
50-foot heated exhaust sample line.
SOURCE: Wade, 1980.

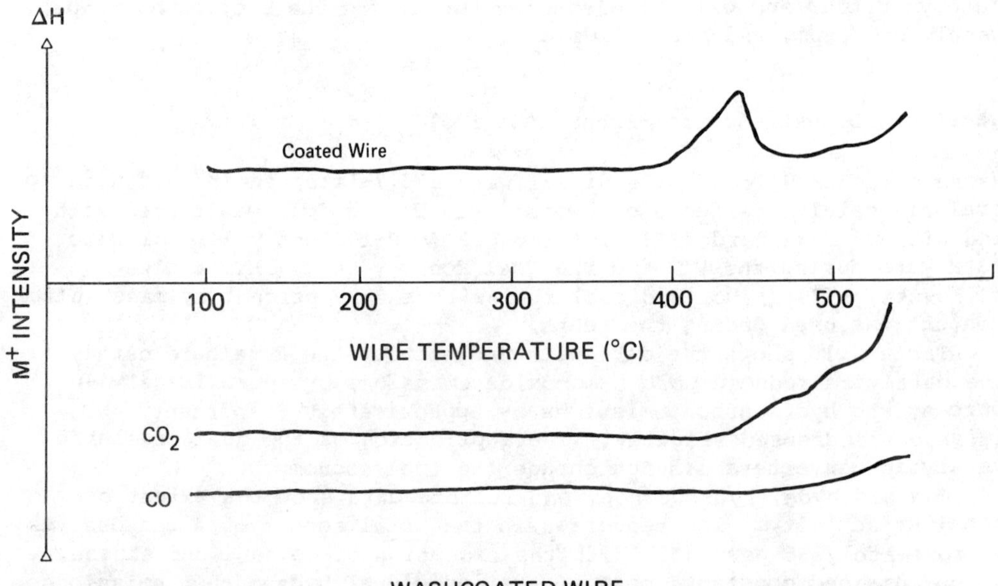

FIGURE 3.44 Data for the Johnson-Matthey washcoated wire catalyst.
SOURCE: Johnson-Matthey, 1980.

emissions by approximately 80 percent and hydrocarbon emissions by
approximately 50 percent. The NO_x emissions increased slightly. The
aged UOP catalyst (UOP-99) exhibited a considerable reduction in
activity. Carbon monoxide and hydrocarbon conversions approximated 40
and 30 percent, respectively. NO_x emissions increased by 14 percent.
Fuel economy was not affected (Bykowski, 1979).

Again, all individual hydrocarbon emission levels, except methane,
were decreased during operation with the fresh catalyst. The aged
catalyst decreased individual hydrocarbon levels, but only about half
as efficiently as the fresh catalyst. Aldehyde and sulfate emissions
for both fresh and aged catalysts did not change appreciably from the
baseline.

Use of Catalyst Substrate in the Exhaust Port

To maintain the highest possible catalyst temperatures, it might be
beneficial to apply the catalyst substrate within the engine exhaust
port. It has shown that a Fecraloy oxidation catalyst, when applied
just outside the port, reduces the hydrocarbons by 60 percent. Within
the port, it might oxidize nearly all the hydrocarbons, which would
provide good control of the soluble organic fraction. Because of the
short residence time, sulfate emissions might be lower than they are
with other catalytic approaches. This approach might be combined with
a particle trap in a total system.

TABLE 3.15 FTP, FET, and Particulate Emissions From 1979 Oldsmobile Diesel Delta 88 With and Without Englehard Catalyst

Test No.	Cycle	Catalyst	Date	Emission Rates, g/km				Fuel Consump., 1/100 km	Sulfate, mg/km	Particulate Emissions, g/km	Calculated Particulate, g/km
				HC	CO	NO_x	CO_2				
1	FTP	None	5-31-79	0.26	1.23	0.93	328.98	10.75	12.888	0.782	—
3	FTP	None	6-01-79	0.17	1.25	0.99	330.58	10.79	7.370	0.770	—
5	FTP	PTX-516	6-06-79	0.09	0.12	1.07	334.69	10.87	52.608	0.811	0.808
7	FTP	PTX-516	6-13-79	0.09	0.14	0.14	333.14	10.82	68.105	0.805	0.808
2	FET	None	5-31-79	0.11	0.71	0.67	228.5	8.55	5.705	0.429	—
4	FET	None	6-01-79	0.09	0.71	0.87	230.7	8.63	6.208	0.441	—
6	FET	PTX-516	6-06-79	0.02	0.04	0.74	228.9	8.52	143.463	0.715	0.559
8	FET	PTX-516	6-13-79	0.05	0.04	0.85	224.7	8.37	180.032	1.029	0.599
8-2	FET	PTX-516	7-26-79	0.01	0.11	0.85	230.0	8.57	47.164	0.515	0.466
8-3	FET	PTX-516	7-27-79	0.01	0.10	0.79	226.0	8.41	46.760	0.525	0.466

Test No.	Cycle	Catalyst	Date	Individual Hydrocarbon Emission Rate, mg/km							
				Methane	Ethylene	Ethane	Acetylene	Propane	Propylene	Benzene	Tolune
1	FTP	None	5-31-79	11.31	20.69	0.88	4.29	0	8.02	9.01	0.81
3	FTP	None	6-01-79	12.16	23.65	1.66	5.06	0	7.18	8.06	1.47
5	FTP	PTX-516	6-06-79	15.42	5.12	1.02	0.06	0	0.26	1.47	0.64
7	FTP	PTX-516	6-13-79	11.00	6.65	2.05	0.06	0.50	0.87	1.37	1.12
2	FET	None	5-31-79	2.30	6.96	0.25	1.43	0	2.36	2.49	0
4	FET	None	6-01-79	2.42	6.52	0.19	1.30	0	2.36	2.61	0.50
6	FET	PTX-516	6-06-79	3.36	0.37	0.19	0	0	0	0	0
8	FET	PTX-516	6-13-79	2.24	0.62	0.44	0	0	0	0	0

Test No.	Cycle	Catalyst	Date	Aldehyde Emission Rates, mg/km								
				Formal- dehyde	Acet- aldehyde	Acetone[a]	Isobutyr- Aldehyde	Methyethyl Ketone	Croton- Aldehyde	Hex- aldehyde	Benzene	Total
1	FTP	None	5-31-79	3.4	1.2	0.1	2.1	0	0	0	0	6.8
3	FTP	None	6-01-79	6.1	0.9	0.8	0	0	0.2	0	0.9	8.9
5	FTP	PTX-516	6-06-79	1.1	0.6	0	0	0	0	0	1.6	3.3
7	FTP	PTX-516	6-13-79	2.4	0.7	0	0	0	0	0	0.4	3.5

[a]Includes acrolein and propanol.

SOURCE: Bykowski, 1979.

TABLE 3.16 FTP, FET, and Particulate Emissions From 1979 Oldsmobile Diesel Delta 88 With and Without UOP Catalyst

| Test No.[a] | Cycle | Catalyst | Date | Emission Rates, g/km | | | | Fuel Consump., 1/100 km | Sulfate, mg/km | Particulate Emissions, g/km | Calculated Particulate, g/km |
				HC	CO	NO_x	CO_2				
9-1	FTP	None	7-06-79	0.27	1.18	1.02	340.41	11.12	23.881	0.782	–
10-1	FTP	UOP-103	7-11-79	0.13	0.18	0.97	324.33	10.54	17.334	0.723	0.747
11-1	FTP	UOP-103	7-12-79	0.04	0.23	1.12	327.70	10.64	14.865	0.605[a]	0.710
12-1	FTP	None	7-13-79	0.28	1.09	1.01	321.18	10.50	21.073	0.756	–
13-1	FTP	UOP-99	7-17-79	0.21	0.66	1.16	329.83	10.75	18.115	0.711	0.760
14-1	FTP	UOP-99	7-18-79	0.18	0.64	1.14	325.61	10.61	15.296	0.670	0.730
11-2[b]	FTP	UOP-103	7.25-79	0.13	0.44	1.11	330.48	10.75	19.897	0.761	0.735
9-1	FET	None	7-06-79	0.14	0.66	0.75	222.0	8.34	19.798	–	–
10-1	FET	UOP-103	7-11-79	0.06	0.06	0.70	226.6	8.44	80.219	0.535	0.507
11-1	FET	UOP-103	7-12-79	0.03	0.08	0.75	202.8	7.55	30.864	0.410	0.442
12-1	FET	None	7-13-79	0.10	0.61	0.75	212.8	7.96	17.098	0.452	–
13-1	FET	UOP-99	7-17-79	0.03	0.20	0.84	222.8	8.30	30.160	0.467	0.449
14-1	FET	UOP-99	7-18-79	0.07	0.28	10.81	224.2	8.36	34.930	0.505	0.462
11-1	FET	UOP-103	7-25-79	0.05	0.11	0.80	227.4	8.47	48.792	0.511	0.464
14-2[b]	FET	UOP-99	7-25-79	0.06	0.31	0.87	235.5	8.78	60.723	0.513	0.487

Test No.	Cycle	Catalyst	Date	Individual Hydrocarbon Emission Rate, mg/km							
				Methane	Ethylene	Ethane	Acetylene	Propane	Propylene	Benzene	Toluene
9-1	FTP	None	7-06-79	12.21	24.03	2.33	4.74	0.0	7.83	10.01	3.01
10-1	FTP	UOP-103	7-11-79	8.57	7.19	0.53	0.17	0.0	1.79	2.23	0.07
11-1	FTP	UOP-103	7-12-79	10.95	9.12	1.09	0.25	0.0	1.40	3.21	0.17
12-1	FTP	None	7-13-79	11.46	22.63	1.44	4.29	0.0	6.87	7.25	0.38
13-1	FTP	UOP-99	7-17-79	11.98	16.62	0.90	1.12	0.93	4.55	5.40	0.30
14-1	FTP	UOP-99	7-18-79	10.94	17.31	1.22	1.31	0.0	4.69	4.70	0.32
9-1	FET	None	7-06-79	4.56	11.45	0.59	2.19	0.0	3.29	4.01	1.56
10-1	FET	UOP-103	7-11-79	3.57	1.61	0.46	0.0	0.0	0.09	0.40	0.00
11-1	FET	UOP-103	7-12-79	4.22	2.32	0.65	0.0	0.0	0.23	0.66	0.23
12-1	FET	None	7-13-79	3.77	7.89	9.24	1.48	0.0	2.50	2.36	0.21
13-1	FET	UOP-99	7-17-79	4.98	4.69	0.50	0.22	0.0	1.21	1.96	0.32
14-1	FET	UOP-99	7-18-79	4.55	5.92	0.50	0.22	0.0	1.34	2.05	0.00

Test No.	Cycle	Catalyst	Date	Aldehyde Emission Rates, mg/km								
				Formal-dehyde	Acet-aldehyde	Acetone	Isobutyr-Aldehyde	Methyethyl Ketone	Croton-Aldehyde	Hex-aldehyde	Benzene	Total
9-1	FET	None	7-06-79	4.2	0.9	0.0	0.0	0.0	0.0	0.0	0.6	5.8
10-1	FET	UOP-103	7-11-79	3.1	0.7	0.0	0.6	0.0	0.0	0.0	0.0	4.4
11-1	FET	UOP-103	7-12-79	2.9	0.9	0.4	0.0	0.0	0.0	0.0	0.0	4.2
12-1	FET	None	7-13-79	4.9	0.1	0.0	0.0	0.0	0.0	0.0	0.0	5.0
13-1	FET	UOP-99	7-17-79	4.5	0.0	0.0	0.0	0.0	0.2	0.0	0.2	4.9
14-1	FET	UOP-99	7-18-79	5.3	0.0	0.0	1.6	0.0	0.0	0.0	1.2	6.5

[a]Piece of filter missing during weighting, unless indicated otherwise.
[b]Includes acrolein and propanol.

SOURCE: Bykowski, 1979.

Using a catalyst should have little effect on engine brake-specific
fuel consumption, performance, or durability. However, the catalyst
itself:

- could have durability problems;
- might require periodic maintenance;
- could cost a significant amount in some engines because it
might require a redesign of the head with insulation and port size
(length and diameter) for high levels of control; and
- could markedly increase the uncontrolled deterioration factor.

Catalysts in Combination with Other Control Technologies

An oxidation catalyst could be used in light-duty diesels to control
the increased hydrocarbon and carbon monoxide emissions resulting from
other control technologies, such as EGR or injection timing retard.

Amano and co-workers (1976) investigated the potential of using
exhaust catalysts to decrease the high hydrocarbon and carbon monoxide
emissions resulting from the use of EGR. Using a simple on/off system
that shut off EGR at 65 percent of the peak load, 66 to 72 percent
reductions in hydrocarbons and 76 percent reductions in carbon monoxide,
with no effect on NO_x, were found. No results for particulates were
reported.

Bassoli and co-workers (1979) studied the feasibility of using
catalysts to control the high hydrocarbon emissions associated with
timing retard and EGR for both turbocharged and naturally aspirated
light-duty diesels (Table 3.17). For the naturally aspirated vehicle
with timing retard, the most effective catalyst lowered hydrocarbon
emissions by 64 percent, to almost the same level as that obtained with
standard timing. Some NO_x formation on the catalyst was noted, but
the exhaust levels were 37 percent lower than the values reported for
standard timing.

The hydrocarbon emissions from the turbocharged vehicle with EGR
and timing retard were 29 percent less than those from the turbocharged
timing retard vehicle and 44 percent less than those from the naturally
aspirated vehicle with standard timing. Again, some NO_x catalyst
formation was noted. Further timing retard lowered NO_x, but the
catalyst did not maintain hydrocarbon emission control at a
sufficiently low level. The particulate data are complicated by the
use of a high-sulfur fuel and the consequent formation of sulfate over
the catalyst. This reemphasizes the necessity of using low-sulfur fuel
if an oxidation catalyst is to be employed.

Traps, Catalyzed Traps, and Trap-Oxidizers

Although a filter is usually thought of as a means for removing
particles from the flowstream, it also can be considered as a method
for concentrating particulate matter. The three main types of filter
materials being developed are metal meshes, ceramic monoliths, and

TABLE 3.17 Effect of Oxidation Catalyst, Injection Timing, EGR, and Turbocharging on Fiat 2.4-Liter Light-Duty Diesel

Type	Static Injection Timing		Rear Axle Ratio	Emissions g/m				Fuel Econ., mpg
				HC	CO	NO_x	Part.	
NA	1	btdc	3.2	0.31	1.68	1.60	0.55	34.6
	5	atdc	3.2	0.99	1.94	0.94	0.68	−
NA, Cat. A	5	atdc	3.2	0.55	0.60	0.99	0.75	−
NA, Cat. B	5	atdc	3.2	0.36	0.33	1.01	1.25[a]	34.5
TC	5	atdc	2.6	0.28	1.33	1.47	0.36	37.7
TC, EGR	5	atdc	2.6	0.31	2.67	1.19	0.53	35.6
TC, EGR, Cat. B	5	atdc	2.6	0.20	2.1	1.28	0.78[b]	35.6
	9	atdc	2.6	0.50	2.1	0.98	0.93[c]	−

[a] Used 0.6 percent by mass-sulfur fuel. Estimated value with low-sulfur fuel: 0.7.
[b] Estimated value with low-sulfur fuel: 0.6.
[c] No estimate for low-sulfur fuel.

SOURCE: Bassoli et al., 1979.

ceramic foams; to a lesser degree, paper and other fiber material also are under investigation. Figure 3.45 shows, in schematic form, two of these materials and the general exhaust gas flow pattern through the material. Table 3.18, from General Motors, shows the durability limits for three types of materials. General Motors has developed the requirements shown in Table 3.19 for its trap material. The Environmental Protection Agency (EPA) summary of the efficiencies of a number of materials is provided in Table 3.20.

The material used in a trap can be impregnated with a catalyst. The concept of a disposable trap is also possible. Because of filter plugging and the low density (0.07 to 0.10 gm/cm^3) of diesel particulate (over a gallon per 1000 miles for a high-efficiency unit), periodic regeneration is probably the most feasible approach, possibly every 100 miles. Rather than a self-contained automatic onboard regeneration system, auxiliary cleaning (at service station) equipment could be used (GM, 1979a). Figure 3.46 shows this concept in schematic form. The particulate disposal problem and the inconvenience and cost to the driver may make this approach impractical.

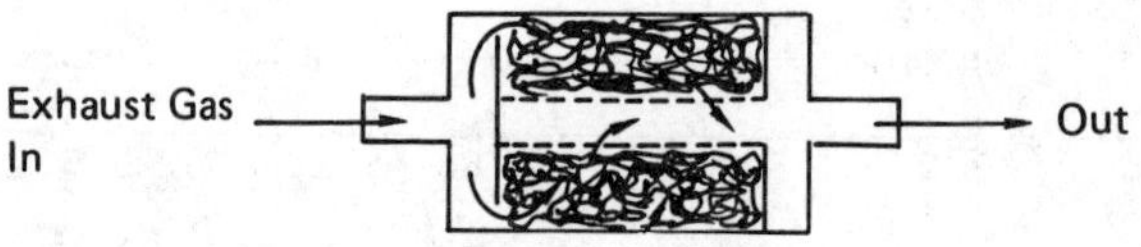

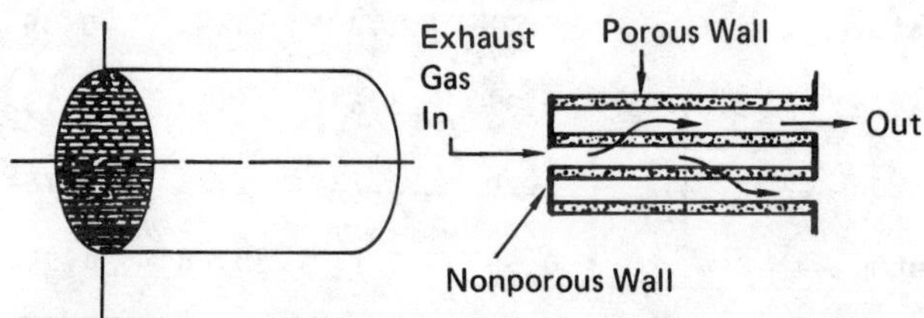

FIGURE 3.45 Schematic: steel wool and
ceramic exhaust aftertreatment filters.
SOURCE: Nissan, 1980.

Regeneration Methods

Currently, the most desirable solution to the trap regeneration problem
appears to be onboard incineration of the particulates. Test data have
shown that with uncatalyzed traps this can be achieved, under certain
conditions, by raising the temperature of the exhaust gases to
approximately 480°C (900°F). Four regeneration methods are being tried
(Murphy, et al., 1979):

1. The trapped particulate is ignited by energy derived from
engine operation--usually from the hot exhaust gases. This has been
accomplished by engine intake throttling, which reduces the air/fuel
ratio and increases the exhaust temperature.
2. The trapped particulate is ignited by an auxiliary heating
device that triggers ignition at programmed times in the engine
operation cycle. One approach has been to use an electric heating
element at low exhaust flow. This conserves energy because it acts on
only a portion of the exhaust flow and provides flexibility in the
placement of the energy generating device. In the second approach, a
dual path trap with two heating elements and a valve routes a fraction
of the exhaust flow to the incineration path (Figure 3.47). Other
approaches, with and without heating elements, including a system that
uses an oil burner, are being investigated (GM, 1979a).
3. The particulate is ignited at temperatures that are substantially lower (800°F (427°C) versus 900°F) than those used in either of

TABLE 3.18 Requirements for Trap Material Durability

Material	Durability Characteristics
Metal mesh	1500°F limit (2100°F low efficiency) 600-mi load-up 12,800 mi. with burn and blow off
Ceramic monolith	2300°F limit 150-mi load-up (difficult to regenerate) No blow off
Paper element	450°F limit 750-mi load-up (no satisfactory cleanup) No blowoff

SOURCE: GM, 1980a.

TABLE 3.19 Requirements for Trap Materials

Characteristics	Specific Requirement of Characteristics
High collection efficiency	50 percent or greater
Minimum engine back pressure rise	Capacity Cleaning cycle time
Heat resistant	Level dependent on application
Particulate must be easily removed	Thermal regeneration Nonthermal cleaning
Filter size and complexity	Cost

SOURCE: GM, 1980a.

the above methods by employing catalytic materials that promote
particulate oxidation.

4. The vapor phase portion of the material, which would later form
the condensed/adsorbed portion of the particulate at lower temperatures,
is catalytically or thermally oxidized at the higher temperatures, and
the remaining particulate is trapped and burned downstream in a trap-
oxidizer unit using any of the above three strategies.

These approaches seem to offer the greatest potential for an
effective particulate aftertreatment device. They would require a

TABLE 3.20 Summary of Trap Material Efficiencies

Engine Type	Material	Efficiency, %
Opel, 2.1-liter engine	Corrugated foil Fecralloy	36
	Chopped Fecralloy	29
	Chromium Alloy Ribbo	37
	Glass-fiber fabric	34-65
	Fiberfax fiber fabric	46
	Aluminum fiber	32-61
	Catalyst beads	10-62
	Ceramic monolith extruded	10-52
	Ceramic-torturous path	39-49
	Ceramic bobbin	42
	Aluminum-coated metal mesh	63
	Metal mesh	28-44
Oldsmobile, 5.7-liter engine	Corrugated Fecralloy	30
	Catalyst beads	56
	Ceramic monolith extruded	30
	Aluminum-coated mesh	65
	Metal wood	60
	Fiberglass	76
	Paper element	90

SOURCE: EPA, 1980b.

microprocessor controller and sensors to properly initiate and control
the trap regeneration process as shown in Figure 3.48. Table 3.21
shows the basic parameters and their effects on the process.

General Motors is investigating the effect of the fuel additive MMT
on emissions with and without regeneration. Figure 3.49 provides
typical short-term durability data on trap efficiency and back pressure
versus time (Roessler et al., 1980) for traps A and B. Trap A is of
particular interest because it has the ability to maintain trapping
efficiency at a better than 50 percent level for 950 miles; a small
increase in back pressure results. In trap B, the efficiency is higher
at zero mileage. A failure in its mild-steel structure prevented
evaluation of performance for 950 miles (Roessler et al., 1980).

Catalyzed Traps

If a catalytic trap material or a catalyst followed by a trap-oxidizer
were used, a lower sulfur fuel might be required to reduce the sulfate
emissions. In addition, the high sulfate emissions during trap
regeneration with either a catalyzed or an uncatalyzed system needs
further investigation. Some increase in nitrogen dioxide could result
when using catalyzed traps or a catalyst in series with a trap-oxidizer.
It would appear that these systems, when properly developed, could

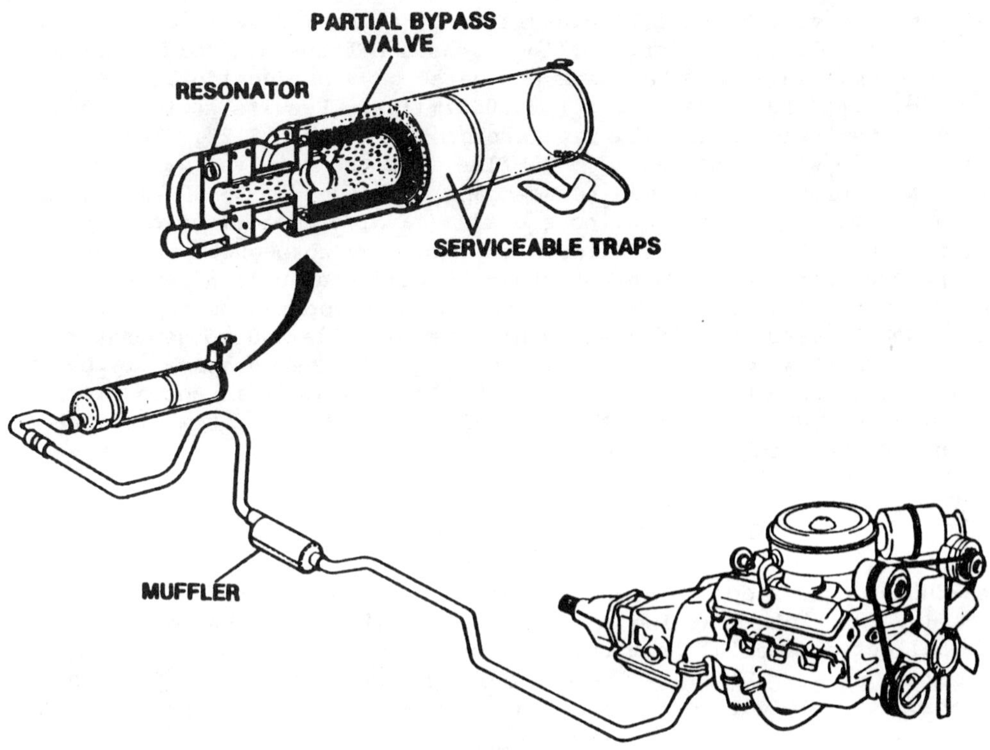

FIGURE 3.46 Serviceable trap system. SOURCE: GM, 1980.

approach 80 percent efficiency; however, their durability has not yet
been proven. They should have minimal effect on brake-specific fuel
consumption, engine performance, and durability. The concept in its
final system form will require active and reliable controls.

The use of catalyzed traps could be costly because some of the
systems will require a microprocessor and sensors in addition to the
container/trap material. Many of these components, such as the microprocessor, sensors, and dual-bed catalysts (oxidation and reduction),
will be used in 1981 gasoline-powered cars. Applying such a control
system to diesel engines will increase the differential cost between
diesel and spark-ignited engine vehicles.

Johnson-Matthey reports that it is developing a precious metal
catalyst that lowers the ignition temperature of the solid particulate
matter. The catalyst support consists of a knitted special alloy-steel
wire formed into compressed rigid blocks. This system gives a high
degree of turbulent exhaust flow with a highly geometric surface area.
A washcoat is deposited onto the support, and the catalyst is formulated
to promote combustion of particulate and gaseous emissions (Johnson-
Matthey, 1980). Sulfate generation and storage has been shown to be a
problem, but Johnson-Matthey believes that the problem is related to
the washcoat and that sulfate generation can be minimized or eliminated

by tailoring the catalyst to the individual vehicle, i.e., by matching
catalyst activity to vehicle exhaust temperature and baseline emissions.

Figure 3.50 is a diagram of the Johnson-Matthey manifold catalyzed
trap system. This trap has been evaluated on a production 2-liter
diesel-powered passenger car for 35,000 miles. Results up to 12,500
miles on the road durability test are shown in Table 3.22. Testing on
this vehicle was halted at 12,500 miles, and the catalyst was stored.
After some time, testing was resumed and continued to 35,000 miles.
The unit suffered damage during storage (catalyst deactivation), but
the catalyst structure maintained its integrity throughout the test.
It appears that the Johnson-Matthey catalyzed trap will also require a
regeneration system. Further research and development on this system
is needed to measure sulfates and nitrogen dioxide and to demonstrate
that the catalyst does indeed lower the ignition temperature of the
solid particulate matter. Tests to show that blowoff is not a problem
also need to be conducted. The type of regeneration system also needs
further definition and development.

Electrostatic Precipitation

The fundamental steps in an electrostatic precipitation process
include: particle charging, particle collection, and the removal of
the collected material from the collection and discharge electrodes
(Faulkner et al., 1979). Particle charging is accomplished through the
creation of an electric field and a corona current by applying a large

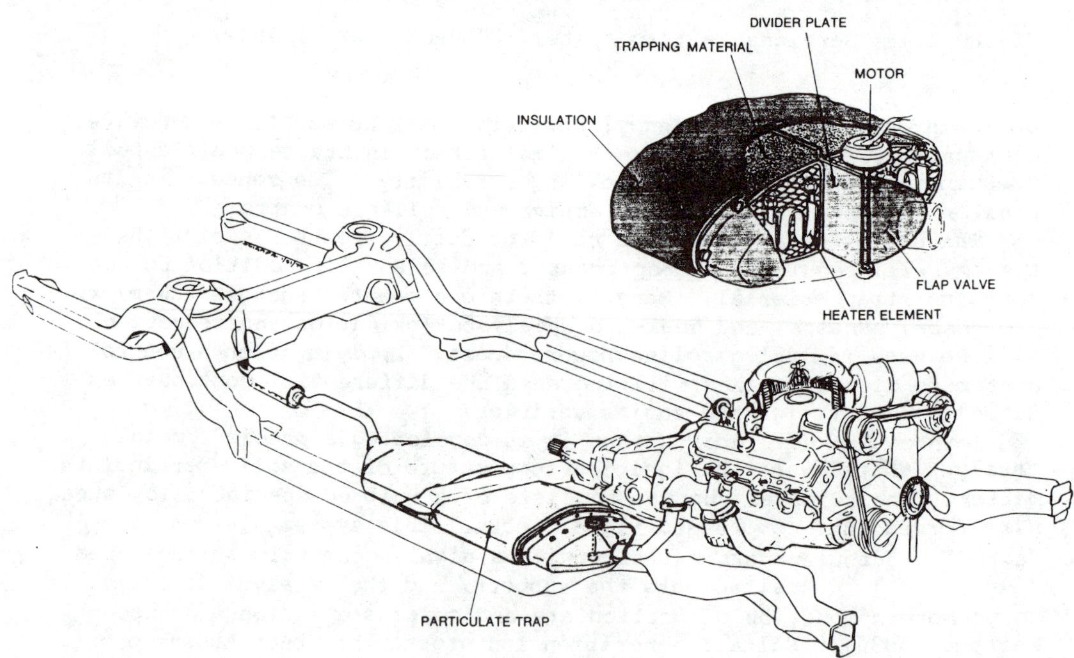

FIGURE 3.47 Dual path trap system. SOURCE: GM, 1980.

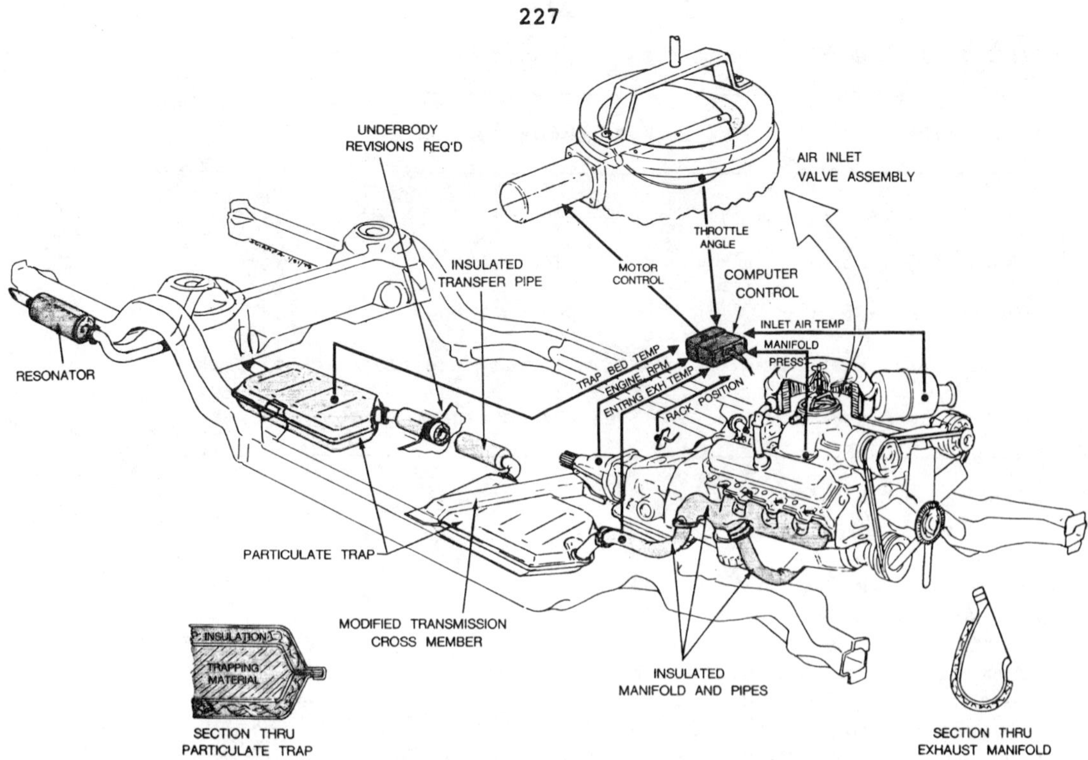

FIGURE 3.48   Diesel exhaust particulate control system.   SOURCE: GM, 1980.

potential difference between a small radius electrode and a much larger
electrode, where the two electrodes are separated by a region of space
containing an insulating gas (Faulkner et al., 1979).

This approach provides high collection efficiency for small
particles with low back pressure and acceptable energy consumption.
However, at least two major development problems must be overcome
before this system can be used in automotive service: particle
reentrainment and current leakage through conductive films on high-
voltage insulators. In addition, a means for removal and disposal of
the collected particulate must be developed.

The collected particulates will be bulky and potentially toxic:
disposal, therefore, could be a major problem. General Motors has
supported work at Battelle Columbus Laboratories on an electric field
to divert particulates into a concentrated portion of the exhaust (GM,
1979c). An arrangement of a charging wire in the center of an exhaust
pipe was used to drive the particulates outward, and a peripheral
collector pipe to gather the concentrated matter. The objectives were
to have 90 percent of the particulates in 10 percent of the flow; but
the best accomplished on an Opel 2100D engine was 50 percent of the
particulates in 16 percent of the flow. The particulates quickly
coated all surfaces exposed to the exhaust gas, causing erratic
operation and several failures of the device. The high voltage and

TABLE 3.21 Control Parameters for Regeneration

Parameter	Effect on Regenerator
Exhaust temperature	Reaction rate
Exhaust O_2 concentration	Reaction rate
Exhaust flow velocity	Convective losses
Trapped particulate mass	Peak temperature, location of peak
Particulate distribution	Peak temperature, location of peak

SOURCE: GM, 1980.

power requirement (on the order of 60 watts at 12,000 volts) was substantial but not prohibitive.

The electrostatic precipitator control efficiency potential for automotive emissions has not been proven sufficiently at this time. The method should have no effect on the other regulated and the unregulated pollutants. It should have no effect on brake-specific fuel consumption, engine performance, and durability. It would be complex, bulky, and expensive when a complete system were developed (including removal and disposal components). It will have to be integrated into the engine design. The technology is in the early development stage and will require several years before its cost, control efficiency, and complexity can be fully assessed.

FUEL MODIFICATIONS

The composition and characteristics of fuel can significantly affect the combustion process and the production of exhaust emissions. Fuel composition is generally described by a number of physical characteristics, including:

- specific gravity, or API gravity number;
- viscosity;
- distillation characteristics;
- cetane number;
- sulfur, nitrogen, and other inorganic contents;
- hydrocarbon family distribution; and
- degree of unsaturation.

These properties are not all independent, making isolation of specific fuel effects difficult. For example, replacing paraffins with aromatics in diesel fuel results in a denser fuel with a lower cetane number. Also, the response to fuel changes is affected by engine design and

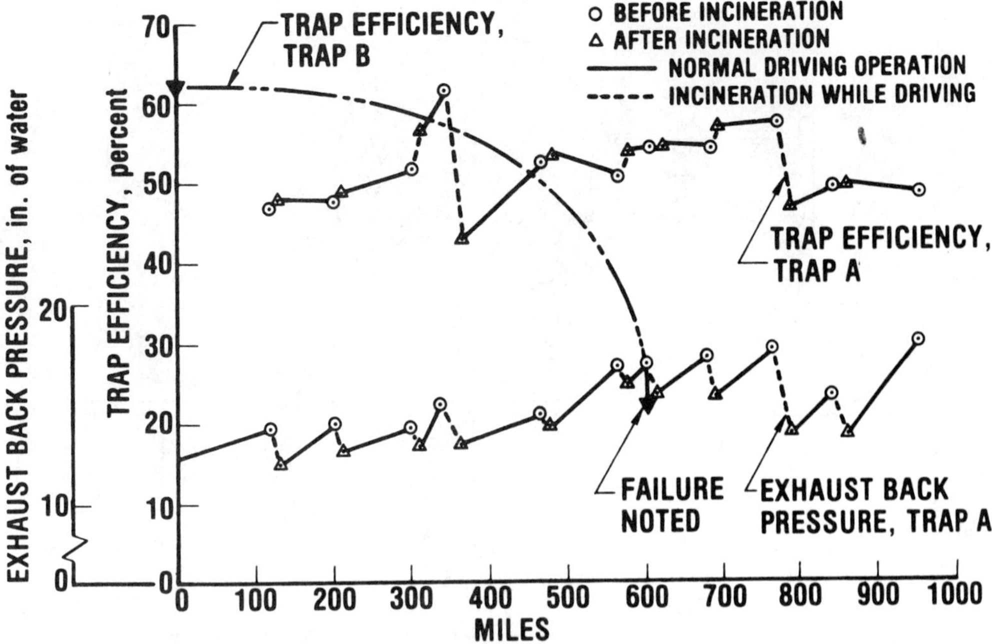

FIGURE 3.49 Short-term durability data: trap efficiency and back pressure versus time, traps A and B. SOURCE: Roessler et al., 1980.

operating conditions; hence engines do not behave uniformly. Nonetheless, a number of general observations and trends have been reported.

FUEL PROPERTIES

The effects of fuel properties on particulate and other gaseous emissions were discussed in the section on effects of fuel properties and will not be repeated in detail here. In brief, reducing aromatic content and decreasing the 90 percent boiling point of diesel fuel generally result in lower particulate, NO_x, and noise emission levels. The NO_x and noise effects are attributable to the decrease in ignition delay and smoother pressure rise rate due to the higher cetane number. Hydrocarbons were found also to be sensitive to fuel cetane number; emissions decrease with higher cetane numbers. Also, fuel sulfur is related to the level of particulate sulfate. Thus, a higher minimum cetane number of 48 and limits on fuel properties (20 percent aromatics maximum, 0.25 percent sulfur maximum, and 90 percent boiling range temperature less than 316°C, or 600°F) are recommended for emission control purposes. A major aspect of this is that new automotive diesel fuel specifications must provide the engine/equipment designer with a better defined fuel, to allow for optimization of the injection system, the combustion chamber design, and control of the combustion process.

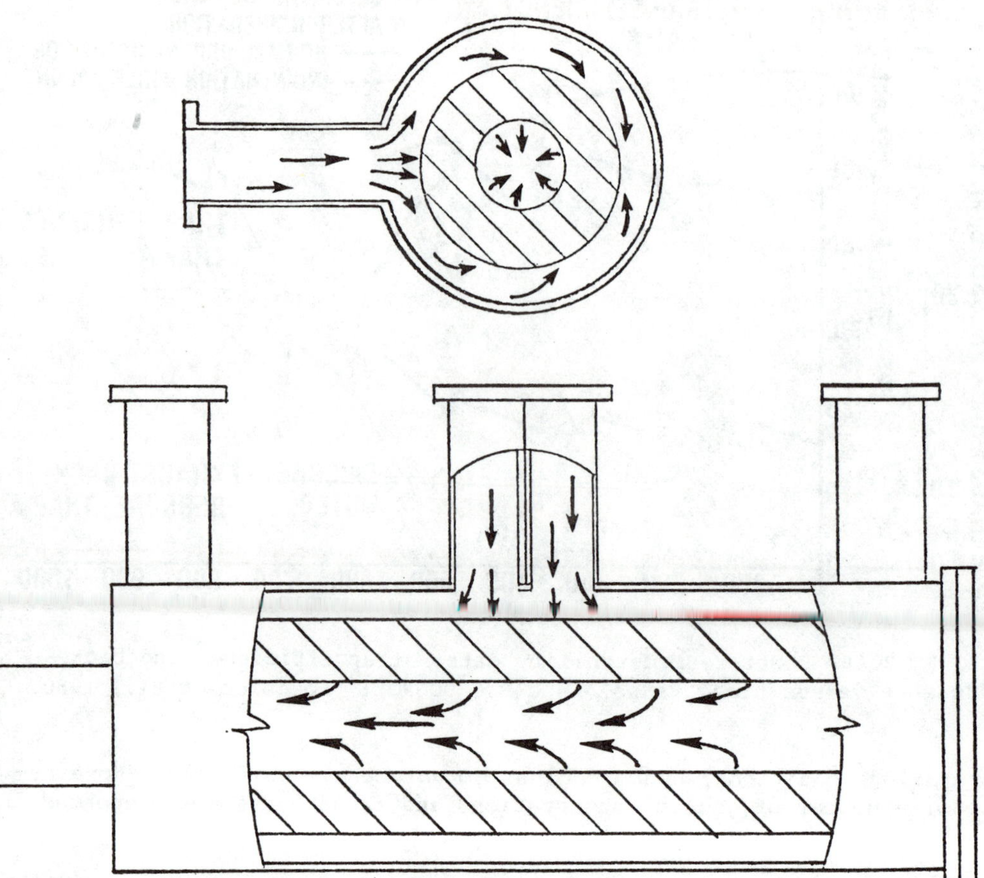

FIGURE 3.50 Diagram of Johnson-Matthey manifold system. SOURCE:
Johnson-Matthey, 1980.

Fuel Properties and Unregulated Emissions

Information related to the effects of fuel properties on unregulated
emissions is less well defined. Significant variations in engine
design and driving cycle have been observed: Thus, an unambiguous
interpretation of the data is difficult. In general, however, minimum
quality diesel fuel (low cetane number, low volatility, and high
aromatic content) generally results in higher emissions of benzo(a)-
pyrene, odor, and aldehydes (Hare and Baines, 1979; Hare, 1979). The
trends for benzo(a)pyrene and aldehydes are illustrated in Figures 3.51
and 3.52; vehicle type and driving cycle effects are also shown. It is
noted that there was little correlation between particulate mass and
particulate benzo(a)pyrene content. This indicates a possible need for
developing additional, component-specific regulations for particulate.

TABLE 3.22 Results From a 12,500-mi Durability Test of the Johnson-Matthey Manifold Catalyzed Trap System

Emission Species	Reduction, %	Deterioration Factor (0 to 12,500 miles)
Particulates	52	1.05
Adsorbed hydrocarbons	76	1.08
Carbon monoxide	58	1.56
Hydrocarbons	61	1.00
Total hydrocarbons	66	1.05

SOURCE: Johnson-Matthey, 1980.

The ramifications of fuel property/emission interactions are summarized by Roessler and co-workers (1980):

> The association of increased emissions with increased aromatic content and 90% boiling point is unfortunate. Aromatic content of diesel fuels is projected to increase as a natural result of catalytic cracking of heavy fuel oil components to increase the yield of diesel fuel relative to residual fuels. Also, the available fraction of diesel fuel from crude oil may be increased by in creasing the allowable 90% boiling point. An increase in the ASTM D-975 D-2 fuel specification maximum value of the 90% boiling point from 640°F to 680°F could provide as much as an additional 4 1/2% distillate fuel. In spite of these factors, based on engine operation and emissions control requirements, the recommended development of an automotive diesel fuel specification is warranted.

Fuel Additives

Fuel additives are developed to modify ignition delay, catalyze the combustion of carbon, and/or improve the fuel-air mixing processes by assisting fuel droplet dispersion and thereby reducing particulate emissions. The last approach generally involves the use of nonconventional fuels or fueling systems (e.g., emulsions).

Effects of Additives

Barium- and calcium-based additives have long been used to reduce smoke opacity. These additives not only catalyze the conversion of carbon particles to carbon dioxide, they also aid in the suppression of free carbon formation during the early stages of the combustion process. As a result, carbonaceous particles are replaced with emissions of oxides and sulfates of barium and calcium, and although smoke opacity was reduced, total particulate emissions frequently increased. Hence, the fate and toxicity of the additive end products must be considered.

A number of experimental and commercially available particulate reduction additives have been screened by General Motors (1979a). As the data in Table 3.23 show, using additives resulted in changes ranging from a 25 percent reduction to a 48 percent increase in the mass of particulates at steady state (40-mph-road-load) operation. Further screening, as depicted in Figure 3.53, showed a load dependence with variable results. One additive, in tests during the FTP driving cycle, resulted in a 16 percent reduction in particulates, with little or no change in gaseous emissions or fuel economy. The data provide no indication of the additive's effects on unregulated emissions, noise, durability, etc. These factors would have to be considered before any of the additives become widely used. Again, the ultimate fate and toxicity of the additive components must be explored.

Cetane Improvers

Commercial cetane improvers are available for diesel fuels. These fuel additives are primarily comprised of mixtures of primary amyl nitrates

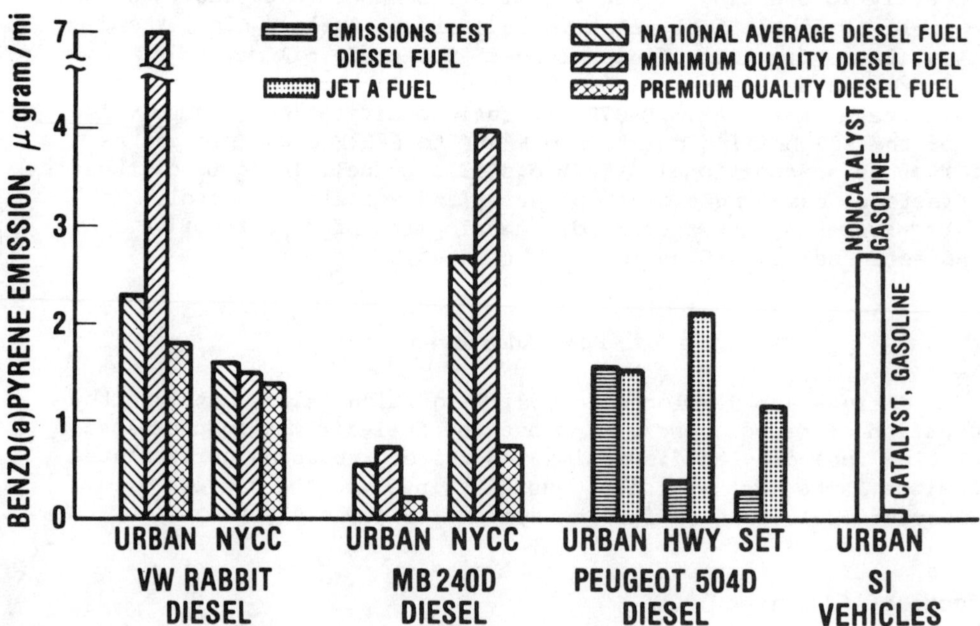

FIGURE 3.51 Benzo(a)pyrene emissions from diesel and spark ignition vehicles. SOURCE: Roessler, 1979.

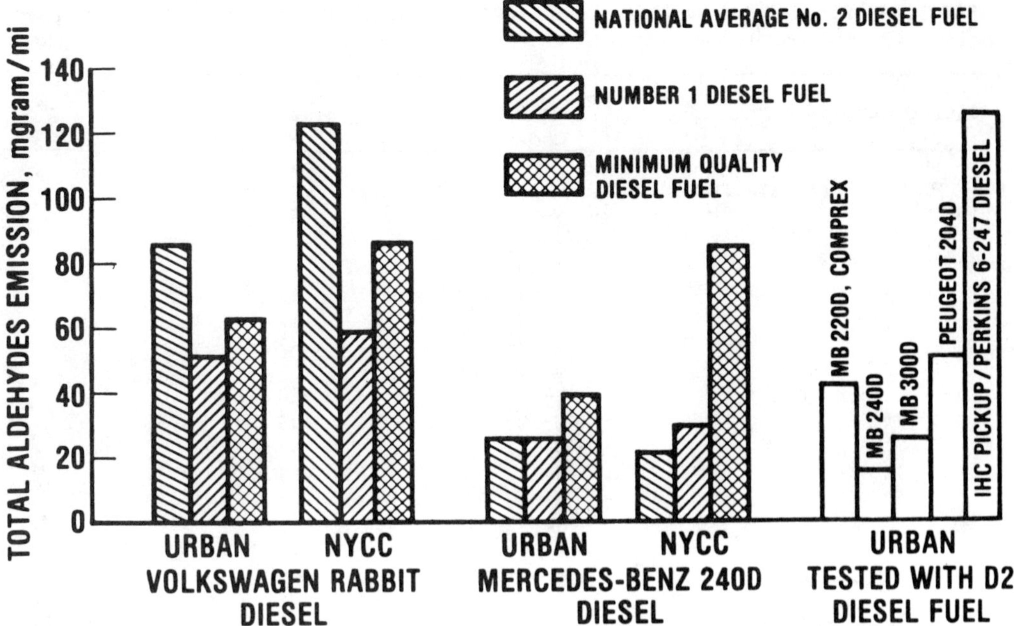

FIGURE 3.52 Aldehyde emission from diesel vehicles. SOURCE: Roessler, 1979.

or primary hexyl nitrates, which in concentrations near 1,000 ppm provide an increase of about four cetane numbers. Organic peroxides are also known to act as cetane improvers. Of course, sensitivity to the cetane improver is fuel-dependent.

Cetane improvers were examined in Burley and Rosebrock's (1979) work on particulate fuel effects. In every case when a cetane improver was added, particulate emissions increased over the base fuel level. Such an increase in particulate emissions, with a reduction in ignition delay and a preignition premixing, is consistent with the particulate formation model of Amano and co-workers (1980).

Based on the reduction of ignition delay, cetane improvers are expected to result in limited reductions in hydrocarbon, carbon monoxide, and odorants (Reading and Greeves, 1980) and in a possible reduction in NO_x (Daimler-Benz, 1979). In addition, reductions in combustion noise are expected. Information on other unregulated emissions is limited and inconclusive. Again, the fate of the additive components (nitrates and nitrites) should be investigated.

Overall, experience with fuel additives for combustion modification and emision control is limited. This area represents a promising area for future investigation. However, additive costs and concerns about the ultimate fate and effect of the additive components may well limit their use.

TABLE 3.23 Effect of Additive Blends on Particulate Emissions From a 5.7-Liter Engine: 40-mph Road Load

Additive Supplier	Blend Code	Change in Particulates, %
Ethyl Corp., Ferndale, Mich.	R-59	+4.4
	R-60	-25.2
	R-61	-23.0
	R-63	-8.8
	R-64	-11.0
	R-66	+12.7
	R-67	-7.5
Texaco, Beacon, N.Y.	U-51	+20.0
	U-52	+12.1
	U-53	-4.6
	U-54	-27.0
	U-55	-23.2
	U-56	+11.5
	U-57	+13.0
	U-58	+48.0
ATI, Milford, Conn.	S-65	-11.0
Econ, Ontario, Canada	T-69	-8.3
Rohm and Haas, Southfield, Mich.	Q-70	+12.6
	Q-71	+21.0

SOURCE: Roessler et al., 1980.

Nonconventional Fuels and Fueling Systems

A number of different nonconventional fuels and fueling systems are being examined for controlling emissions and extending engine operations. These approaches include: introducing a portion of the fuel into the combustion chamber prior to the main injection (pilot injection and fumigation); introducing water into the combustion chamber, either through direct cylinder injection, or through manifold injection, or as an emulsion; and using alcohols and other fuels either directly, or as a blend, or as an emulsion.

Fumigation

Supplying a portion of the engine's fuel requirement with the intake air (fumigation) improved NO_x, smoke, and noise emissions from a single-cylinder indirect injection research engine operated at midspeed; a small reduction in fuel economy also occurred. However, when the engine was operated at rated speed and load, the emission improvements diminished, but fuel consumption increased. At both speeds, hydrocarbon emissions increased. No particulate measurements were reported. It was concluded that at full load some oxidation products were produced during the compression stroke, which reduced ignition delay and accelerated combustion of the main charge (Russell, 1977).

The principal disadvantage of fumigation is the increased level of hydrocarbon emissions that results, particularly at light load. At light load the injected portion of the fuel engages a smaller portion of the very lean fumigated air-fuel charge within the combustion chamber, which results in incomplete combustion of the fumigated air-fuel charge. A potential advantage of fumigation is that it provides the capability to use fuels other than diesel for a portion of the fuel supply, particularly products that can be derived from nonpetroleum sources such as ethanol or methanol.

Overall, because of system complexity and light load hydrocarbon emissions, use of fumigation for light-duty diesel engines is not likely.

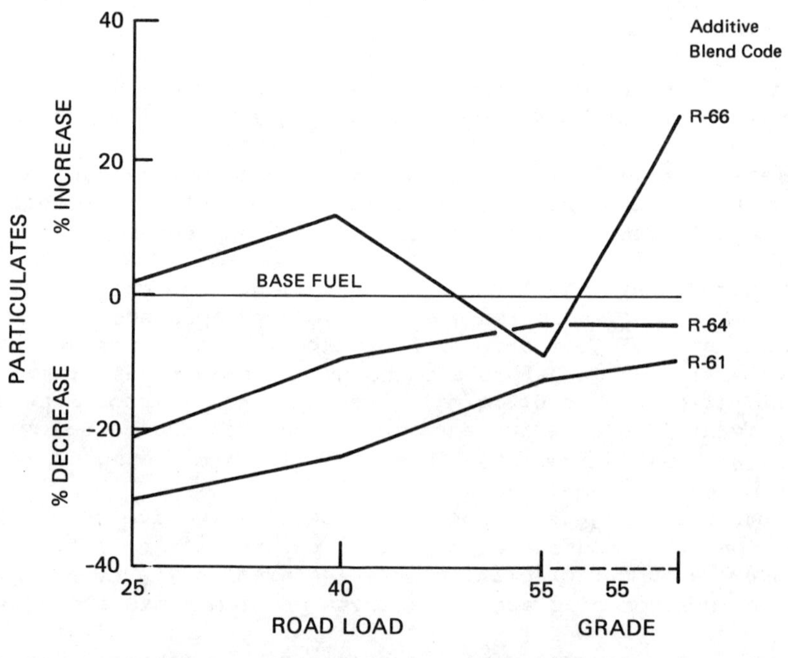

FIGURE 3.53 Effect of vehicle speed on particulate emissions for additive blends. SOURCE: Roessler _et al._ 1980.

Pilot Injection

With pilot injection a portion of the engine's fuel requirement is directly injected into the cylinder, prior to injection of the main charge. Pilot injection generally results in emission trends and behavior similar to fumigation. Thus, there is a problem with low load hydrocarbon emissions. Also, using conventional fuel injection equipment, pilot injection requires a second injection system; hence, it is very expensive. With the introduction of electronic injection systems, pilot injection could be achieved with a single injector, which also could provide better control and operational matching. As such, advanced pilot injection systems may have the potential for better hydrocarbon performance and should be explored.

Water Injection

The presence of water during the combustion process can cause changes in fuel consumption, output power, and emissions (Roessler et al., 1980). Water introduced as an aerosol with the intake air acts as a charge diluent and, because of its high heat of vaporization, reduces temperature and pressures at the end of the compression stroke and at the start of combustion. The consequent reduction in peak pressures, temperatures, and oxygen availability during combustion decreases NO_x formation rates.

The injection of water into the intake manifold or directly into the cylinder, either separately or as a fuel-water mixture, has been shown to be an effective NO_x control technique. Limited test data indicate a 50 percent reduction in NO_x in experimental diesel engines and moderate increases or reductions in hydrocarbon, carbon monoxide, and smoke emissions. There are insufficient data to characterize the particulate emissions accurately, and additional research in this area is needed.

Tests with water injection into the cylinder indicate a potential for even greater NO_x reductions, with moderate improvements in carbon monoxide and smoke. Test data from an experimental system using both EGR and water injection before a turbocharger show a very large reduction in NO_x (from 1.5 to 0.3 g/mi), a fuel economy improvement of about 7 percent, and some increase in hydrocarbon and carbon monoxide emissions. Representative water injection data are presented in Table 3.24.

Although water injection appears to be an effective method for NO_x reduction, with considerable promise for stationary engine applications, a number of critical problem areas may prevent its application to automobile engines. These problem areas include:

- potential corrosion of all system parts in contact with water or water vapor,
- added equipment and complexity,
- packaging and space requirements of a water storage tank, and
- freeze protection.

TABLE 3.24 Effect of Water Injection on Emissions and Fuel Economy

Water Addition System	Aspiration	Inertia Weight, lbs	Configuration	Emissions, g/mi			Comb. Fuel Econ., mpg
				HC[a]	CO	NO_x	
Carburetor	NA	3,500	Baseline with EGR	0.08	0.78	0.88	31.30
			EGR + water	0.35	2.66	0.36	29.11
Injection before turbo	TC	4,000	Baseline	0.06	0.76	1.49	28.40
			EGR + water	0.16	1.93	0.29	30.00

[a]Nonheated lines to FID.

SOURCE: Roessler et al., 1980.

Alcohols

The use of alcohols, either ethanol or methanol, as extenders to
available supplies of diesel fuel has been under study recently
(Roessler et al., 1980). A major problem with the use of alcohols as
diesel fuel is their high rates of vaporization, which result in
mixture cooling and large ignition delays. To overcome this
characteristic, higher compression ratios (on the order of 25:1 to
30:1) and charge preheating may be required to achieve satisfactory
combustion (Roessler et al., 1978). Additionally, the stoichiometric
mixture ratio of alcohols is considerably lower than that of diesel
fuel. Therefore, because stoichiometric mixtures of these fuels have
roughly equivalent energy content, the diesel fuel injection system
must be modified to deliver larger amounts of alcohol fuels for the
same power output. Alcohol and diesel fuel are not readily miscible;
however, through the use of surfactants, stable mixtures can be
achieved if the fuels are emulsified.

Neat alcohols have not been extensively tested in diesel engines
because of their poor ignition qualities. Investigations of alcohol as
a neat fuel for diesel engines have resorted to the addition of a spark
ignition system or have used alcohols in an impure state, with large
amounts (16 to 20 percent) of ignition modifying compounds such as
hexyl nitrate or cetonox added (Roessler et al., 1980). Engine tests
of an alcohol-hexyl nitrate mixture indicated an increase in hydrocarbon, carbon monoxide, and NO_x emissions. In other tests using
neat ethanol, the Bosch Smoke Number was close to zero at all loads
(Bandel, 1978). The ability to combust without forming particulates,
which appears to occur with ethanol and has also been observed with
methanol, makes neat alcohols an attractive fuel for use in meeting
particulate standards (Adelman, 1979). However, there are a number of
equipment and operational problems that may limit the widespread use of
alcohols in automotive applications. These include: solving materials
compatibility and size problems, developing feasible combustion and
fuel injection systems, securing sources of supplies, and developing a

segregated distribution system. Such problems are more readily overcome in stationary applications.

Emulsions

Alcohol or water introduced with the fuel in the form of an emulsion causes changes often associated with the so-called "microexplosion" phenomenon. With the microexplosion phenomenon, small droplets of water or alcohol emulsified within a larger droplet of fuel are subject to early vaporization within the droplet's lifetime. This early vaporization can shatter the larger droplet into many small fragments. The resulting increase in evaporative surface area, and improved atomization, increases burn rate and reduces the formation of sooty residues (Dryer, 1977).

Long-term stability has not been attained with emulsions of water or alcohol in diesel fuel: Some success has been achieved with water where evident phase separation from the diesel fuel may require weeks. Methanol/diesel fuel emulsions prepared to date separate after periods ranging from several seconds to 30 minutes--depending on water concentration, temperature, nature and quantity of the emulsifying agent, and effectiveness of the emulsification unit.

Emulsions of low stability prepared on board the vehicle appear practical. At high loads and speeds, with proper injection advance to compensate for increased ignition delay and reduced flame speed, limited data on heavy-duty direct injection and indirect injection engines indicate that diesel fuel consumption can be reduced (7 percent for water; 11 to 18 percent for methanol) at the expense of maximum power output (less energy content of the emulsion) and with possible increases in one or more pollutant species (50 percent higher hydrocarbons with water and 120 percent higher hydrocarbon and NO_x with methanol were observed). However, hydrocarbon and NO_x emissions were not design criteria in this work, and, hence, hydrocarbon and NO_x optimization was not attempted. Also, considerable variations and differences in results are obtained with different engine systems. In general, low-load and low speed operations with emulsions have resulted in higher than normal diesel fuel consumption and emissions (Lawson and Last, 1979a,b).

In addition to possible high increases in hydrocarbon emissions, problems with emulsions include startability, poor low load operation, materials compatibility, and cost. These problems are readily overcome in stationary applications but may become limiting in automotive applications. The problems of poor engine starting and operation at low loads can be avoided by reserving emulsions for use at high loads. However, in this case, system complexity might pose problems. Transient engine operation could also pose problems with mixing systems. Additional engine testing, vehicle design, compatibility investigation, system analysis, and cost analyses are necessary to evaluate the overall attractiveness of diesel emulsions in automotive applications.

DIRECT INJECTION ENGINES

Direct injection diesel engines are 10 to 20 percent more fuel efficient than indirect injection diesels. Their development for use in automobiles is, therefore, being pursued vigorously by a number of automobile manufacturers and other organizations. To date, however, all light-duty diesels employ indirect injection engines because direct injection engines generally:

* have lower speed and brake-mean-effective-power output capabilities;
* emit higher levels of hydrocarbons, NO_x, noise, and odor;
* are heavier; and
* have higher peak cylinder pressures.

A more specific comparison between indirect and direct injection engines is provided in Table 3.25.

Before direct injection engines become commercially available for light-duty applications, most of the current drawbacks will need to be minimized--for both vehicle certification purposes and consumer acceptability. At a minimum, the engine's power output must be increased and the emission levels lowered.

Manufacturers believe that most of the problems with the direct injection engine can be corrected through improvements in the combustion process. Other improvements will include the addition of emission control devices.

Combustion Improvements

Efforts are in progress to increase the rate of combustion and thereby the power output of the direct injection engine. Table 3.26 shows the combustion system design variables that AVL is testing. Attempts to lower hydrocarbon, odor, and noise emissions, by increasing the turbulence level, retarding injection timing, minimizing the injection nozzle sac volume, and increasing the compression ratio, are under way.

Further efforts are underway to improve the direct injection engine fuel system. Manufacturers are trying to determine the feasibility of unit injectors because they offer the advantage of increased injector pressures and the resulting higher rates of injection necessary for low particulate and NO_x emissions. Figure 3.54 provides data on injection characteristics with in-line pump and unit injector systems. Electronic fuel injection, as discussed earlier, is also an important development that should improve the direct injection engine performance.

Turbochargers and Exhaust Gas Recirculation

Turbocharging and exhaust gas recirculation (EGR) are both being tested in direct injection engines. It appears that turbocharging is needed to reduce the smoke limited brake mean effective pressure (Ricardo,

TABLE 3.25 Evaluation of Direct Injection in Relation to Indirect Injection Engines

	Better	Equal	Inferior
Fuel economy	15-20%		
Heat rejection to coolant	-30%		
Starting	___		
Suitable for turbocharging	___		
Speed range		___	
Brake-mean-effective-pressure potential		___	
Smoke, full load			___
Smoke, part load		___	
Low NO_x/HC potential (USA only)			___
Particulates		___	
Noise (combustion)			
Full load			___
Idle		___	
Durability		___	
Production cost		___	

SOURCE: Cartellieri, 1980.

1980). Tests and computer simulation data from experimental direct injection engines, using vehicles in the 2,250- to 3,000-pound inertia weight range, operated during the Federal Test Procedure (FTP), resulted in hydrocarbon, carbon monoxide, and NO_x emissions of about 0.35, 1.1, and 1.4 g/mi, respectively, without EGR, and of 0.3, 1.6, and 0.7 g/mi, respectively, with EGR. As illustrated in Figure 3.55, these emission levels are comparable to those from similar indirect injection vehicles. Without EGR the observed fuel economy advantage of the direct injection vehicle is about 15 percent, and with the indirect injection engine it is about 20 percent with EGR. Figure 3.56 shows AVL data for particulates, hydrocarbons, and NO_x for a number of engines or vehicles operated with and without EGR (Cartellieri, 1980). Figure 3.57 shows the overall emission trade-off for a 4,000-pound-inertia-weight car with a direct injection turbocharged 3.2-liter

TABLE 3.26 Direct Injection Combustion System Design Variables Being Tested by AVL

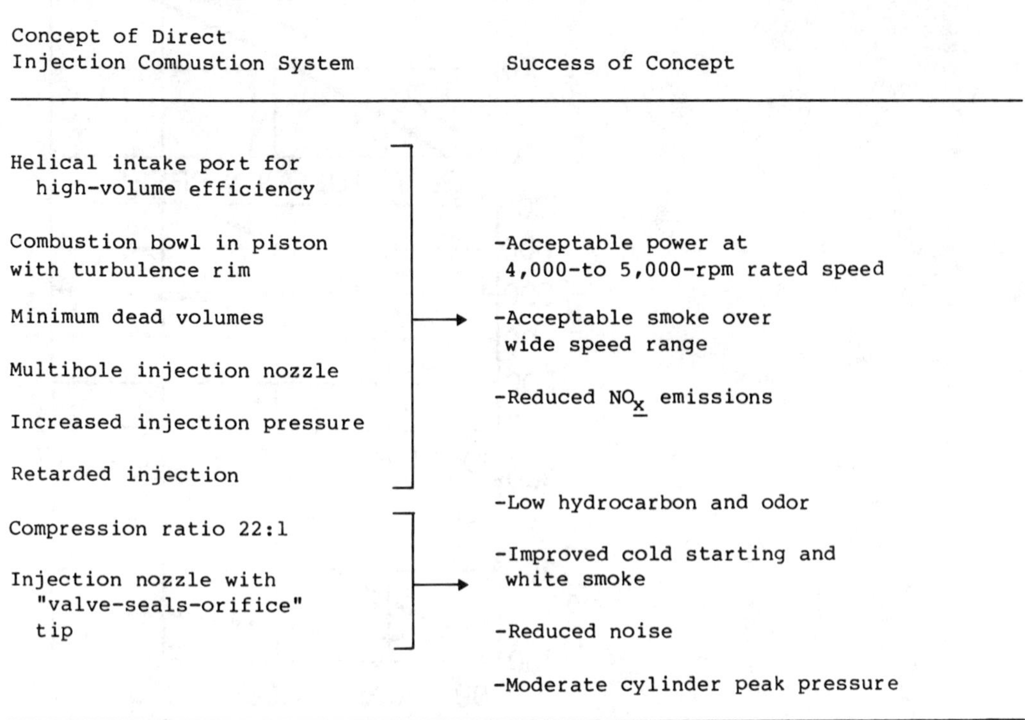

SOURCE: Cartellieri, 1980.

engine. If manufacturers attempted to achieve a production goal of 0.7 g/mi NO$_x$, which will be required to meet the federal 1.0 g/mi standard in 1985, resulting hydrocarbon and particulate emission levels could exceed 0.41 and 0.6 g/mi, respectively. Trap technology therefore would be needed to achieve a 0.2 g/mi emission level for particulates. Compression ratios of 23:1 or 24:1 might also be needed to reduce hydrocarbons further.

Future Prospects for the Direct Injection Diesel

Prospects for a light-duty direct injection passenger car engine are reasonable. Considerable progress in its development has been made during the past few years and continued progress is expected. Successful production of durable injection nozzles with "valve-seals-orifice" tips (which would reduce the sac volume to less than 0.3 mm^3 so that the hydrocarbon standard can be met) is important for the future development of the light-duty direct injection engine. Changes in the hydrocarbon measurement procedure, as proposed in the section on

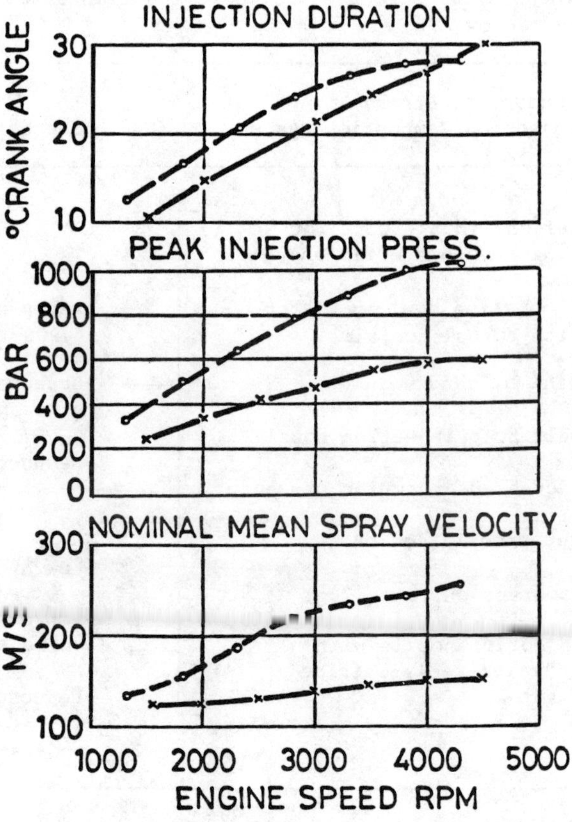

FIGURE 3.54 Injection system characteristics. SOURCE: Cartellieri, 1980.

federal particulate and gaseous emission measurement procedures, would also affect its future prospects.

Introduction of a direct injection engine is expected between 1985 and 1990. This would probably first occur in Europe, where emission standards are less stringent and fuel economy is more important. Figure 3.58 provides a fuel economy comparison of a number of vehicles with different weights and engine types operating on the European city driving cycle (Cartellieri, 1980). The clear fuel economy advantage of the direct injection turbocharged engine is emphasized in this comparison.

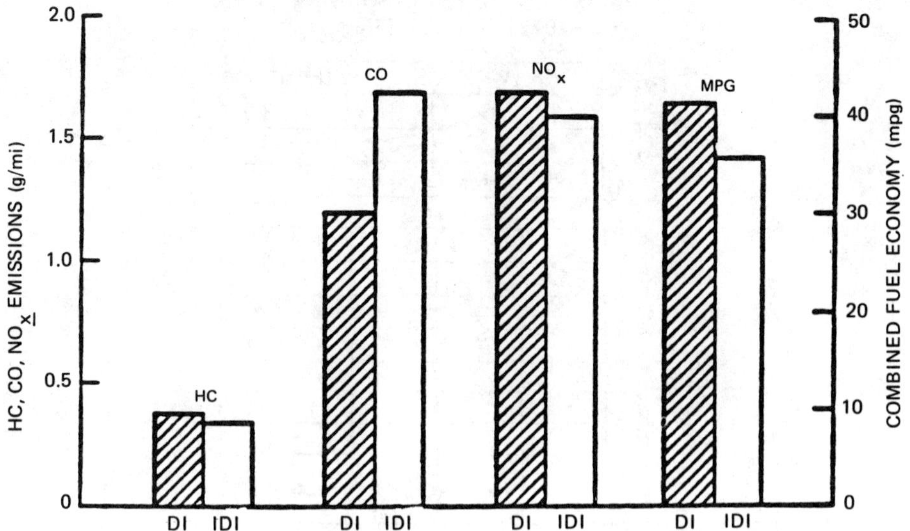

FIGURE 3.55 Exhaust emissions and fuel economy of a direct and indirect injection diesel car. SOURCE: Roessler, 1979.

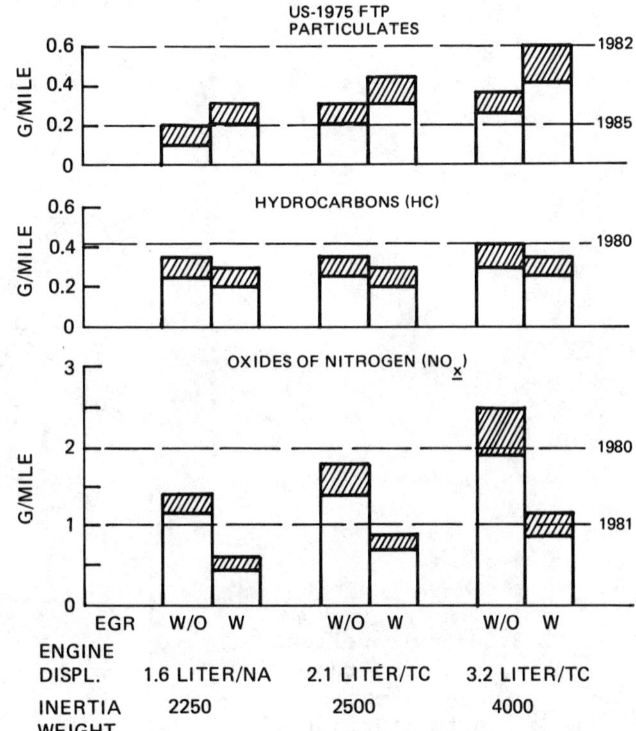

FIGURE 3.56 Low emission potential of AVL direct injection diesel engine. SOURCE: Cartellieri, 1980.

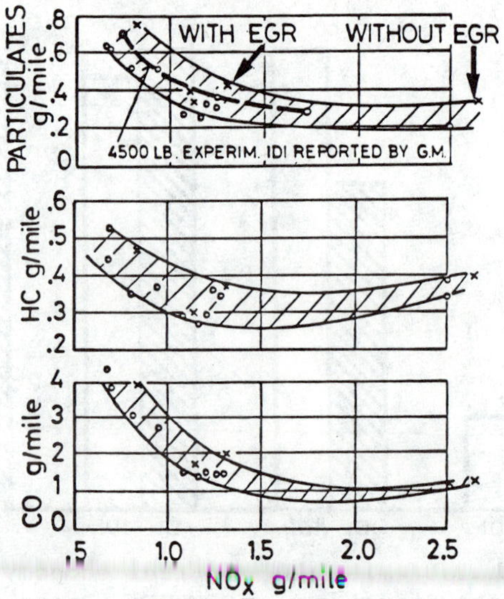

FIGURE 3.57 Emission trade-offs.
SOURCE: Cartellieri, 1980.

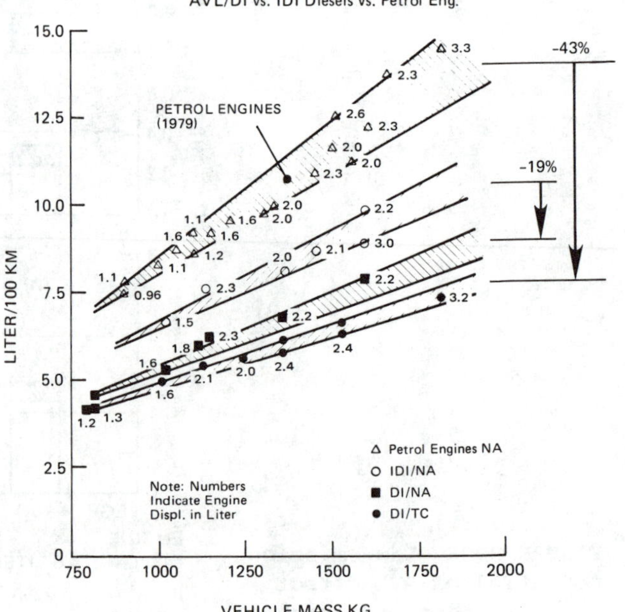

FIGURE 3.58 Average fuel
consumption during the
European driving cycle.
SOURCE: Cartellieri, 1980.

4

FUELS

FUEL REFINING

In a typical United States refinery, a substantial part of the No. 2
diesel fuel yield is obtained by distilling crude oil and separating
the portion with the proper boiling range. The amount of fuel obtained
varies according to crude oil type. Older crude oil sources, such as
Mid-Continent and Arabian Light crudes, contain about 35 percent by
volume of petroleum within the diesel fuel boiling range. The newer
crude oil sources, such as Arabian Heavy and Alaska North Slope crudes,
contain about 25 to 30 percent by volume of petroleum within the diesel
fuel boiling range.

As is shown in Figure 4.1, simple distillation of crude oil
typically yields less gasoline and more heavy material than are needed
to meet current U.S. demand. Several processes are commonly used to
upgrade the heavier materials into salable products and to improve the
quality of refinery streams. The processes include: reforming,
coking, catalytic cracking, and alkylation. Figure 4.2 is a simplified
schematic of a modern refinery. The schematic does not show facilities
for process utilities, desulfurization, sweetening (mercaptan removal),
chemicals, and lubricants manufacture.

Crude oil is separated into nine major streams in the distillation
unit. As they come from the crude still, light virgin naphtha and
light and heavy virgin distillates have the proper octane or cetane
quality and volatility for blending directly into final products.
Heavy virgin naphtha also has proper volatility, but its octane number
must be upgraded by catalytic reforming before it can be blended into
gasoline. Some is also used directly in jet fuel and diesel or burner
fuels. Reduced crude boiling above 1000°F is converted into naphthas,
distillates, and gas oils in the coker. Virgin and coker gas oils are
combined with heavy distillate for feed to the catalytic cracker. They
are used primarily to obtain gasoline and distillate components. Light
gases (butylenes) from the catalytic cracker are combined by alkylation
with isobutane from the crude still and reformer to increase gasoline
volume and octane number. Other processes include isomerization and
gas oil refractionation. Isomerization is used to convert low octane
light napthas into high-octane gasoline components. Gas oil refractionation is used to separate the No. 1 and No. 2 distillates in gas oil by
vacuum distillation.

245

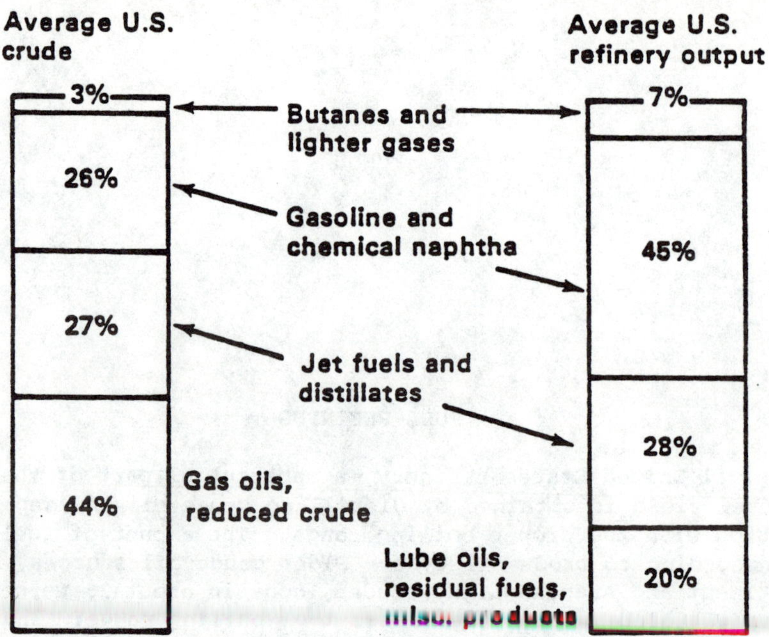

FIGURE 4.1 Crude composition and refinery output, 1980. SOURCE: Amoco Oil Co., 1980.

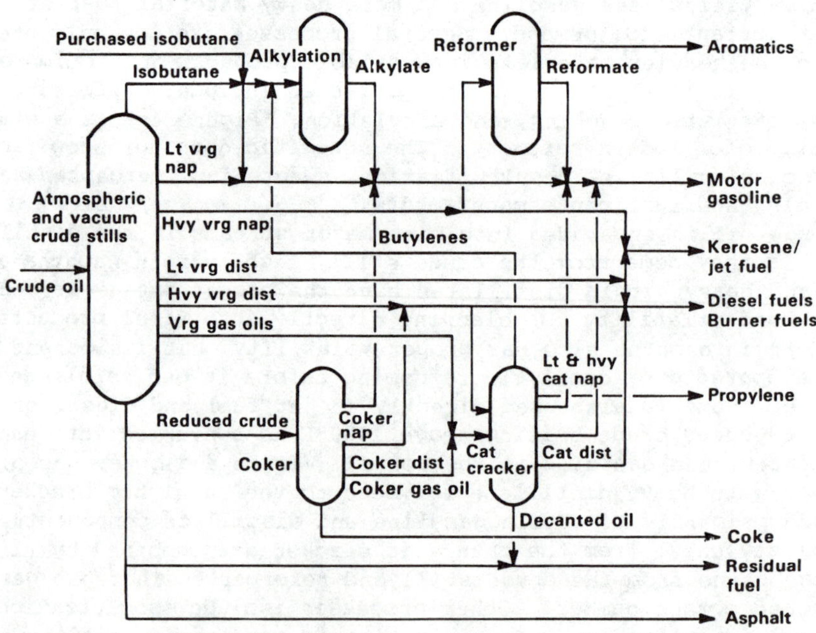

FIGURE 4.2 Simplified refinery processing.
SOURCE: Amoco Oil Co., 1980.

To meet changes in demand, the volume of gasoline and distillates from the refinery can be adjusted by changing temperature fractionation points of the crude still and by varying operating conditions of the catalytic cracker. This flexibility is very important because the relative demand for gasoline and distillate varies according to seasons and geographical areas. Thus, market demands at any one refinery may be markedly different from the U.S. average.

The hydrocarbon composition of diesel fuel distilled directly from crude oil depends entirely on the crude oil composition. The aromatic content ranges between about 12 and 30 percent by volume. A highly paraffinic crude such as Arabian Light will produce a low-aromatic, high-paraffin-content distillate with a high cetane number; North Slope or typical California crudes will produce distillates having higher aromatic and naphthene contents. In a modern refinery, additional diesel fuel is obtained by catalytic cracking of higher boiling crude oil fractions to produce lower boiling materials in the gasoline and diesel fuel range. The diesel fuel obtained from this type of processing is relatively high in aromatic content. When blended with the diesel fuel component distilled directly from crude oil, a diesel fuel containing from 30 to 35 percent aromatics by volume results.

Hydrocracking is another process used to produce gasoline and diesel fuel from higher-boiling-range hydrocarbons. The process produces a higher cetane number, lower-aromatic-content cracked diesel fuel component; however, hydrocracking capacity is presently limited worldwide.

SUPPLY AND DEMAND FOR CURRENT FUELS
AND ALTERNATIVES FOR THE FUTURE

In 1979, total demand for all distillate diesel fuels amounted to 2,000,000 barrels per day (2,000 TBD). Demand for distillate heating oils amounted to 1,300 TBD for a total U.S. petroleum product distillate demand of 3,300 TBD. Highway use of diesel fuel was 850 TBD, only 1 to 2 percent being used by automobiles and the balance by trucks and buses.

Table 4.1 shows estimated passenger car diesel fuel requirements for 1985 and 1990 for various degrees of diesel sales penetration, starting from an assumed 4.5 percent of new-car sales. Assuming that penetration grew to 25 percent by 1990, the total demand for passenger cars represents only 13.3 percent of the 1979 consumption of all middle-distillate-type fuels and approximately 12.2 percent of the total middle distillate market projected for 1990 (3,610 TBD). Even if diesel sales penetration reached 50 percent by 1990, passenger car requirements would represent only 20 percent of the total distillate market in 1990.

Several oil companies were canvassed by the committee for projections of passenger car diesel penetration used in their planning programs. These projections ranged from a low of 6 percent to a high of 20 percent of new car sales in 1990. The highest forecast of 20 percent would mean that less than 10 percent of total distillate fuel production would be used by passenger cars.

TABLE 4.1 Estimated Diesel Fuel Requirements (in thousands of barrels per day) for 1985 and 1990.

Projected Diesel Penetration	1985		1990	
	Diesel Car Sales, %	Diesel Fuel Required, TBD	Diesel Car Sales, %	Diesel Fuel Required, TBD
Case 1	6.0	92	6.0	145
Case 2	7.5	104	12.0	219
Case 3	15.0	179	20.0	286
Case 4	15.0	179	30.0	482
Case 5	25.0	244	50.0	773

SOURCE: Standard Oil Company of California, 1980.

Some projections call for sharply increased dieselization of medium-duty trucks. This, together with almost 100 percent use of diesel engines in heavy-duty trucks, results in forecasts as reported to the committee of large increases in diesel fuel demand for highway trucks. Typical oil industry forecasts call for an increase from about 900 TBD in 1980 to 1,700 TBD in 1990. Thus, the indicated demands for passenger car diesel fuel during the next 10 years are relatively small in the context of total middle distillate demand and diesel fuel requirements for highway trucks.

Kerosene type jet fuel (kero jet), the principal fuel for commercial aviation, is derived from the same part of the crude petroleum barrel as diesel fuel. The composition and properties of kero jet are similar to those of No. 1 diesel fuel. Any analysis of future diesel fuel availability must also consider the requirements for kero jet. Kero jet consumption amounted to 870 TBD in 1979. Demand for this product is expected to grow at an annual rate of 2 percent through 1990; the forecasted consumption in that year is about 1,080 TBD.

A study of gasoline demand recently released by the Petroleum Industry Research Foundation predicts that gasoline consumption will drop by 20 percent between 1979 and 1990. The reduction will be caused primarily by a switch to smaller and more efficient cars. To a lesser extent, gasoline consumption will also be reduced by the growth in sales of cars with diesel engines. In 1979, total gasoline consumption was 7,070 TBD. By 1990, consumption is forecasted to be 5,650 TBD--a drop of 1,420 TBD from 1979--representing an average annual decrease of 2.0 percent. The same study predicts an average annual increase in total distillate fuel demand of 0.8 percent (3,300 TBD in 1979 to 3,610 TBD in 1990). The Petroleum Industry Research Foundation forecasts 10

percent diesel passenger car sales in 1985, and it further states that
the share could be twice as high in 1990.

Another projection being developed from a survey by the National
Petroleum Council is shown in Table 4.2. This projection shows
somewhat higher volumes of middle distillate for 1990 than the
Petroleum Industry Research Foundation predictions.

Petroleum Industry Ability to Meet Increased Diesel Fuel Demands

During the past two decades, a large portion of petroleum refinery
industry investments were focused on increasing the yield of gasoline
from the crude oils processed, increasing the octane number of gasoline
to satisfy engine knock requirements and the phase down of tetraethyl
lead usage, and lowering the sulfur content of fuel oils. The industry
currently faces a declining demand for gasoline, and the octane
upgrading program is nearing completion.

Major new investments are now required to accommodate the processing
of heavier crudes and a forecasted decline in residual fuel demand.
Although the types of processes selected for heavy crude upgrading may
depend on the relative demands of gasoline and distillate fuels, the
magnitude of the required investments will not be greatly influenced by
these relative demands. If anything, decreasing the production ratio

TABLE 4.2 Domestic Petroleum Product Demand (in TBD)

Fuel Use	Actual 1978	Projected 1990
Motor gasoline	7,412	6,124
Jet fuel		
Naphtha type	199	124
Kerosene type	858	1,203
Distillates		
Automotive diesel[a]	988	1,988
Other[b]	2,619	2,287
Residual fuel oil	3,023	2,344
Other[c]	3,748	4,827

[a] Includes off-highway diesels.
[b] Includes heating oils for households, electric utilities,
railroads, vessels, military, and miscellaneous.
[c] Includes aviation gasoline, napthas, liquefied gases, petrochemical
feed stocks, lubricants, waxes, coke, asphalts, road oils, still gases
for fuel, and miscellaneous.

SOURCE: National Petroleum Council preliminary unpublished data.

of gasoline to distillate will tend to lower both investment and operating costs.

A survey of the petroleum industry undertaken by the committee indicates that there will be few problems supplying increased demands of current specification diesel fuel during the next 10 years, even if passenger car diesel sales penetration reaches 50 percent in 1990. Basically, this will be accomplished by reducing the conversion of distillates to gasoline and adding a limited amount of refinery equipment.

The industry reported to the committee that maintaining current cetane number levels will be difficult in the future because of the expected reduction in the quality of crude oils and the increasing demand for diesel fuel. Maintaining an adequately low cloud point temperature will also be difficult. The industry is already experiencing difficulties in providing sufficient No. 1 diesel fuel for colder climate requirements.

Current petroleum product demands require that the typical U.S. refinery produce gasolines and distillates (G/D)* at a ratio of approximately 1.6. As demands for middle distillates increase, with additional demands for diesel fuel and kero jet, which will be offset slightly by decreasing demands for middle distillate heating oils, the typical refinery will be required to operate at a G/D ratio of 1.2. Most refiners conclude that their existing refineries can be adjusted to a G/D ratio as low as 1.0 without large new investments and major operating problems. One study, using a computer model to simulate operation of a modern refinery, shows the feasibility of operating at an even lower G/D ratio of 0.7 (Lawrence et al., 1980). The G/D ratio of 0.7 represents a shift of 30 percent of total motor fuel from gasoline to diesel fuel. ASTM-D975 specifications were used as a basis for defining the diesel fuel specifications. Lawrence found that although some existing process units had to be modified, the only major change to attain the G/D ratio of 0.7 was the addition of a gas oil refractionator. Using the automotive diesel fuel property limits outlined in the section on effect of fuel properties would probably have affected some of the conclusions of this study.

Another published study, using linear program simulation of a typical U.S. refinery, demonstrates that it is possible to shift from a G/D ratio of 1.6 to a ratio of 0.6 without shutting down existing processing facilities or constructing major new facilities (Barry et al., 1977). In this study, diesel fuel production was increased, at the expense of gasoline, by using portions of distillates normally charged to the catalytic cracking unit, and heavy naphtha normally reformed into gasoline, as diesel fuel components. In addition, catalytic cracking unit processing conditions were changed to maximize distillate fuel production. Distillate desulfurization capacity was

---

*G/D ratio is the volume of motor gasoline divided by the volume of total distillates--diesel fuel, distillate fuel oils, kerosene, and jet fuels. It is commonly used to describe refining operations. Although U.S. refineries currently average a G/D ratio of about 1.6, the ratio varies among refineries and according to season.

expanded to avoid an increase in diesel sulfur content. As the yield
of diesel fuel is increased in this fashion, the distillation range
becomes broader (the 10 percent point decreases and the 90 percent
point increases), the cloud point tends to increase, and the cetane
number tends to increase. Despite these trends there was no problem
meeting ASTM-D975 specifications.

Gasoline to distillate ratios lower than about 0.6 will be difficult
to achieve. At that point essentially all of the components falling in
the diesel fuel boiling range that occur naturally in crude oil or that
are created by conventional refining conversion processes are fully
used for distillate products. Production of additional diesel fuel
will require changes in boiling range limits or the development of new
refining processes. Below the G/D ratio of 0.6, the costs of producing
additional diesel fuel would be expected to increase sharply.

If dieselization of passenger cars reaches a high level by 1990 and
remains at a high level, the demand for diesel fuel in the 1990s could
exceed the availability, on the basis of existing refining technology.
If that level is reached, the advantage of further dieselization will
probably be offset by increased fuel costs. It might be expected that
an equilibrium between gasoline- and diesel-powered vehicles would be
reached because of the relative availability and cost of the two fuels.

It should be recognized that the U.S. refineries vary greatly in
size and degree of processing capability. Some refineries with little
or no conversion capability may have little flexibility to increase
diesel fuel output. The typical refinery referred to in this report
represents a reasonably modern refinery of about 150 TBD of crude oil
processing capacity and conversion capability to maximize gasoline
production and minimize residual fuel output.

Distribution

The survey of industry representatives that was undertaken by the
committee indicates that no major problems in distributing current
specification automotive diesel fuel exist. As demand for diesel fuel
rises, it will be substituted gradually for leaded gasoline without
major changes in pipeline, terminal, or service station facilities. In
general, existing tanks and pumps will be switched from one grade of
gasoline to diesel fuel as the demand for diesel fuel increases and the
demand for leaded gasoline declines.

The cost of delivering a given volume of diesel fuel from the
supply terminal to the service station will be higher than that for
gasoline. Tank truck payloads could be reduced as much as 15 percent
for a full load delivery because diesel fuel is heavier and there are
weight limits. Payload for split loads would be reduced in proportion
to the amount of gasoline in the load. The law requires that when
gasoline and distillates are transported in the same vehicle, double
bulkheads must separate the compartments. Retrofitting an existing
vehicle with double bulkheads costs about $1,500: the incremented cost
on a new vehicle is $275.

It is estimated that changing existing tanks and pumps at a retail
service station from one grade of gasoline to diesel fuel will cost

$500. If an additional dispensing system is required to accommodate diesel fuel, the cost is estimated to be $14,000 per station. This additional system is based on an 8,000-gallon tank, piping, and two pumps (Mobil Oil Corp., 1980).

In 1979 there were about 160,000 service stations in the United States. Estimates of the number of service stations requiring a new system for handling diesel fuel are not available, but the percentage is thought to be small. The cost impact will be the greatest in the situation where only two grades of gasoline are sold and the service station elects to add diesel fuel without eliminating a grade of gasoline.

Revised Product Specifications

In view of the evidence that the emission of particulate matter from a diesel engine is influenced by the boiling range and aromatic content of the fuel (Burley and Rosebrock, 1979), the petroleum industry was canvassed by the committee concerning the effect on availability and cost if the boiling range or aromatic content of diesel fuel for passenger cars were reduced. In a qualitative sense, the response from the petroleum industry was negative. Such a diesel fuel would be very costly to produce; the supply would be limited; and an expensive, segregated system would be required for distribution and marketing.

Few data are available to quantify the cost and availability of a low boiling range, low aromatic content diesel fuel. The American Petroleum Institute has recently initiated a study that will provide quantitative information on this subject. When available, this information will provide necessary input for a cost-benefit analysis of modified diesel fuel specifications.

Most of the existing U.S. refineries have very limited potential for producing a low end-point, low aromatic diesel fuel. The production of this fuel would have to compete with jet fuel, which is in increasing demand. Such a fuel would be based on light straight-run distillates, which vary in aromatic content from about 12 to 30 percent, depending on the crude oil source. Reductions in aromatic content below the 12 to 30 percent range would require development of new technologies, major capital expenditures for installation of refining facilities, and increased energy consumption in the refining process.

A brief study made in 1979 by one petroleum company indicated that reductions of about 100°F in distillation temperatures and reduction of aromatic content from 20 percent to 5 percent would raise the cost of diesel fuel by 10 cents per gallon (Amoco, 1980). Part of the cost was due to capital charges and operating costs. Most of the cost, however, was attributed to downgrading components to lower-valued products.

Product Costs and Pricing

The cost of crude oil is the major factor determining the cost of producing diesel fuel. Typically the crude oil raw material cost may represent 85 to 90 percent of the cost of producing diesel fuel. Although a minor part of the total, the other significant element is the refining cost.

In an absolute sense, therefore, future diesel fuel prices will be largely determined by crude oil costs. With the uncertainties of crude oil availability and producing country pricing policies, no attempt will be made here to forecast the impact of raw material costs on diesel fuel prices.

The relative costs of diesel fuel and gasoline are an important consideration. Table 4.3 shows the average annual prices of regular grade gasoline and No. 2 home heating oil for the years 1967 through 1979. Because No. 2 home heating oil and No. 2 diesel fuel are similar products, and in most instances common products, trends in the prices of home heating oil should be indicative of trends in diesel fuel

TABLE 4.3   Retail Prices of Regular Grade Gasoline and No. 2 Home Heating Oil

	Current Cents			1967 Constant Cents		
Year	Gasoline, Service Station, Excl. Taxes	Home Heating Oil	Diff.	Gasoline, Service Station, Excl. Taxes	Home Heating Oil	Diff.
1967	22.55	16.91	5.64	22.55	16.91	5.64
1968	22.93	17.44	5.49	22.01	16.74	5.72
1969	23.85	17.82	6.03	21.72	16.23	5.49
1970	24.55	18.48	6.07	21.11	15.89	5.22
1971	25.20	19.63	5.57	20.77	16.18	4.59
1972	24.46	19.72	4.74	19.52	15.74	3.78
1973	26.88	22.75	4.13	20.19	17.09	3.10
1974	40.41	36.01	4.40	27.36	24.38	2.98
1975	45.44	38.98	6.46	28.19	24.18	4.01
1976	47.44	41.80	5.64	27.82	24.52	3.30
1977	50.70	47.37	3.33	27.93	26.10	1.83
1978	53.09	50.80	2.29	27.17	26.00	1.17
1979	74.76	72.80	1.96	34.38	33.49	0.89

SOURCE: American Petroleum Institute, 1980.

price. On a current cost basis, No. 2 heating oil sold at a
differential in the general range of 4 to 6 cents per gallon below
regular gasoline from 1967 to 1976. Since 1976 the differential has
dropped to about 2 to 3 cents per gallon. Since 1974, price control on
these products has probably affected these differentials, and the
prices may not represent what would have happened under free market
conditions. The main question now is, What will happen to the diesel
fuel price in the future as demand increases and gasoline demand
decreases?

During the past several decades, maximizing gasoline yield has been
the main emphasis in the fuel refineries of the United States. Gasoline
demand grew steadily during this period, and refinery expansions and
modifications were largely directed toward the production of greater
gasoline volumes. This was accomplished through increased crude oil
processing, conversion of heavier distillates, and alkylation of lighter
hydrocarbons. Under these conditions gasoline was priced to bear a
large proportion of refinery operating and capital costs and to provide
a large share of refining profits. With the prospect of declining
gasoline demand and increasing distillate demand, the petroleum
companies surveyed by the committee predict a narrowing of the
differential between the price of gasoline and distillate. If diesel
fuel price does not increase as production is shifted from gasoline to
diesel, refining revenues and net refining profit will decrease because
of the lower diesel fuel price and only minor changes in refinery
operating costs. The refiner will attempt to maintain revenues by
increasing the diesel fuel price. Assuming free market conditions,
some refiners predict that sometime between 1985 and 1990 the price of
a gallon of No. 2 diesel fuel and a gallon of unleaded regular gasoline
will be the same. In the same time period, kerosene-type jet fuel
prices are expected to rise above the price of gasoline. To maintain
refinery profit margins, the refiner will shift a larger proportion of
total refining costs onto the increasing middle distillate production.

In a study published in 1977, two pricing options to maintain
refinery profitability as diesel fuel demand increases were examined
(Wagner, 1977). In option A, the price of gasoline was assumed to
remain constant as its volume declines, and the price of diesel fuel
was permitted to rise in order to maintain refinery revenue the same as
at its historic 1.7 gasoline to distillate ratio. At a ratio near 1.1
(a ratio frequently predicted for the late 1980s) the price of diesel
would equal that of gasoline.

In option B, the price of diesel was assumed to rise in proportion
to its incremented manufacturing cost, with the price of gasoline
adjusted to maintain refinery revenue. In this option, diesel and
gasoline prices would equate at a gasoline-to-distillate production
ratio of about 1.4. Below this ratio, diesel prices would rise
rapidly, and gasoline prices would decline sharply.

Any pricing strategy the industry might choose would likely fall
between these two options. No matter which strategy is chosen, in a
free market, as more diesel fuel is produced, its price will rise
compared to that of gasoline. Government taxing and price control

policies will also affect the market price of diesel and gasoline fuels in the future.

Total Energy Considerations

In considering the national benefits of the use of diesel engines in the automotive fleet, it is important to evaluate the crude oil required, and energy consumed, to produce additional diesel fuel. A study published by Amoco assesses these energy requirements (Lawrence et al., 1980).

This study used a computer model that simulated a modern fuel refinery and identified the process requirements to meet varying product demands at minimum cost. The base refinery had a capacity of 150 TBD, which was considered representative of the refineries that produce most petroleum products in the United States. The process configuration was chosen to meet 1980 product demands economically. The refinery was self-sufficient in all energy needs, which made it possible to determine accurately the total energy requirements for producing varying product slates. In the model, product shifts were accommodated by changing processing conditions, modifying existing equipment, or adding new process units where economical. The alternative product slates were centered in projected demands in 1995. In the study the total amount of gasoline plus diesel fuel was held constant at 76 TBD, while the relative amounts of gasoline and automotive diesel fuel were varied.

Starting at a G/D ratio of 1.6, diesel fuel production was successively increased from 5 to 28 TBD (0.7 G/D), which represents a change of 30 percent of total automotive fuel to diesel. Below this G/D ratio of 0.7 there is no practical way to make additional diesel fuel.

Refinery product yields in thousands of barrels per day (TBD) for the various G/D cases are provided in Table 4.4. The refinery energy balances at various G/D ratios while producing 86 Research and Motor (R&M) pool octane number gasoline are shown in Table 4.5. This information demonstrates that a typical modern-fuel-type refinery will consume less energy as diesel fuel production is increased. In this case a 12.5 percent reduction in refinery energy consumption results, with most of the reduction in refinery energy consumption occurring in the catalytic cracking process as diesel production is increased.

Although the volume of crude oil input required to make a given total volume of gasoline and diesel fuel increases as diesel fuel production increases, the volume of butane input required to meet gasoline vapor pressure requirements decreases. The net change in the total volume of raw material input is therefore small. In the cases shown above, crude input increased from 150 to 152 TBD when the G/D ratio went from 1.6 to 0.7. At the same time butane input decreased from 3 to zero TBD, for a net drop in volume of total input of 1 TBD.

Other petroleum companies surveyed by the committee generally agreed that refinery energy consumption as a percent of energy input would decrease as diesel fuel production is increased and gasoline production

TABLE 4.4 Refinery Product Yields at Four G/D Ratios, 150-TBD Capacity (in TBD)

Refiner Product Yields	G/D Ratio			
	1.6	1.3	1.0	0.7
Motor gasoline	71	66	58	48
Diesel fuel	5	10	18	28
Kerosene and jet fuel	----------	15.0	----------	
No. 2 fuel oil	----------	25.0	----------	
Aromatics	----------	3.0	----------	
Asphalt	----------	6.2	----------	
Residual fuel	----------	6.6	----------	
Special naphthas	----------	7.2	----------	

SOURCE: Lawrence et al., 1980.

is decreased. Several made reference to the Amoco study described above as a reasonable appraisal of the energy balance effects.

Alternative Sources of Diesel Fuel Shale Oil

Raw shale oil is notable for its high nitrogen content, large amounts of unsaturated hydrocarbons, and metallic contaminants. Advanced commercial state-of-the-art refining technology is capable of refining shale oil into fuel products that meet current specification requirements. The key to successful shale oil refining is an effective initial hydrotreating step. Following this step, conventional refining processes may be used. Pilot plant studies have produced high quality transportation fuels, including diesel fuel, using this processing approach (Sullivan, 1979).

Coal Liquids

In addition to high nitrogen content, raw coal liquids are high in aromatic content and low in hydrogen content. Typical petroleum crude oils contain 5 to 25 percent aromatics; raw coal liquids contain 34 to 49 percent aromatics. With this hydrogen deficiency and high aromatic content it currently appears unlikely that an acceptable automotive diesel fuel can be refined from coal liquids at a reasonable cost.

TABLE 4.5 Refinery Energy Balance

Energy Dispersal Elements	G/D Ratio			
	1.6	1.3	1.0	0.7
Energy in refinery input, 10^9 Btu/day	829	830	826	826
Energy output in products, 10^9 Btu/day	763	766	765	768
Energy consumed in refinery, 10^9 Btu/day	66	64	61	58
Energy in product output as percentage of energy input	92.0	92.3	92.6	93.0

SOURCE: Lawrence et al., 1980.

Advances in indirect coal liquefaction processes may provide a higher-quality diesel fuel.

Alcohols

Although spark ignition and gas turbine engines can use alcohol fuels, the diesel engine, thus far, can not. In the diesel engine, the low cetane ratings prevent autoignition. Alcohols have been used in unmodified diesel engines only with the addition of large amounts of ignition accelerators. Because alcohols have not been suitable compression ignition fuels when used alone, their use in conjunction with diesel fuel has been extensively studied. Used in combination with diesel fuel, alcohol will reduce engine power output, increase consumption per unit of work, delay combustion, and increase engine noise. In view of these problems it seems unlikely that there will be any significant use of alcohol as a fuel for automotive diesel engines (Adelman, 1979; Strait et al., 1979).

Introduction of Alternative Fuels

In view of the potential availability of shale oil, coal liquids, and alcohols as alternative raw material sources for transportation fuels, the Department of Energy is sponsoring a major study to determine the requirements for optimum future use of these materials. This project involves the Southwest Research Institute, Bonner & Moore Associates, and Standard Oil Company of Ohio as subcontractors. Substantial shale and coal oil production in the Rocky Mountain, Mid-Continent, and Great Lakes regions is expected by 1995, the base period for the study's productions.

In this study, comparisons of alternative raw materials will be made by using a baseline projection of 1995 conventional crudes and products. Conventional product slates, a maximum diesel operation, and a broad-cut fuel operation will be evaluated. The study has not yet progressed sufficiently for results to be used to assess the impact of these alternative sources of raw materials on diesel fuel availability and quality.

CONCLUSIONS

* The petroleum industry will be able to produce and distribute to motorists the needed amount of current specification diesel fuel for passenger cars during the next decade, even with a passenger car diesel sales penetration reaching 50 percent in 1990.
* Diminishing availability of high-quality crude oils and increasing demand will result in a general trend toward lower-quality diesel fuel. Cetane number will tend to decrease, and aromatic content will tend to increase. However, current ASTM specifications will generally be met.
* Although quantifying studies are not currently available, oil industry representatives surveyed by the committee believe the production of a lower aromatic-content, lower-boiling-range diesel fuel for passenger cars would have substantial cost and energy disadvantages.
* Sulfur content (at the 0.3 percent level) can be controlled with existing technology at moderate cost.
* As diesel fuel demand increases and gasoline demand decreases, diesel fuel prices will increase in a free market. Sometime between 1985 and 1990, gasoline and diesel fuel prices are expected to become equal.
* The amount of energy consumed to refine a given volume of gasoline and diesel fuel will decrease by approximately 10 percent as diesel fuel production is increased and gasoline is decreased to a G/D ratio of approximately 0.7.
* There is a need to study the sulfur removal costs in the 0.3 to 0.05 percent by mass in diesel automotive fuels because particulate control being developed for the post-1985 period might require a low-sulfur (less than 0.15 percent) fuel. Local air basin requirements may also require a low-sulfur diesel fuel.

5

RESEARCH AND DEVELOPMENT OPPORTUNITIES

The need to improve the fuel economy of light-duty vehicles while simultaneously meeting increasingly stringent pollution regulations presents numerous research and development opportunities. More basic research is necessary to understand the fundamental physical and chemical phenomena of combustion and pollutant formation and control. The vehicle itself needs to be optimized to reduce aerodynamic drag, rolling resistance, and vehicle weight. Indirect injection diesels, spark ignition engines, and control systems in current use require further development. Investigation of alternative power plants, including advanced diesels and advanced spark ignition engines, is needed, as is research on the use of alternative fuels in traditional and advanced power plants. These research and development opportunities and needs are discussed in the following sections.

Obviously, any potential engine/vehicle system will have to compete in the marketplace not only on the basis of fuel economy but also on the basis of other criteria, including:

* regulated gaseous and particulate emission levels and control technologies;
* unregulated pollutant emission levels, noise, and odor;
* drivability;
* engine performance and durability;
* complexity and maintenance requirements;
* relative cost; and
* possibly, fuel flexibility.

BASIC RESEARCH

Technological improvements in the diesel engine, particularly improvements in efficiency, drivability, and manufacturing processes, as well as reductions in pollutant, odorant, and irritant emissions; costs; and noise, will be assisted by a better understanding of related fundamental processes. This understanding of related fundamental processes will be enhanced by the encouragement of basic research in areas critical to the continuing development and improvement of the diesel engine.

The Department of Transportation has recently assembled several working groups to draft a "technical framework" for its evolving "Cooperative Automotive Research Program" (DOT, 1980). Many of the research topics identified in this document are relevant to the future of the diesel engine. Of particular and specific relevance are the areas of combustion, fluid mechanics and heat transfer, materials science and processing, control systems, tribology, acoustics, and catalysis.

The combustion process controls energy release and the formation of all pollutant species. Just how pollutants are formed in diesel combustion, especially particulates, remains vaguely understood. An understanding of the mechanisms of droplet spray combustion at the high pressure and temperature of diesel engines is essential to the design of low emission systems. The prediction of the behavior of lower-quality and alternative diesel fuels similarly requires an understanding of the fundamentals of the combustion process. The pyrolysis and oxidation chemistry of hydrocarbon fuels is an important component of this research.

The fluid mechanics of intake, prechamber, main chamber, and exhaust flows is important to the improvement of the efficiency of operating of these individual components of the diesel engine. The fluid mechanics and aerodynamics of compressors and turbines are important to the design of new turbochargers that will have high efficiencies at low engine power levels. Heat transfer plays an important role in the development of prechambers, the eventual development of ceramic "adiabatic" engine concepts, and the development of higher-efficiency more durable trap-oxidizer particulate control systems.

Materials research is required in the area of oxidizing traps to improve durability. Ceramic technology promises higher component operating temperatures, greater durability, and reduced manufacturing costs. Satisfying the need for injector systems with more precise delivery capabilities will in turn depend largely upon the development of manufacturing and material processing methods of greater precision that are still suitable to mass production application.

Onboard computer based technology promises the ability to operate the engine, its emission control systems, and the transmission in a manner to optimize dynamically engine parameters for the simultaneous improvement of efficiency, reduction of pollutants and noise, and improvement of drivability and performance. Research in control theory, sensors, and optimization methodology is required.

Tribology, the science of friction, wear, and lubrication, is particularly critical to the diesel engine, which is plagued by lubrication and wear problems related to high particulate levels. Improvements in thermodynamic efficiency at high compression ratio are degraded by frictional losses. The chemical and physical behavior of wear surfaces in the harsh environment of the diesel engine is an important area for fundamental research.

Acoustics research is important to the understanding of combustion-generated noise and its transmission through engine components to the engine surroundings. Current understanding does not allow an early consideration of acoustics implications during design.

The reduction of particulate emissions through trap-oxidizer systems may be improved through the use of catalytic materials. Little understanding exists of the surface catalysis and oxidation of carbonaceous particulate or of the potential use of catalysts as fuel additives.

The continued development of the diesel engine would progress more rapidly if backed by a better understanding of fundamental processes and phenomena relevant to critical diesel engine components and processes. Conducting basic research in support of diesel engine development is considered essential and fruitful.

VEHICLE OPTIMIZATION

Figure 3.58 showed the effect of vehicle weight on fuel consumption for light-duty vehicles with various types and sizes and engines. This figure graphically illustrates the well known fact that mileage per gallon of fuel increases with decreasing vehicle weight, a fact that is responsible for the recent "downsizing" of domestic passenger cars. Reducing vehicle weight has also a beneficial effect on particulate and NO_x emissions from light-duty diesels, as shown in Figure 5.1 and in Appendix D. Continuing efforts to further reduce vehicle weight include not only downsizing but also the increasing use of lighter components. Examples of advanced technologies are the use of composites for vehicle bodies and engine parts, including pistons, rods, and rocker arms (Holtzberg, 1979). Current opinion is that safety considerations mandate a practical lower vehicle weight limit of approximately 2,000 pounds. Thus, research on improved safety systems is necessary if crashworthy vehicles of less than 2,000 pounds are to be developed.

Figure 5.2 illustrates the steady state fuel economies of a production light-duty diesel and a production light-duty gasoline-powered vehicle. It is evident that fuel consumption increases as

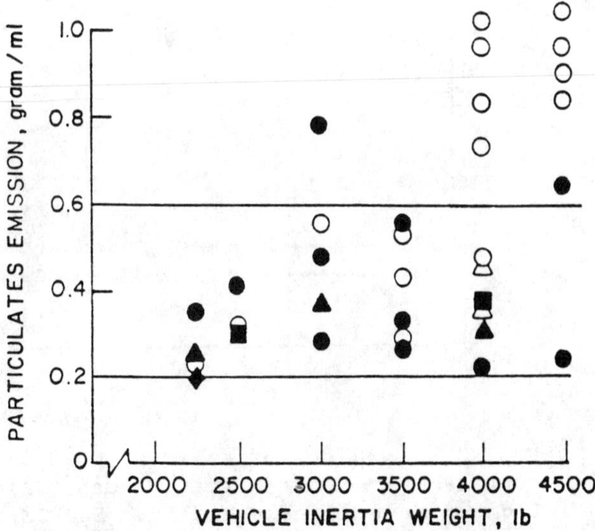

FIGURE 5.1 Effect of vehicle weight on particulate emissions from light-duty diesels. Open symbols, production vehicles; closed symbols, development vehicles; circles, naturally aspirated indirect injection; triangles, turbocharged indirect injection. Diamond, naturally aspirated direct injection; and squares, turbocharged direct injection. SOURCE: Adapted from Roessler et al. (1980) with additional data from Cartellieri (1980).

vehicle speed increases. (The reason that the fuel economy of the diesel degrades more rapidly than that of the spark ignition engine vehicle is discussed in a later portion of this chapter.) This trend results from aerodynamic drag, increasing approximately with the square of vehicle speed and tire rolling resistance, increasing approximately linearly with speed. Figure 5.3 illustrates these trends. At speeds ranging from 35 to 40 mph, for the newer vehicles, about half of the power required by the vehicle is needed to overcome rolling resistance; the other half is used to overcome aerodynamic drag (Sovran, 1980). At higher vehicle speeds, the aerodynamic drag dominates. Thus, rolling resistance is dominant over aerodynamic drag during most of the Highway Fuel Economy Test (FET), which EPA discontinued using in 1979 as a basis for fuel economy figures, retaining only the EPA urban estimate (Federal Register, 43, 21412, 1978). However, improvement in rolling resistance has outpaced improvement in aerodynamic design, and both factors merit further research. Aerodynamic drag has a greater influence at low speeds for small cars than for large cars, and, therefore, aerodynamics will continue to increase in importance as the vehicle weight of the on-the-road fleet decreases. There has been little basic research on the physics of flow fields around vehicles (White, 1980). Such research could lead to quantitative methods for reducing aerodynamic drag for vehicles subjected to head winds

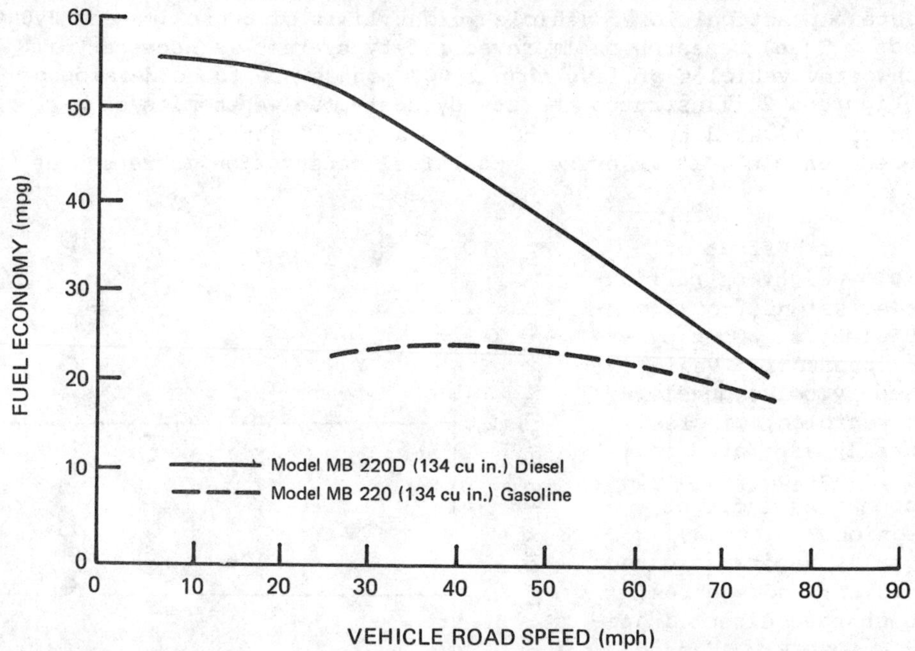

FIGURE 5.2 Effect of vehicle speed on the fuel economy of comparable 1972 model year Mercedes light-duty diesel and light-duty gasoline vehicles. SOURCE: EPA, 1980a,b.

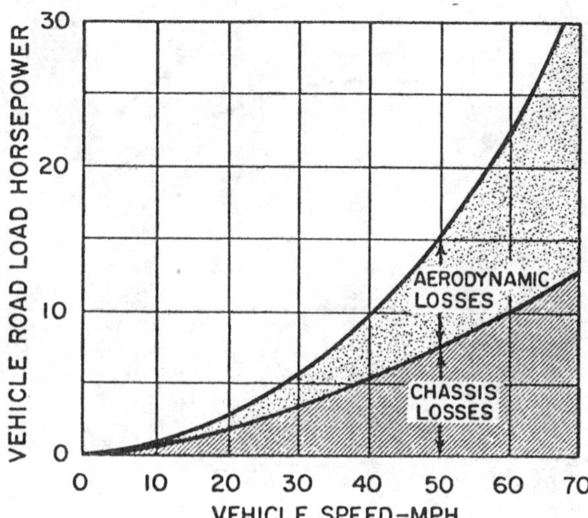

FIGURE 5.3 Effect of vehicle speed on vehicle aerodynamic drag and chassis losses (tire rolling resistance and rotational losses in drive train). SOURCE: Tenniswood and Graetzel, 1967.

(currently trial and error techniques are used) and ways of decreasing the sensitivity of drag to side winds (Figure 5.4).

Both drag and rolling resistance increase with road roughness (White, 1980). Resurfacing rough roads to increase fuel economy is probably not a viable policy because of the fragmented responsibility (federal, state, municipal, county, etc.) for road maintenance.

INDIRECT INJECTION DIESELS, ADVANCED DIESELS, AND CONTROL TECHNOLOGIES

The Indirect Injection Diesel Engine

Current indirect injection diesels offer significant fuel economy advantages in comparison with current gasoline-powered light-duty vehicles. It should be noted that indirect injection light-duty diesel technology is a relatively new technology and improvements in the indirect injection light-duty diesel should occur more rapidly than improvements in older light-duty gasoline engine technology.

Emission standards may, however, limit the future use of the light-duty diesel; emission control technology must be developed during a very compressed time schedule. Emission control technologies currently under investigation have already been discussed (Chapter 3). The technologies that are considered to have the greatest potential and for which development must be expanded and accelerated include trap-oxidizers and catalytic trap-oxidizers, exhaust catalysts, modulated exhaust gas recirculation systems (EGR), and electronically controlled rotary fuel injection pumps.

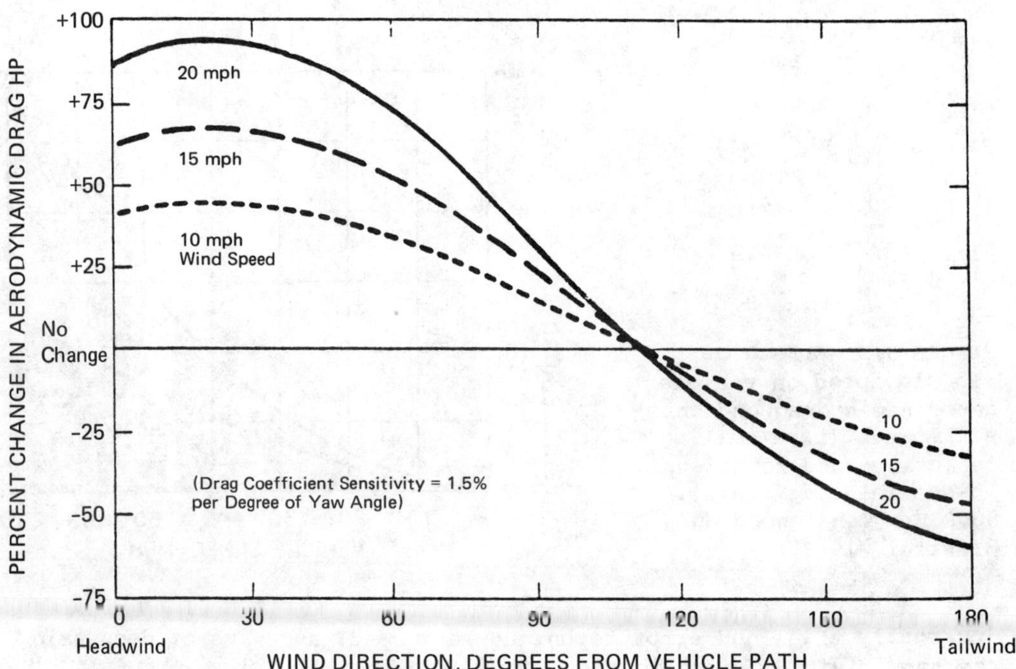

FIGURE 5.4 Effect of wind and wind direction on vehicle aerodynamic
drag at 55 mph. SOURCE: EPA, 1980a,b.

In addition to these advances in emission control technology, other
improvements in the indirect and direct injection light-duty diesel
engine system should be the subject of new or expanded research
programs. These areas for improvement are discussed in the following
paragraphs.

Compression Ratio Optimization

The optimum compression ratio for the light-duty diesel for fuel
economy is about 17:1. Above this value, increases in the thermal
efficiency are offset by greater increases in frictional and heat
transfer losses. Current indirect and direct injection light-duty
diesels have compression ratios of about 22:1 because of cold-start
difficulties at lower compression ratios. Alternative strategies
should be investigated, such as retractable glow plugs (Horwitz, 1980)
or spark-assisted ignition.

Lubricating Oil

Particulate contamination of the engine lubricating oil is a
characteristic of diesels. As was discussed in the section on engine

modifications, exhaust gas recirculation (EGR) accelerates particulate contamination of the oil. These oil-suspended particulates adsorb zinc dithiophosphate (or its decomposition products), the antiwear agent used in engine oils. Depletion of the antiwear agent results in decreased oil and filter change intervals and necessitates using more expensive engine component construction materials that have higher resistance to wear. Development of an additive "package" that would not be adsorbed by the particulates is highly desirable. Improved EGR technology might also lessen this problem.

Between 16 and 80 percent of the particulates emitted by the light-duty diesel have been traced to the lubricating oil (Mayer et al., 1980). The mechanism of this contamination process is currently not understood; thus, research in this area is warranted.

Turbochargers and Superchargers

Improvements in turbocharger technology may be expected to result in improvements in fuel economy and emissions. Current turbochargers would be improved by electronic rather than mechanical control of the exhaust waste gate. Microprocessor logic could be used to keep the exhaust waste gate open during engine operating conditions where the exhaust back pressure would otherwise be greater than the intake boost pressure. This may result in lower particulate emissions and fuel consumption. Also, microprocessor control could be used to open the waste gate under high boost conditions not only in response to boost pressure but also in response to intake air density (this requires the development of an air density sensor or the use of pressure and temperature sensors) and possibly other engine operating conditions. This would improve the peak power that is possible with turbocharged engines that are subjected to a wide range of ambient conditions.

Current turbochargers suffer from poor efficiency during low-engine-speed operation. Development of light weight ceramic impellers will help inertia but not necessarily efficiency. Another means of improving low speed boost would be through the use of a variable area turbo-charger. Research related to this technology is ongoing, but variable area turbochargers promise to be more expensive than traditional fixed geometry turbochargers. Alternatively, resonant manifolds such as those used on high performance spark ignition engines might be used to improve torque during transients (Brands, 1980). Because superchargers do not have poor low speed efficiency, more research is warranted in this area. The high cost of Roots superchargers eliminates them from consideration for production vehicles. However, Brown Boveri is developing the Comprex pressure wave supercharger, and Bendix is investigating a vane-type supercharger. Much more research needs to be done before a durable, reasonably priced high-efficiency-at-low-engine-speed turbocharger or supercharger can be made available.

Because increasing the engine boost pressure also increases NO_x emissions, an aftercooler may be necessary. However, aftercoolers that decrease the intake air temperature through heat exchange with cooling water from the engine's radiator may actually increase the intake air temperature under some operating conditions. Therefore, separate system aftercoolers (SSAC) may be necessary. SSACs require the use of a separate heat exchanger for the aftercooler and are, therefore, even more expensive than traditional aftercoolers. Because of the difficulty of NO_x control of the diesel, especially the direct injection engine, development of effective inexpensive aftercoolers may be highly attractive.

Turbocompounding offers promise as an exhaust energy recovery technology. This concept uses a second turbine downstream from the turbocharger. The second turbine is mechanically coupled to the engine output shaft. A 15 percent fuel economy improvement has been noted during testing of this system in long haul truck application (Brands, 1980). Unfortunately, because of the limited energy available in the exhaust gases of the lightly loaded light-duty diesel, this technology will probably not be suited for this application. However, turbo-compounding might prove to be applicable to the light-duty "adiabatic" diesel (discussed in a later section).

Advanced Fuel Injection Systems

Precise control of fuel injection timing, rate, and duration is necessary to optimize the fuel economy, emissions, and performance of the light-duty diesel. Combinations of these three factors produce other parameters of interest such as quantity of fuel per injection stroke, maximum rate of delivery, and rate of change in rate of delivery (rate tailoring). In current turbocharged indirect injection light-duty diesel engine systems, the maximum deliverable fuel rate is limited until the boost pressure reaches a preselected level. However, timing and quantity are not controlled as precisely as desired, because of mechanical tolerance, limitations of mechanical controls, and cycle-to-cycle and cylinder-to-cylinder variations. Current systems deliver fuel too rapidly at some engine operating conditions and too slowly at others (Montanari et al., 1975; Parker, 1976).

The electronically controlled rotary pump systems now under development will provide much more accurate control of timing in response to engine operating conditions as a function of sensor signals for EGR rate, coolant temperature, engine speed, and throttle position. Such systems can be expanded to provide even more favorable timing control if sensors for torque, air flow rate, fuel flow rate, air density, smoke, and NO_x are added (e.g., Eisle, 1980).

The development of inexpensive and reliable sensors and actuators is a significant need. However, even the most advanced electronically controlled rotary pump systems envisioned do not incorporate improvements in rate or quantity control, and they have no capability for rate tailoring. Rate tailoring is attractive because high rates near the beginning of injection will promote good mixing of fuel and air and lower rates near the end of injection will promote complete oxidation, thereby simultaneously controlling NO_x, hydrocarbons, and particulates. Research is needed to investigate the advantages of precise rate control, precise quantity control, and rate tailoring.

Precise control of fuel injection parameters, in response to ambient conditions, should help alleviate cold-start difficulties (possibly enough to allow a decrease in the compression ratio). The same sensors, microprocessor, and actuating systems should allow the idle speed to be minimized in response to ambient conditions to minimize fuel consumption and emissions during idle running.

Valve Timing

Engine valve timing characteristics dictate air induction rate and, therefore, the volumetric efficiency of the engine as a function of engine speed. Hence, valve timing capability constrains fuel economy, acceleration (drivability), and maximum speed of the vehicle. Because valves are currently driven by a camshaft with fixed geometry, valve opening and closing times and amount of valve overlap are constant and cannot be controlled in response to engine speed. This has a major effect on the volumetric efficiency, exhaust residual, and thermal efficiency at each increment of engine speed. Also, the valves must be opened and closed relatively slowly to avoid excessive wear and valve bounce. This constraint introduces additional fluid dynamic losses. Most of the recent research in the area of valve timing has been conducted on spark ignition engines, as will be discussed in a later section. Many of the gains attainable from adjustable valve timing on spark ignition engines are also attainable for the diesel.

The Direct Injection Diesel

As is shown in Figure 3.58, the direct injection light-duty diesel has a significant fuel economy advantage in comparison to current light-duty indirect injection diesels and gasoline engines. For the higher inertia weight vehicles, the turbocharged direct injection diesel engine has a 43 percent fuel consumption advantage over the light-duty gasoline engine (this comparison is based on current developmental stages of these engines operated during the European ECE 15 cycle). Direct injection diesel-powered vehicles enjoy a 10 to 20 percent fuel economy advantage during the Federal Test Procedure (FTP) over the indirect injection diesel. Recent Department of Transportation (DOT) calculations (Miller, 1980) reveal that a three-cylinder direct injection diesel in a 1,700- to 2,000-pound vehicle would get 80 to 90 mpg. Many of the European vehicle manufacturers are developing direct injection light-duty diesel engines for introduction into the European market-place. Ford, General Motors, and BMW are considering offering direct injection engines in the United States, but this possibility is uncertain because of emission constraints. Compared with indirect injection engines, the direct injection diesel generally has higher NO_x, hydrocarbon, and odor emissions, and a more severe full load smoke problem. Additionally, the direct injection diesel is noisier, has a limited engine speed range, lower power/weight ratio, and lower power output capability than a comparable indirect injection diesel.

Much of the emission research necessary for the indirect injection diesel will also be applicable to direct injection models, such as development of trap-oxidizer and modulated EGR systems. However, as discussed in the section on direct injection engines, research that is specifically related to direct injection applications is needed. This includes the development of an effective zero sac volume or minimum sac volume nozzle and development of techniques to improve the engine speed range. As with the indirect injection engine, the fuel injection system plays a primary role in determining the emission levels. For this reason, unit injectors are currently being studied because of the higher injection pressures obtainable with them. Recent research and development has improved the emissions and speed/power range of the direct injection considerably (Middelmiss, 1978; Brandl, 1979; Cartellieri, 1980), but much more research is necessary before a direct injection diesel-powered light-duty vehicle can be introduced into the American market.

The Adiabatic Diesel

In conventional diesel engines as much as 40 percent of the energy liberated by the combustion process is rejected to the cooling and lubricating systems. For this reason there is much current research interest concerning the possibility of building insulated "adiabatic" diesel engines that can recover a significant portion of that lost energy and convert it to increases in power and fuel economy. Most of the current research in this field has been aimed at heavy-duty diesel engines, but much of the technology that might be developed from these investigations should be directly transferrable to the light-duty vehicle. Results from testing a single-cylinder direct injection engine showed decreases in carbon monoxide and NO_x, but increases in hydrocarbons and specific fuel consumption and a decrease in power output (Murray, 1980). It is believed that these results are anomalous and due to absorption of the impinging fuel into the porous ceramic coating. Design improvements or use of an indirect injection combustion chamber is expected to minimize this difficulty and yield the expected performance and hydrocarbon emission improvements. Amann (1980) has calculated that this concept will only improve the thermal efficiency by a few percent, with up to 85 percent of the conserved energy being rejected in the exhaust. Therefore, the adiabatic diesel will be a prime candidate for exhaust energy recovery technologies such as turbo-compounding. This will also improve the potential of using an exhaust catalyst, an application that is currently difficult because of the low exhaust temperatures of the conventional light-duty diesel. The difficulties of long-term bonding of the insulating ceramic to the engine components are also the subject of ongoing research. Solutions to these problems observed in the development of the adiabatic diesel require materials science research combined with fundamental research into heterogeneous ignition and combustion.

The Catalytic Diesel

The probability that performance of the adiabatic diesel is affected by surface reactions has led to an interest in in-cylinder catalysis in diesels. If the catalytic reactions can be understood and controlled, the development of an adiabatic diesel will be significantly enhanced. Surface catalysis in either an insulated or a traditional diesel would be expected to reduce engine knock by decreasing the ignition delay of the fuel, improve the engine's fuel tolerance, increase the rate of combustion, and decrease emissions. In turn, improved fuel economy of the diesel would be expected. Interpretation of the results of Murray (1980) in a catalytic adiabatic direct injection diesel is complicated by the fuel absorption difficulty mentioned in the previous paragraph. Sapienza and co-workers (1980) found improvements in performance and a 45 percent decrease in particulate emissions from an uninsulated catalytic diesel. However, only a single engine operating condition was observed, and the experimental measurements were neither extensive nor sophisticated. Further basic and applied catalytic combustion research is needed to determine the practicality of this concept.

SPARK IGNITION ENGINES

The spark ignition engine has undergone continual development as a light-duty-vehicle power plant since the inception of the automobile. However, emissions and fuel economy constraints have been important developmental issues only since the early 1970s. Therefore, further improvements in both the emission characteristics and the fuel economy of light-duty gasoline vehicles can be expected.

Figure 5.2 compares the steady state fuel economies of a production model light-duty gasoline engine and a comparable production model indirect injection light-duty diesel. The fuel economy of the vehicle powered by the spark ignition engine is lower for several reasons. The most important reasons are the lower energy density of gasoline in comparison to the energy density of diesel fuel, the lower compression ratio of the spark ignition engine, the relatively constant spark ignition engine fuel/air ratio versus the variable diesel air/fuel ratio, and the significant throttling losses associated with spark ignition operation. Because these throttling losses decrease with increasing vehicle speed (due to higher loads required to overcome the higher power required to move the vehicle), the fuel economy of the light-duty gasoline vehicle comes closer to the diesel with increasing vehicle speed. Also, the diesel fuel economy decreases (due to higher loads required) with increasing engine speed and approaches that of the spark ignition engine. The fuel economy advantage of the light-duty diesel is as high as 100 percent under low-speed driving conditions but is only about 15 percent for high-speed driving (which is approximately the fuel energy difference), as compared to the fuel economy of the light-duty gasoline vehicle. The fuel economy of the light-duty gasoline-powered vehicle can be improved by investigating methods to decrease throttling losses and increase the compression ratio. The thermal efficiency of the spark ignition engine can be improved by insulation to decrease the energy loss to the engine's cooling and lubricating systems (the "adiabatic" spark ignition engine). As with the diesel, microprocessor control of the fueling system of the spark ignition engine could be used to minimize the idle speed in response to ambient conditions, to decrease fuel consumption during the idling mode of operations. The frictional and accessory losses of the spark ignition engine should also be minimized, but these factors will not be discussed further. The optimized spark ignition engine will probably not be as fuel efficient as the optimized diesel because of the lower energy density of gasoline and because of the lower thermal efficiency of the actual spark ignition engine as compared to the actual thermodynamic cycle on which the diesel operates. However, the diesel does not operate under optimum conditions (such as compression ratio) either, and emission controls on the diesel may decrease its fuel economy. Also, without extensive modifications, refineries must produce a product mix containing a substantial gasoline fraction (about 40 percent gasoline, 60 percent distillate fuels). Even with refinery modifications, a significant fraction of the crude barrel can only be used to produce gasoline. Therefore, methods for improving the fuel economy of both gasoline-powered and diesel-powered light-duty vehicles must take a high research priority.

Throttling Losses

Throttling losses in the spark ignition engine may be minimized or eliminated by valve timing modifications, improvements in transmission technology, lean fuel/air operation, or use of turbochargers or superchargers.

Valve Timing

The effects of valve timing for diesel engines were briefly discussed in the previous section. Because the volumetric efficiency of the spark ignition engine is lower than that of the diesel engine (due to throttling of the intake air of the spark ignition engine), even more improvement is expected for the spark ignition engine from optimizing valve timing. For this reason, valve timing has been the subject of much recent interest.

An extreme example of the use of valve timing is found in the 1981 Cadillac V8-6-4. This system uses microprocessor actuated solenoids to stop operation of the intake and exhaust valves of two or four cylinders under certain engine operating conditions. This forces the remaining operational cylinders to do more work by requiring that the throttle be opened farther, with a corresponding and proportionate decrease in throttling losses as the throttle is opened. Flegl and co-workers (1979) estimated that fuel consumption at idle speed could be reduced by 50 percent in a V8 engine by using a partial cylinder cutoff technique, but this approach is not used in the Cadillac.

Tuttle (1980a) has recently investigated the use of late intake-valve closing (LIVC) as a means of decreasing pumping losses. In this concept, the intake valve closes very late during the compression stroke. Thus, some of the fresh charge is pushed back into the intake manifold, raising the manifold pressure and decreasing the throttling losses. The LIVC engine would have up to a 6.5 percent lower net specific fuel consumption with lower NO_x and equivalent hydrocarbon emissions. Preliminary results on early valve timing indicate that this technique has even more potential (Tuttle, 1980b).

Microprocessor control of the intake and exhaust valves would allow use of the late and early valve timing approaches being investigated by Tuttle and use of the cylinder subtraction method currently offered as an option on the Cadillac. Such a system would have other advantages as well. Current valve timing is fixed because of the mechanical coupling of the valves and the camshaft. The camshaft (and therefore valve timing) that is chosen must be a compromise for all engine speed conditions. Thus, a high performance engine will have longer duration and greater valve overlap than the comparable production engine, which yields higher power at high engine speeds but less power at low engine speeds. Microprocessor control of valve timing would allow optimization of valve timing for all engine operating conditions. This requires assessment of the benefits from this concept and the development of electronic valve timing.

Transmissions

Flegl and co-workers (1979) have investigated the benefits of optimizing the transmission and using microprocessor control of the gearbox shift points, as shown in Figure 5.5. This technique allows the attainment of minimum engine speed operation in response to driver demand while maintaining satisfactory drivability.

Another approach is to use an infinitely variable gear ratio transmission to minimize specific fuel consumption. This concept maintains engine operation at or near wide open throttle for all vehicle driving conditions. Such systems are currently under development (e.g., Lindsley, 1980).

Continuing effort is required in this area. Current automatic transmissions generally exhibit greater fuel consumption than manual transmissions for similar vehicles. More research is needed to decrease these losses and to take advantage of the fuel economy benefits available from microprocessor control of an optimized automatic transmission and from infinitely variable transmissions.

Lean-Burn Engines

The lean-burn engine, in one form or another, has been the subject of research for almost 30 years. The first production model lean burn engine, the divided chamber Honda CVCC, was introduced primarily because of its low-emission-level characteristic. Combustion of

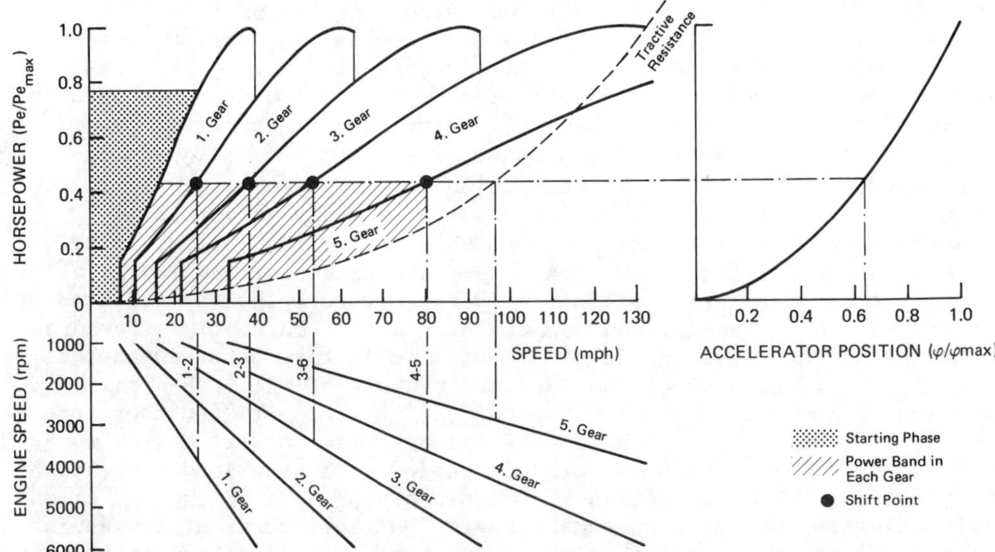

FIGURE 5.5 Example of microprocessor optimization of shift points to improve fuel economy of a vehicle with an automatic transmission. SOURCE: Flegl et al., 1979.

increasingly leaner mixtures results in increasingly lower emissions of NO_x, carbon monoxide, and hydrocarbons up to the point where the lean flammability limit is encountered. A benefit of more recent interest is the fact that combustion of leaner mixtures allows the use of wider throttle openings and thereby decreases throttling losses.

However, it should be noted that to meet NO_x emission standards of 1.0 g/mi, spark ignition engines require the use of a dual or three-way catalyst to control NO_x. This approach is not possible with an engine that does not operate near stoichiometric air/fuel ratios. Thus, the attainment of extremely lean combustion is necessary to meet increasingly stringent NO_x standards with a lean-burn engine.

The divided chamber spark ignition engine or similar form of lean-burn engine also has a decreased knock tendency because the mixture strength of the end gases is low due to the lean fuel/air ratios used. Thus, the lean burn engine should be able to operate at higher compression ratios than the conventional spark ignition engine, especially if microprocessor controlled spark retardation is used.

The lean flammability limit can be extended by increasing the turbulence in the combustion chamber. This in turn increases the internal heat feedback and the flame speed (thus the terminology of "fast-burn engine"). This technique has been shown to be viable to meet earlier emission standards while improving fuel economy of the divided chamber spark ignition engine (Noguchi et al., 1976), but considering the present state of development regarding chamber turbulence, compliance with future standards does not appear to be possible.

The use of fuels or fuel supplements that provide very lean flammability limits is another method for extending the lean burn concept. Alcohols are examples of fuels that have leaner flammability limits than gasoline has. Methanol has a lean flammability limit that is about 0.2 equivalence ratios lower than that of isooctane (Ebersole and Manning, 1972). A blend of 5 percent methanol and 95 percent gasoline would lower the lean flammability limit by about 0.04 equivalence ratios (Pefley, 1979). Gasohol (10 percent ethanol, 90 percent gasoline) would have characteristics somewhere in this range. Engines are not currently designed to take advantage of the leaner flammability limits available from alcohol/gasoline blends because of engineering problems associated with more than 10 percent alcohol in the fuel. Restriction to only 10 percent alcohol constrains the ability to use leaner mixtures to a range that would not be in compliance with the emission regulations. Although these developmental problems have potential engineering solutions, there is presently not sufficient alcohol production capacity available to provide the quantities necessary for widespread use of fuels containing more than 10 percent alcohol. Also, current federal regulations would not permit an engine designed specifically to operate on alcohol/gasoline blends to use this fuel during certification.

Hydrogen (H_2) is a fuel that has a much leaner flammability limit than is the case for either gasoline or pure alcohols. Use of pure hydrogen as a fuel has the added advantage that there is no opportunity for the formation of hydrocarbon and carbon monoxide exhaust emissions. However, the lean combustion limits for H_2 have not been well established, especially on multicylinder engines (Grosshandler, 1980). Because of flashback difficulties associated with use of hydrogen in carbureted engines, it appears that direct injection into the combustion chamber will be necessary. Furuhama and Azuma (1979) applied this technique to a two-stroke spark ignition engine. This is an especially interesting application because the two-stroke spark ignition engine is disappearing because of its normally high hydrocarbon emissions.

Handling and storage are serious problems to be overcome before the advantages of use of H_2 can be implemented. Storage difficulties may be solved by use of metal hydrides rather than gaseous or liquid H_2 storage, but this area requires further investigation. Because of the low energy density of hydrogen, current H_2-fueled vehicles are limited in range: This is probably the most serious problem facing those interested in conversion to hydrogen fuels.

An alternative concept is to catalytically convert a portion of the fuel gasoline to gaseous components including H_2 or to add "small" amounts of H_2 from a separate source (Rupe, 1973; Stebar and Parks, 1974; Houseman and Cerini, 1974; Hoehn and Dowdy, 1974; Houseman and Hoehn, 1974; Cichanowicz and Sawyer, 1976). Such use of hydrogen as a fuel supplement rather than as a pure fuel overcomes most of the handling, storage, and range problems otherwise encountered. It remains to be determined whether this technique will allow lean enough operation for compliance with future NO_x emission standards. Increased hydrocarbon emissions for very lean operation have been noted with this system. Although this concept has not been the subject of recent investigation, it is possible that it should be reevaluated in the light of current fuel economy concerns.

Direct Injection Stratified Change (DISC)

The direct injection stratified change (DISC) engine is another method that may be used to minimize the throttling losses and to achieve variable air/fuel operation in spark ignition engines. Examples of DISC engines currently under development are the Ford PROCO for light-duty vehicles and the Texaco TCCS for light- to heavy-duty vehicles. Because the DISC engine uses direct fuel injection into the combustion chamber, load control is achieved by varying the fuel/air ratio as in the diesel, rather than by limiting the quantity of fuel-air mixture entering the combustion chamber as in the traditional spark ignition engine. For this reason, the DISC engine can eliminate throttling losses. However, the PROCO uses limited throttling for emission control and improved drivability.

The DISC engine has several other advantages, including good fuel tolerance and the ability to operate at compression ratios that are higher than those of the conventional spark ignition engine. In the DISC engine, the end gases are primarily comprised of air because of the charge stratification. Thus, the PROCO, for example, can operate at compression ratios up to almost 13:1, which is another advantage of this type of engine. Table 5.1 shows about a 50 percent improvement in steady state fuel economy for the TCCS operating on diesel fuel as compared to the fuel economy of the conventional spark ignition engine operating on gasoline in a four-cylinder 2,750-pound AMC Gremlin. Acceleration was also improved with the DISC engine. Steady state fuel economy of the gasoline DISC engine was increased by about 35 percent

TABLE 5.1 Steady State Fuel Economy Comparison of a Four-Cylinder Spark Ignition Engine With a Comparable Four-Cylinder DISC Engine Operating on Gasoline and Diesel Fuel

Vehicle Speed, mph	L-141 Gasoline, mpg	L-163STCCS Gasoline, mpg	Diesel Fuel, mpg
20	24.1	33.1	36.2
30	21.3	29.2	31.9
40	18.6	25.2	27.7
50	16.0	21.2	23.5
Idle	0.463 gal/hr	0.179 gal/hr	0.158 gal/hr

SOURCE: Lewis and Tierney, 1980.

(Lewis and Tierney, 1980). Table 5.2 compares the emissions and fuel economy of a carbureted four-cylinder engine in a 2,700-pound vehicle with those of a 141-cubic-inch PROCO engine in the same vehicle. The DISC engine designed for best fuel economy has a 43 percent advantage over the carbureted engine, with neither engine incorporating emission controls. The catalyst-equipped PROCO with EGR offers an 18 percent fuel economy advantage over the uncontrolled spark ignition engine (Simko et al., 1972). Ford is currently redirecting its PROCO research effort away from the large V8 toward the in-line six-, four-, and three-cylinder engines in compliance with the needs of smaller vehicles designed for upcoming years. The V8 PROCO was in the final development stages, but development of the smaller engines is not as advanced. The PROCO loses some advantage in vehicles with a low power/weight ratio because these vehicles do not have throttling losses that are as significant as those of vehicles with higher power/weight ratios. Getting proper swirl and flow patterns in the smaller engines appears to be the most important objective in PROCO engine design. The small PROCO is expected to be about the energy equivalent of the small diesel but would suffer a 13 percent mile-per-gallon disadvantage because of the lower energy density of gasoline. The PROCO does not appear to have as severe a particulate emission problem as the diesel, even when operating on diesel fuel. The PROCO technology is very promising and should be pursued.

TABLE 5.2 Comparison of 1972 CVS Procedure Emissions and Fuel Economy of Conventional Four-Cylinder Spark Ignition Engine with a Comparable DISC Engine Designed for Best Fuel Economy and a Comparable PROCO Engine With and Without an Exhaust Catalyst

Vehicle	Emissions, g/mi			Fuel Economy, mpg	No. Test Averaged
	HC	CO	NO_x		
Carbureted spark ignition without emission controls	5.65	46.24	4.47	16.6	3
Stratified, best economy	4.96	7.75	3.85	23.8	2
PROCO without catalyst	3.10	13.75	0.33	21.2	1
PROCO with catalyst	0.54	1.18	0.37	19.6	4

SOURCE: Simko et al., 1972.

Turbocharging and Supercharging

Turbocharging or supercharging can also be used to minimize the throttling losses of the spark ignition engine. This technology has the same difficulties discussed earlier in connection with turbocharging the diesel: poor low speed turbocharger efficiency, transient response lag, and increased NO_x emissions. Also, turbocharging the spark ignition engine increases its knock tendency; turbocharging the diesel decreases its tendency for combustion knock. Production model turbocharged light-duty gasoline-powered vehicles incorporate a knock sensor that retards the spark timing at incipient knock. Injection of water or alcohol is a better, but less practical, technique of preventing knock. Present turbochargers do not exhibit a fuel economy advantage over nonturbocharged engines because of poor turbocharger efficiency during the FTP driving schedule. The improvements in this technology discussed in the section pertaining to diesel engines may result in a fuel economy advantage for advanced turbochargers or superchargers.

Compression Ratio Optimization

Similar to the diesel, the optimum compression ratio for the spark ignition engine is about 17:1. The maximum compression ratio of the spark ignition is limited by the need to avoid combustion knock. Knock occurs in the spark ignition engine when the mixture near the periphery of the combustion chamber ignites spontaneously after the spark has occurred, but before the flame front has reached these end gases. Knock can be inhibited by using a fuel with a longer chemical ignition delay time (higher octane rating), by decreasing the chemical reactivity of the mixture, or by decreasing the residence time between the spark initiation and arrival of the flame front at the end gas region. Current production engines have compression ratios of only about 8.5:1 compared with earlier spark ignition engines, which had compression ratios as high as about 12:1. This change is due to the decreasing octane level of gasoline as produced by the refinery and as a partial remedy for NO_x production. However, NO_x is more effectively controlled (from a fuel economy perspective) through use of a three-way catalyst, which is generally used in conjunction with exhaust gas recirculation. Increasing the compression ratio will improve power and fuel economy and may be attained in several ways.

Fuels

The octane rating of gasoline is expected to continue to decline due to the declining quality of available crude. Tetraethyl lead (TEL), the additive traditionally used to boost octane ratings, is being phased out due to lead air pollution problems and lead poisoning of exhaust catalysts. The most promising replacement for TEL, methyl cyclopentadienyl manganese tricarbonyl (MMT), also exhibits catalyst poisoning and is more expensive than TEL. Most other octane boosters, used primarily in racing applications, are much more expensive than TEL. Gasohol has a higher octane rating than regular gasoline; higher alcohol/gasoline blends have even higher octane ratings. However, several problems are associated with use of alcohol/gasoline blends with more than 10 percent alcohol: attack by the alcohol on many metallic, plastic, and rubber components in the fuel system; phase separation of the alcohol and gasoline in the presence of water; etc. Most of these problems have engineering solutions. Alcohol/gasoline blends have a lower energy density than pure gasoline. This fuel economy (mpg) disadvantage might be offset through engine design optimization (such as increasing the compression ratio) to take advantage of the characteristics of alcohol. There is currently not sufficient capacity to produce enough alcohol even for widespread use of gasohol, much less higher alcohol/gasoline blends. The FTP currently prescribes the fuel to be used by certification light-duty gasoline vehicles as indolene clear, and a vehicle designed specifically for alcohol or alcohol/gasoline blends would probably not operate satisfactorily on indolene clear during certification. This EPA policy should be reevaluated, if the advantages of gasohol are to be optimized.

Mixture Reactivity

An alternative method for inhibiting knock and allowing higher compression ratios is to decrease the reactivity of the end gases. This is commonly accomplished on nonproduction vehicles by water injection or alcohol injection. However, this technique is not believed to be practical for vehicles maintained by the general public.

Retarding the spark timing will result in lower temperatures and pressures of the end gases after the initiation of combustion because the piston is receding. In this way the mixture reactivity is decreased without decreasing the mixture strength. Spark retardation generally produces a loss in fuel economy, but if retardation is only used when necessary to prevent knock and if the compression ratio is also increased, a net fuel economy gain will result. Chevrolet used this concept on a 1981 six-cylinder light-duty truck to increase the compression ratio from 8.4:1 to 9.2:1. A knock sensor coupled to microprocessor control of the spark timing was used so that spark retardation was only used when necessary to avoid knock. This technology should be extended and its wider application considered.

Decreasing Residence Time

The end gas residence time can be decreased by increasing the flame speed or by decreasing the distance between the flame source and the end gases. The flame speed can be increased by increasing the turbulence in the chamber. It can be increased by using a fuel or fuel supplement the flame speed of which is higher than that of gasoline, such as hydrogen, which was discussed in an earlier section. Because the flame speed is dependent on temperature, it is expected that the flame speed would be higher in an "abiabatic" spark ignition engine. The adiabatic spark ignition engine is discussed in the following subsection.

Decreasing the size of the combustion chamber to decrease the end gas residence time is limited by the dependence of hydrocarbon emissions on the surface/volume ratio of the combustion chamber. An alternate method is the use of multiple spark plugs. There might be a fuel savings versus initial engine cost advantage in using multiple spark-plugs in a high compression ratio engine that operates on relatively low octane gasoline. Use of a catalyst on the cylinder head and piston top could also accomplish this goal with additional advantages as well. The catalytic spark ignition engine will be discussed in the following subsection.

Catalytic and Adiabatic Engines

The use of a catalytic coating in the combustion chamber of a spark ignition engine was first studied in 1925, when Dr. Edward Sokal successfully used a cerium dioxide coating to inhibit knock. The introduction of tetraethyl lead as a gasoline antiknock additive halted further research in this area--until recently. Haskell and Legate (1972) used a platinum coated Vycor disc located in the center of the piston crown to try to decrease hydrocarbon emissions, but they found increased hydrocarbon emissions and degraded performance. This catalyst was not located in a position to afford knock inhibition, but the observed results were unexpected. A properly designed catalytic engine should inhibit knock by initiating combustion at the walls almost immediately after the spark initiates the usual flame front. This can be approximately envisioned as having a very large number of spark plugs located all around the combustion chamber. Because of the resulting knock inhibition, it should be possible to increase the compression ratio and improve performance and fuel economy. Further-more, hydrocarbons that are normally not burned, because of wall quenching, should be consumed by the catalytic reactions; thus, hydrocarbon emissions should also be lower. However, it is necessary to tailor the catalyst properties precisely and thus avoid preflame reactions during the intake and compression strokes which might significantly alter the reactant characteristics before the normal initiation of combustion by the spark. The rate of catalyst-induced activity must be high enough to initiate reaction during the time scale of the ignition process. Wojciki and co-workers (1980) used a coil of platinum wire in the prechamber of a divided chamber engine and found performance improvements.

Like the diesel engine, the spark ignition engine rejects up to 40 percent of the energy liberated by combustion to the cooling system and lubricating oil. A significant portion of this loss could be prevented by using suitable insulating materials. In contrast to the adiabatic diesel, little research has been done on the adiabatic spark ignition engine. Murray (1980) found that a zirconium coating promoted knock for compression ratios greater than 10:1. This means that the effect of a higher initial temperature of the end gases was more important than the presumably higher flame speed. The observed decrease in ignition delay of the fuel, even for compression ratios below 10:1, supports this argument. However, at lower compression ratios there was a significant improvement in power and fuel consumption and in reduction of carbon monoxide and hydrocarbon emissions. As expected, NO_x emissions were higher, and the exhaust temperature was increased up to 25°K.

Murray also investigated the catalytic adiabatic spark ignition engine. Cerium was found to decrease the ignition delay, but platinum, a very active catalyst, actually increased the ignition delay. Emissions, performance, and fuel economy were either comparable or degraded in these catalytic adiabatic engines. Knowledge of this complex system is so limited that a satisfactory explanation of these results is not available. It is possibly related to the requirements for an optimum catalyst, as discussed previously. It is possible that the high temperatures of the insulated spark ignition engine will render many conventional catalysts too active for this application.

The catalytic and adiabatic engines are a great departure from the conventional design. However, the potential of these concepts dictates the need for further research in these areas.

ALTERNATIVE POWER PLANTS

The spark ignition engine has been the dominant power plant for light-duty vehicles for the past 70 years. The diesel has been used in light-duty vehicles for almost that long. During the infancy of the automobile, the steam-powered car and the electric vehicle were also in contention. Before the turn of the century and the development of oil refining technology that eventually led to the popularity of the spark ignition engine, the Stirling engine was a relatively widely used power plant. The late 1940's saw the introduction of the gas turbine as a potential power plant for light-duty vehicles. These alternatives to the spark ignition and diesel engines are the subject of renewed interest because of their potential for higher fuel economy and lower exhaust emissions.

External Combustion Engines

External combustion engines derive their energy from an external heat source rather than from a chemical reaction inside the combustion chamber. Thus, they could operate on nuclear or solar power, for example, but for automotive applications, only combustion of gaseous or liquid fuels is more practical. The combined cycle fuel economy of recent developmental external-combustion-engine-powered vehicles is exhibited in Figure 5.6.

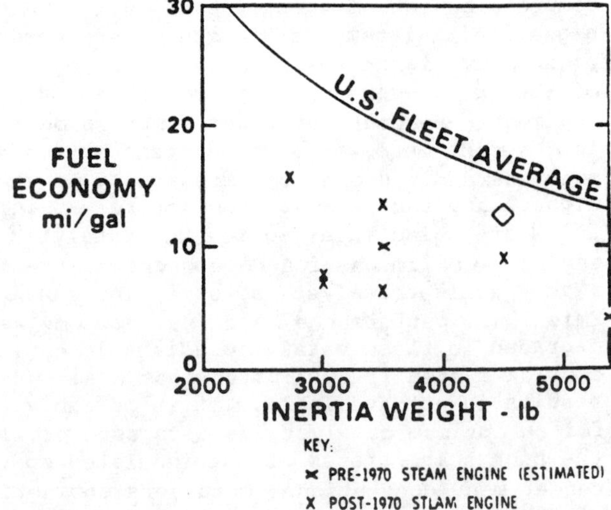

FIGURE 5.6 Fuel economy on the federal composite driving schedule for experimental cars with closed-cycle external-combustion engines compared to average for production cars sold in the United States. SOURCE: Amann et al., 1980.

Stirling Engines

The Stirling engine, invented 150 year ago by an English preacher, enjoyed considerable use during the latter half of the 19th century before it was supplanted by the Otto cycle engine. The efficiency of the Stirling cycle with ideal regeneration is equivalent to that of the Carnot cycle. Because it is an external combustion engine, it is inherently quiet. Also, it has few moving parts, and it is safe and reliable. Stirling engines suffer from a low power/weight ratio. The working fluid requires a high thermal conductivity and low viscosity, and hydrogen or helium are normally used. Use of hydrogen, the optimum working fluid, results in embrittlement problems of materials under high temperatures. Sealing problems are severe with both hydrogen and helium. Because the efficiency of the cycle increases with increasing pressure, pressures up to 2,000 atmospheres are used. This results in a "mobile pressure vessel" which must meet the appropriate safety codes, which is a primary cause of the low power/weight ratio. Adequate control and transmission of the generated power are two areas that are currently receiving less attention than they deserve. Exhaust emissions from the Stirling are higher than necessary because the combustion systems used have not been optimized. Most surprisingly, almost no regenerator research has been conducted recently, although optimum regenerators are needed to approach Carnot efficiency. The fuel economy of the present Stirling engine is only 15 to 25 mpg, less than that of an efficient light-duty gasoline engine and much less than that of a light-duty diesel. The potential of this engine justifies more research. The variety of problems requires that research funding not be directed primarily at a few large scale projects, as is presently the case. Dowdy and Burke (1979) predict that the fuel economy of the Stirling may be as much as 40 percent greater than that of the spark ignition vehicle when both are subjected to either a 1.0 g/mi or a 0.4 g/mi NO_x regulation.

Steam Engines

The steam engine was surpassed by the early reliability and efficiency of the spark ignition engine. Practical steam engines operate on the Rankine cycle. Amann (1980) has reviewed recent research on steam-powered vehicles and concluded that although the most stringent expected emission standards may be met with continuing development, the extrinsic losses inherent in the steam power plant package will prevent it from ever realizing fuel economy values equivalent to those of an average spark ignition engine. The Rankine cycle engine also suffers from an inability to operate at high ambient temperatures without a massive condenser or significant steam rejection. Amann (1980) claims that even if these problems are solved, the steam-powered vehicle will still not be competitive. The prognosis for the steam power plant is not very promising.

The Gas Turbine

The gas turbine is unlike either the continuous external combustion Stirling and Rankine engines or the intermittent internal combustion diesel and spark ignition engines. The gas turbine is a continuous internal combustion engine that operates on the Brayton cycle. The efficiency of the ideal Brayton cycle is equivalent to that of the ideal Otto cycle. Thus, renewed interest in the gas turbine was prompted not by fuel efficiency but by emission considerations.

The gas turbine, as used in the jet engine applications, has inherently low carbon monoxide and hydrocarbon emissions. However, the automotive gas turbine has a load factor different from that of the jet engine, and it appears necessary to use either a lean premixed prevapo-rizing combustor or a catalytic combustor for emission control. The lean premixed prevaporized combustor must remain above 1100°C to complete carbon monoxide oxidation and below 1750°C to inhibit NO_x formation. The catalytic combustor appears to have several advantages over the premixed system but is farther behind in development. The catalyst promotes oxidation of carbon monoxide even at temperatures below 1100°C. Also, the catalytic combustor has wider flammability limits than the premixed combustor and can therefore maintain stable combustion at leaner fuel/air ratios. This can be interpreted to reveal that the NO_x levels for a driving cycle will be significantly lower if the gas turbine vehicle has a catalytic combustor than if it had any other type of combustor. However, catalytic combustion is a relatively new technology, and very little research has been done on transient characteristics. From the automotive perspective, research on transient characteristics and the development of high-temperature ceramic substrates are the primary research needs in the catalytic combustion field.

It should be noted that the maximum allowable burner temperature for NO_x control (1750°C) is much higher than the current allowable turbine inlet temperature (1050°C). Although the efficiency of the ideal Brayton cycle is independent of peak combustion temperature, the efficiency of the actual gas turbine cycle increases with increasing turbine inlet temperature. Thus, development of ceramic turbine vanes and rotors is crucial to the fuel economy of the gas turbine (Dowdy and Burke, 1979). Other areas of materials research necessary for the gas turbine to exhibit superior fuel economy include the development of ceramic combustion liners, gasifier inlet plenums, regenerators, and exhaust diffusers (Herschfeld, 1978). Using ceramic materials for the turbomachinery has the additional advantage of lower rotational inertia, thus allowing better acceleration--poor acceleration has been a drivability problem with single-shaft gas turbine vehicles (Amann, 1980).

In fact, it appears that if a single shaft (single turbine) gas turbine is ever to be marketed, a continuously variable transmission with 800 recharges (thought to be good for the 100,000 mile life of the car) and low rotational kinetic energy is necessary (Amann et al., 1980). These power transmission difficulties can be relieved by using a free-shaft (two-turbine) or KTT (three-turbine) system. However, additional turbines imply additional costs but not necessarily additional volume. The free-shaft gas turbine may require the development of variable geometry vanes for power control.

Improvements in turbomachinery efficiency and downsizing of the passenger car fleet indicate the need for smaller compressors. Scaling of compressors is not straightforward and will require significant gas dynamic research (Amann et al., 1980).

Other than the potential for low emissions, the gas turbine enjoys other advantages over spark ignition and diesel engines. One is a higher power/weight ratio. A second is the ability to burn almost any fuel. Thus, the fuel economy of a gas turbine designed to burn diesel fuel or kerosene could be higher than that of a spark ignition engine, on a miles-per-gallon basis, because of the higher volumetric energy density of the fuel. Dowdy and Burke (1979) conclude that a full sized gas turbine powered vehicle could get up to 30 percent better fuel economy than either a 1978 gasoline vehicle at 1.0 g/mi NO_x or a 1985 light-duty gasoline vehicle at 0.4 g/mi NO_x. This conclusion was valid only for full-sized vehicles. The gas turbine has no significant advantage for small vehicles. The fuel economy of current gas-turbine vehicles is shown in Figure 5.7. The 3,500-pound vehicle was operated on gasoline, unlike the other two vehicles, which were operated with diesel fuel.

Electric Vehicles

Electric vehicles fall into two categories: the battery-powered vehicle and the electric hybrid vehicle. Battery-powered vehicles were investigated as early as 1830, and hybrids date back to 1899. During the decade following the turn of the century, there were over 100 manufacturers of electric vehicles, including Studebaker and Oldsmobile

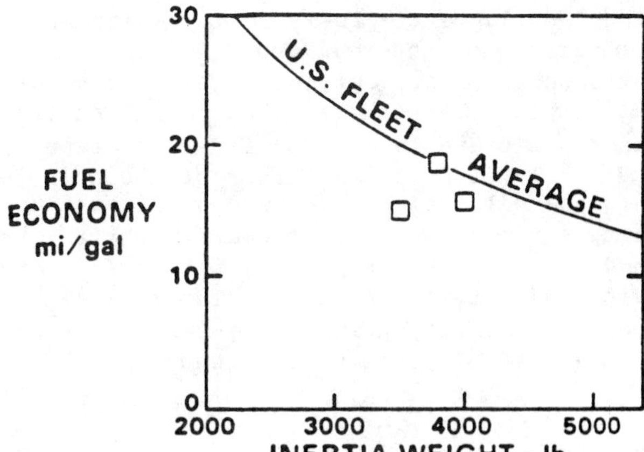

FIGURE 5.7 Fuel economy on the federal composite driving schedule for gas-turbine-powered passenger cars. SOURCE: Amann et al., 1980.

(Kirk and Davis, 1980). In spite of the advantages of the electric vehicle, it was eventually displaced by the gasoline vehicle. Kirk and Davis cite the reasons for the demise of the electric vehicle as: limited performance and range, high initial and operating costs, development of assembly line methods for production of low-cost spark ignition engines, and conservation in development and advertising by the electric vehicle industry. As was the case with the gas turbine, renewed interest in the electric vehicle was prompted by environmental considerations. This interest has since been compounded by concern of the energy situation.

Battery-Powered Vehicles

The Department of Energy program in battery-powered vehicles is aimed at instigating commercial production by the mid-1980s. Present battery-powered vehicles have a range limited to only about 40 miles, but the goal is 100 miles. It is anticipated that for success in the commercial market, the battery on the vehicle must have a specific energy of 56 W-hr/kg and a specific power of 104 W/kg. These goals are thought to be obtainable in the near future. More problematic is the attainment of a battery cycle life and cost that would make the cost per mile equivalent for the battery-powered vehicle and the light-duty gasoline-powered vehicle. The 20 percent initial higher cost of the battery-powered vehicle would be offset by the fuel and operating costs of the light-duty gasoline-powered vehicle (Savitz, 1980).
Watson (1977) has concluded that the limited performance of the battery-powered vehicle is a serious safety constraint. Other difficulties include short range per charge and the 8- to 16-hour recharge time. These difficulties can be overcome by having an auxiliary power supply on board the vehicle. Such an auxiliary power supply could be a high-power-density battery, a fuel cell, or some means of hydraulic or mechanical energy storage. The primary difficulty with the high-power-density battery is probably cost rather

than development. Fuels cells, which are used for the direct
conversion of chemical energy in the fuel to electricity, are very
attractive in concept but require much more research and development.
Burrows and co-workers (1980) conclude that the flywheel is superior to
energy storage in a hydraulic accumulator because of much higher power
per unit weight (by a factor of 100 or more). However, flywheel energy
storage will require the use of a continuously variable transmission to
couple the flywheel to the drive shaft. Thus, flywheel energy storage
has been selected for three out of four of DOE's Advanced Electric
Propulsion System Studies. The flywheel reduces the life cycle
operating costs by about one-third compared to the alternate choice, a
battery-only system with regenerative braking (DOE, 1980a).

Hybrid Vehicles

The performance and range of the electric vehicle can be extended to
attain equivalence with conventional vehicles by using an internal
combustion engine (either spark ignition or diesel) in conjunction with
the electric motor and batteries. The internal combustion engine would
be used for acceleration and would also be used exclusively beyond the
range of the electric power plant. Thus, the electric motor would be
used primarily in urban driving. A flywheel could be used in
conjunction with the electric motor to provide power for acceleration
instead of relying on the internal combustion engine for this purpose.
Again, this would require the development of a continuously variable
transmission.

The hybrid vehicle can be designed in either a series or a parallel
configuration. The series design uses the internal combustion engine
to drive the motor continuously, thereby overcoming the need for
batteries. However, analysis has shown that a parallel system will
provide better performance and efficiency. Figure 5.8 shows two
possible configurations of the parallel system. It has been concluded
that the internal combustion engine should provide about 60 percent of
the overall power, based primarily upon acceleration requirements. In
the short run, the hybrid should realize a 40 to 70 percent petroleum
savings (DOE, 1980a,b) and should cost $2,000 to $4,700 more than
conventional vehicles. In the long run, however, hybrids will cost
$1,800 to $2,500 more than comparable conventional vehicles. This
higher initial cost will be offset by lower operating and fuel costs.
Therefore, the life cycle cost comparison is primarily dependent upon
the price of gasoline, with equivalence expected when gasoline prices
reach $2 to $3 per gallon. Surber (1980) concludes that the "critical
technical areas where money can be most usefully spent are: (1) system
design and development—it remains that . . . hybrid designs can be
built in mass producible and driveable forms; (2) development of low
cost, long lived batteries—even at the expense of specific power and
specific energy; and (3) development of low cost controllers."

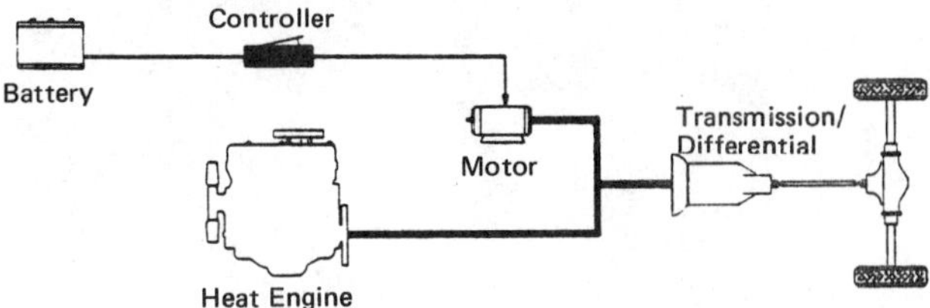

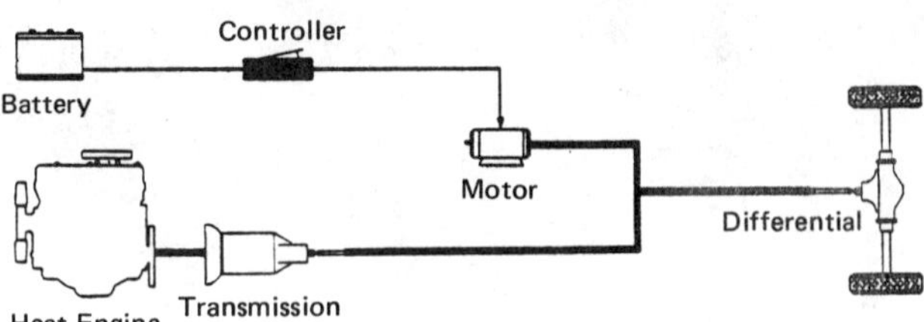

FIGURE 5.8 Potential hybrid vehicle parallel propulsion system configurations. SOURCE: DOE, 1980b.

CONCLUSIONS

As was discussed here and in the other chapters of this report, there remain substantial opportunities for further improvements to the diesel engine. Gasoline- and diesel-powered vehicles will dominate as light-duty power plants for the remainder of this century. Use of alternative power plants, with the exception of the electric vehicle, is unlikely:

* the gas turbine will not be competitive with the diesel in light-duty applications (from an efficiency standpoint) but is likely to be used in heavy-duty vehicles;
* the steam engine is too inefficient; and
* the Stirling engine is too complicated and too expensive to manufacture for light-duty use.

Electric vehicles may be used in special purpose applications--as commuter or delivery vehicles--but will not compete directly with gasoline or diesel engines in most light-duty vehicle applications. Hence, the major competition for automotive power plants for the remainder of the century will be between the gasoline and diesel engine. Since the diesel currently appears to be the lesser developed, opportunities for greater improvements exist for the diesel.

REFERENCES

Adelman, H. (1979). Alcohols in diesel engines--A review. *SAE Tech. Pap.* 790956.

Aerospace Corporation (1978). Diesel engine research and development: Status and needs. Rep. SAN-2184-1.

AiResearch (Garrett) (1980). Personal communication with C. McInerney, A. Shadbourne, W. D. Burley, and G. P. Kanelas.

Amann, C. A. (1980). Why not a new engine? *SAE Tech. Pap.* 801428.

Amann, C. A., D. L. Stivender, S. L. Plee, and J. S. MacDonald (1980). Some rudiments of diesel particulate emissions. *SAE Tech. Pap.* 800251.

Amano, M., H. Sami, S. Nakagawa, and H. Yoshizaki (1976). Approaches to low emission levels for light-duty diesel vehicles. *SAE Tech. Pap.* 760211.

American Petroleum Institute (1980). *Basic Petroleum Data Book.* Washington, D.C.

American Society for Testing and Materials (1978). *Annual Book of ASTM Standards*, part 23. Philadelphia, Pa.

Ames, B. N. (1971). The detection of chemical mutagens with enteric bacteria. In *Chemical Mutagens: Principles and Methods for Their Detection*, vol. I. Plenum Press, New York.

Ames, B. N., F. D. Lee, and W. E. Durston (1973). *Proc. Natl. Acad. Sci.* 70:782-86.

Ames, B. N., J. McCann, and E. Yamasaki (1975). *Mutat. Res.* 31:347-64.

Amoco Oil Company (1980). Communication to David Christison, Consultant to Technology Panel, Diesel Impacts Study Committee, National Academy of Sciences.

Andon, J., H. M. Siegel, J. H. Johnson, D. C. Leddy, S. Smaby, J. N. Pitts, A. M. Winer, K. A. Van Cowenberghe (1979). An initial assessment of the literature on the measurement, control, transport, transformation, and health effects of unregulated diesel engine emissions. Rep. DOT HS-804010.

ANSI (1971). Method for the physical measurement of sound. Rep. ANSI S1.2-1962(R1971).

Aoyagi, Y., T. Kamimoto, Y. Matsui, and S. Matsuoka (1980). A gas sampling study on the formation processes of soot and NO in a DI diesel engine. *SAE Tech. Pap.* 800254.

Appleton, J. P. (1973). Soot oxidation kinetics at combustion temperatures. Paper presented at Meeting on Atmospheric Pollution by Aircraft Engines, AGARD Propulsion and Energetics Panel, London.

Argonne National Laboratory (1979). _Environmental Development Plan for Light-Duty Diesel Vehicles_. U.S. Department of Energy.

Ball, R. T., and J. B. Howard (1971). Electric charge of carbon particles in flames. _Symp. Int. Combust. Proc._ 13th, p. 353.

Bandel, W. (1978). Problems in the application of ethanol as fuel for utility vehicles. CONF-T 11175. In _Proceedings of the International Symposium on Alcohol Fuel Technology_. U.S. Department of Energy.

Barnes, G. J. (1968). Relation of lean combustion limits in diesel engines to exhaust odor intensity. _SAE Tech. Pap. 680445_.

Barry, E. G., F. J. Hills, and L. J. McCabe (1979). Diesel fuel-- Availability, trends and performance. _SAE Tech. Pap. 790921_.

Barry, E. G., A. Ramella, and R. B. Smith (1977). Potential passenger car demand for diesel fuel and refining implications. _SAE Tech. Pap. 770315_.

Barth, E. B. (1980). Environmental Protection Agency, Ann Arbor. Personal communication.

Bassoli, C., G. M. Cornetti, and G. Levizzari (1977). Combustion noise and ignition delay in diesel engines. _SAE Tech. Pap. 770012_.

Bassoli, C., G. M. Cornetti, G. Biaggini, and A. Di Lorenzo (1979). Exhaust emissions from a European light duty turbocharged diesel. _SAE Tech. Pap. 790316_.

Benson, R. S., and N. D. Whitehouse (1979). _Internal Combustion Engines_, vol. 1. Pergamon Press Ltd., New York.

Bittner, J. D., and J. B. Howard (1977). Alternative hydrocarbon fuels: Combustion and chemical kinetics. _Prog. Astronaut. Aeronaut._ 62.

Bittner, J. D., and J. B. Howard (1980). Composition profiles and reaction mechanisms in a near-sooting premixed benzene/oxygen/argon flame. Paper presented at the Eighteenth Symposium (International) on Combustion, Waterloo, Canada.

Black, F., and L. High (1979). Methodology for determining particulate and gaseous diesel hydrocarbon emissions. _SAE Tech. Pap. 790422_.

Blyholder, G., J. S. Binford, and H. Eyring (1958). Kinetic theory for the oxidation of carbonized filaments. _J. Phys. Chem._ 62:263.

Boisdron, Y., and J. R. Brock (1972). Particle growth processes and initial particle size distributions. In _Assessment of Airborne Particles_, p. 129. C. C. Thomas, Springfield, Ill.

Bonne, U., K. H. Homann, and H. G. G. Wagner (1965). Carbon formation in premixed flames. _Symp. Int. Combust. Proc._ 10th, p. 503.

Braddock, J. N., and R. L. Bradow (1975). Emission patterns of Diesel-powered passenger cars. _SAE Tech. Pap. 750682_.

Braddock, J. N., and P. A. Gabele (1977). Emission patterns of diesel-powered passenger cars, part II. _SAE Tech. Pap. 770168_.

Bragdon, C. R. (1970). _Noise Pollution: The Unquiet Crisis_. University of Pennsylvania Press, Philadelphia, Pa.

Brandl, F., I. Reverencic, W. Cartellieri, and J. C. Dent (1979). Turbulent air flow in the combustion bowl of a D.I. diesel engine and its effect on engine performance. _SAE Tech. Pap. 790040_.

Brands, M. C. (1979). Helmholtz tuned induction system for turbocharged diesel engine. SAE Tech. Pap. 790069.

Brands, M. C. (1980). Vehicle testing of a turbocompound heavy-duty diesel engine. Paper presented at the DOE 5th International Symposium on Automotive Propulsion, April 14-18.

Broome, D. and I. M. Khan (1971). The mechanism of soot release from Combustion of hydrocarbon fuels with particular reference to the diesel engine. In Air Pollution Control in Transport Engine, p. 185. Institution of Mechanical Engineers, London.

Bugliarello, G., A. Alexandre, J. Barnes, and C. Wakstein (1976). The Impact of Noise Pollution. Pergamon Press, New York.

Burley, H. A., and T. L. Rosebrock (1979). Automotive diesel engines-- Fuel composition vs particulates. SAE Tech. Pap. 790923.

Burrows, C. R., G. Price, and F. G. Perry (1980). An assessment of flywheel energy storage in electric vehicles. SAE Tech. Pap. 800885.

Bykowski, B. B. (1979). Gasohol, TBA, MTBE effects on light-duty emissions. Final Report of Southwest Research Institute to EPA.

Cadle, R. D. (1971). Particulate matter in the lower stratosphere. In Chemistry of the Lower Stratosphere. Plenum Press New York.

Cadle, S. H., G. J. Nebel, and R. L. Williams (1979). Measurements of unregulated emissions from General Motors light-duty vehicles. SAE Tech. Pap. 790694.

Cain, W. S., and M. R. Garcia-Medina (1980). Possible adverse biological effects of odor pollution. APCA Pap. 80-23.2.

California Air Resources Board, Engineering Evaluation Section (1979a). Emission Tests of Four 3/4 Ton Dodge Pickups Converted from Gasoline to Diesel Engines, p. 4.

California Air Resources Board, In-Use Vehicle Surveillance Section (1979b). Accomplishments for 1975-1979.

Carpenter, K., and J. H. Johnson (1979). Analysis of the physical characteristics of diesel particulate matter using transmission electron microscope techniques. SAE Pap. No. 796815.

Cartellieri, W. (1980). Direct Injection For Light Duty Diesel Engines. Manuscript. La Societe des Ingenieurs de l'Automobile.

Cavender, J. H., D. S. Kircher, and A. F. Hoffman (1973). Nationwide air pollutant emission trends 1940-1970. Environ. Prot. Agency U.S. Publ., AP Ser. 115.

Cermak, G. W. (1979a). Exploratory laboratory studies of the relative aversiveness of traffic sounds. J. Acoust. Soc. Am. 65(1):112-123.

Cermak, G. W. (1979b). Laboratory experiments on traffic noise annoyance. ASTM Spec. Tech. Publ. 692.

Cermak, G. W., and P. C. Cornillon (1976). Multidimensional analyses of judgments about traffic noise. J. Acoust. Soc. Am. 59(6):1412-1420.

Cermak, G. W., and C. R. von Buseck (1978). Quantitative studies of traffic noise annoyance. SAE Tech. Pap. 780390.

Cermak, G. W., C. R. von Buseck, R. D. Blanchard, and J. G. Cervenak (1979). Analysis of traffic noise abatement strategies. Res. Lab. Rep. GMR-3136, General Motors

Cernansky, N. P., C. W. Savery, I. H. Suffet, and R. S. Cohen (1978). Diesel odor sampling and analysis using the diesel odor analysis system (DOAS). SAE Tech. Pap. 780223.

Cernansky, N. P., C. W. Savery, I. H. Suffet, and R. S. Cohen (1980). Physical and chemical phenomena responsible for odor formation in diesel engines. Interim Progress Report. Drexel Univ., Philadelphia, Pa.

Chakraborty, B. B., and R. Long (1968). The formation of soot and polycyclic aromatic hydrocarbons in diffusion flames, part 1. Combust Flame 12:226.

Cichanowicz, J. E., and R. F. Sawyer (1976) Rotary engine combustion with hydrogen addition. SAE Tech. Pap. 760611. Chemical and Engineering News (1979). October 22, p. 71.

Chigier, N. A. (1976). The atomization and burning of liquid sprays. Prog. Energ. Combust. Sci. 2:97.

Cohen, R. (1980). Acoustics and vibration research for automotive vehicles. Report to CARP Planning Staff, April 25.

Colket, M. B., D. W. Naegeli, F. L. Dreyer, and F. Glassman (1974). Flame ionization detection of carbon oxides and hydrocarbon oxygenates. Environ. Sci. Technol. 8:(1):43-46.

Commins, B. T. (1962). Nat. Cancer Inst. Monogr. 9:225.

Compton, W. D. (1980) Automotive challenges for R & D. Mech. Eng. September: 32-41.

Coordinating Research Council (1979). Development and evaluation of a method for measurement of diesel exhaust odor using a diesel odor analysis system (DOAS). CRC Rep. No. 507.

Coordinating Research Council, Inc. (1980). Informational report: 1979 progress of the chemical characterization of the composition of diesel exhaust project and results of particulate extraction round robin. CRC Rep. No. 516, March 1980.

Cornetti, G. M. (1979a). April and May progress report. Report of Centro Ricerche, FIAT S.p.A. to DOT.

Cornetti, G. M. (1979b). Fuel economy study of the FIAT swirl-chamber diesel engine family. Report under Contract DOT-TSC-1424.

Cornetti, G. M. (1979c). June, July and September progress report. Fiat Report to DOT-TSC, DOT-TSC-1424.

Courtney, R. L., and H. K. Newhall (1979). Automotive fuels for the 1980s. SAE Tech. Pap. 790809.

Cronan, C. S., and C. L. Schofield (1979). Aluminum leaching response to acid precipitation: Effects on high-elevation watersheds in the northeast. Science 204:304-306.

Cullis, C. F. (1976). Role of acetylene species in carbon formation. Petroleum Derived Carbons. ACS Symp. Ser. 21:348.

Daimler-Benz, AG (1979). Statement before the U.S. Environmental Protection Agency. Washington, D.C., March.

Daudel, H. L., M. J. Fieber, D. M. Murthy, and H. O. Hardenberg (1979). Detection and instrumental analysis of diesel engine exhaust gas odorants--A new approach to an old problem. SAE Tech. Pap. 790489.

Davis, H. (1957). The hearing mechanism. In Handbook of Noise Control, chapter 4. McGraw-Hill, New York.

Degobert, P. (1980). Diesel exhaust offensive effect: True odor or irritancy. *SAE Tech. Pap. 800423*.

Delfan, J. L., P. Michaud, and A. Barassin (1979). Formation of small and large positive ions in rich and sooting low-pressure ethylene and acetylene premixed flames. *Combust. Sci. Technol.* 20:165.

Department of Energy (1979). *The Impact of Increased Diesel Fuel Use on the Petroleum Industry and on Crude Oil Consumption*.

Department of Energy (1980a). *Electric and Hybrid Vehicle Program Quarterly Report*, May 1980, DOE/CS-0026-10.

Department of Energy (1980b). *Electric and Hybrid Vehicle Program Quarterly Report*, February 1980, DOE/CS-0026-9.

Department of Transportation (1980) and the National Science Foundation. Cooperative automotive research program, technical framework. Draft for Review, September 1980.

Dietzman, H. E., and R. L. Bradow (1975). Protocol to characterize gaseous emissions as a function of fuel-additive composition. Rep. EPA-600/2-75-048.

Dimick, D. L. (1980). Prospects for diesel passenger cars--and the need for an improved fuel. Paper presented to API Automotive and Industry Forum, January 23, 1980.

Dowdy, M., and A. Burke (1979). Automotive technology projections. *SAE Tech. Pap. 790021*.

Dravnieks, A. (1978). Fundamental considerations and methods for measuring air pollution odors. In *Proceedings of the Sixth International Symposium in Olfaction and Taste*, pp. 429-436. Information Retrieval, Ltd., London.

Dravnieks, A., J. Stockham, and A. O'Donnel (1969-1970). Chemical species in engine exhaust and their contribution to exhaust odor. IIT Res. Inst. Rep. No. IITRI C8150-5, June, 1969; IITRI C6183-5, November, 1970.

Drexle, K. (1979). Presentation of Daimler-Benz to Diesel Impacts Study Committee, National Academcy of Sciences, November 15.

Dryer, F. L. (1977). Water addition to practical combustion systems--concepts and applications. *Symp. Int. Combust. Proc. 16th*.

Ebersole, G. D., and F. S. Manning (1972). Engine performance and exhaust emissions: Methanol versus isooctane. *SAE Tech. Pap. 720692*.

EIC, Inc. (1978). *Environment Regulation Handbook*, vol. 3. EIC, Inc., New York.

Eisele, H. (1980). Electronic control of diesel passenger cars. *SAE Tech. Pap. 800167*.

Eldred, K. M. (1977). Community noise. In *Noise and Fluids Engineering*. American Society of Mechanical Engineers, New York.

Energy Research and Development Administration (1975). *The Impact of Automotive Fuel Changes on the U.S. Refining Industry*.

Engine Manufacturers Association Alternate Fuels Committee (1979). Alternate portable fuels for internal combustion engines. *SAE Tech. Pap. 790426*.

Environmental Protection Agency (1972). *Report to the President and Congress on Noise*. U.S. Government Printing Office, Washington, D.C.

Environmental Protection Agency (1975). Compilation of air pollutant emission factors. PHS Rep. No. 999 AP-42, Suppl. 5, Rep. PB-249 526.

Environmental Protection Agency (1977a). 1978 Gas Mileage Guide.

Environmental Protection Agency (1977b). Precautionary notice on laboratory handling of exhaust products from diesel engines. Memorandum from the Acting Assistant Administrator for Research and Development, November 4.

Environmental Protection Agency (1977c). Recommended Practice for Measurement of Exhaust Sulfate Emissions from Light Duty Vehicles and Trucks.

Environmental Protection Agency (1978). OMSAPC Advisory Circ. No. 76, June.

Environmental Protection Agency (1979a). Draft light vehicle noise test procedures and industry comment. Docket #79-02.

Environmental Protection Agency (1979b). 1980 Gas Mileage Guide.

Environmental Protection Agency (1979c). Research summary: acid rain. EPA-600/8-79-028.

Environmental Protection Agency (1979d). Summary and Analysis of Comments on the Notice of Proposed Rulemaking for the Control of Light-Duty Diesel Particulate Emissions From 1981 and Later Model Year Vehicles.

Environmental Protection Agency (1980a). 1980 Gas Mileage Guide, 2nd. ed.

Environmental Protection Agency (1980b). Passenger car fuel economy: EPA and road. Rep. EPA 460/3-80-010.

Environmental Protection Agency (1980c). Regulatory Analysis of the Light-Duty Diesel Particulate Regulations for 1982 and Later Model Year Light-Duty Diesel Vehicles.

Environmental Protection Agency (1980d). Standard for emission of particulate regulations for the 1982 and later model year light-duty diesel vehicles and light trucks. Fed. Regist. 45 (Late February).

Escher, W. J. D., and E. E. Ecklund (1976). Recent progress in the hydrogen engine. SAE Tech. Pap. 760571.

Faulkner, M. G., E. B. Dismukes, J. R. McDonald, D. H. Pontius, and A. H. Dean (1979). Assessment of diesel particulate control: Fillers, scrubbers, and precipitators. EPA Rep. No. 600/7-79-232a.

Federal Register (1972). EPA-new motor vehicles and new motor vehicle engines--Control of air pollution, vol. 37, (221).

Federal Register (1978). Vol. 43:21412.

Federal Register (1979). Vol. 40:(140):42487.

Fenimore, C. P., and G. W. Jones (1968). Comparative yields of soot from premixed hydrocarbon flames. Combust. Flame 12:196.

Fenimore, C. P., and G. W. Jones (1969). Coagulation of soot to smoke in hydrocarbon flames. Combust. Flame 13:303.

Flegl, H., D. Gruden, I. Szodfridt, and R. Wust (1979). The Porsche 995 research car power train--A contribution to reducing the fuel consumption. Paper presented at First International Automotive Fuel Economy Research Conference, Washington, D.C.

Flower, W. L., and T. M. Dyer (1980). Time- and space-resolved measurements of particulate formation during premixed constant volume combustion. Pap. No. CSS/CI 80-01, presented at Technical Meeting, Central States Section, The Combustion Institute, Baton Rouge, La.

Ford, H. S., Jr., D. F. Merrion, and R. J. Hames (1970). Reducing hydrocarbons and odor in diesel exhaust by fuel injector design. SAE Tech. Pap. 700734.

Ford Motor Company (1979a). Ford diesel emissions review. Presentation to National Academy of Sciences, Diesel Impacts Study Committee, January.

Ford Motor Company (1979b). Ford nonregulated emissions program status report. Report to EPA, September 28.

Forrest, L., W. B. Lee, W. M. Smalley, and J. C. Sturm (1979). Impacts of light duty vehicle dieselization on urban air quality. APCA Pap. No. 79-7.6.

Forrest, L., W. B. Lee, and W. M. Smalley (1980). Assessment of environmental impacts of light-duty vehicle dieselization. Rep. DOT-TSC-NHTSA-80-5. Aerosp. Corp.

Frakes, R. G. (1979). General Motors, Fuel Economy Activities. Letter to Dr. H. Rowen, September 13.

French, C. C. J., and D. A. Pike (1979). Diesel engine, light duty vehicles for an emission controlled environment. SAE Tech. Pap. 790761.

Friedlander, S. K. (1977). Smoke, Dust and Haze, p. 175. John Wiley and Sons, New York

Frisch, L. E. (1978). The effect of exhaust dilution and fuel properties on the character of diesel particulate emissions. M.S. Thesis, Michigan Technical University.

Frisch, L. E., J. H. Johnson, D. G. Leddy (1978). Effect of fuels in dilution ratio on diesel particulate emissions. In Measurement in Control of Diesel Particulate Emissions. SAE Progress in Technology 17.

Friswell, N. J. (1979). The influence of fuel composition on smoke emissions from gas-turbine-type combustors: Effect of combustor design and operating conditions. Combust. Sci. Technol. 19:119.

Fujii, N., and T. Asaba (1973). Shock tube study of the reaction of rich mixtures of benzene and oxygen. Symp. Int. Combust Proc. 14th, p. 443.

Funkenbusch, E. F., D. G. Leddy, and J. H. Johnson (1979). The characterization of the soluble organic fraction of diesel particulate matter. SAE Tech. Pap. 790418.

Furuhama, S. and H. Azuma (1979). Hydrogen injection: 2-stroke spark ignition engine. The Hydrogen Energy Systems, T. N. Veziroglu, ed. Pergamon Press, New York.

Gabele, P. A., J. N. Braddock, and R. L. Bradow (1977a). A characterization of exhaust emisions from lean burn, rotary, and stratified charge engines. SAE Tech. Pap. 770301.

Gabele, P. A., J. N. Braddock, F. M. Black, F. D. Stump, and R. B. Zweidinger (1977b). A characterization of exhaust emissions from a dual catalyst vehicle. Rep. EPA-600/2-77-068.

General Motors (1979a). Report to NAS-DISC.

General Motors (1979b). _Response to EPA Notice of Proposed Rulemaking on Particulate Regulation for Light-Duty Diesel Vehicles_. Submitted to EPA April 19.

General Motors (1979c). _Application for Waiver of the 1981-1984 NO_x Emission Standard for Light-Duty Diesel Engines_. Submitted to EPA May 1.

General Motors (1980). Report to NAS-DISC, January.

General Motors Environmental Research Staff (1978). _1977 Annual Report of General Motors on Advanced Emission Control System Development Progress_, vol. II. Submitted to EPA January 16.

Gibbs, R. E., G. P. Wotzak, B. J. Hill, T. M. Hussey, R. E. Johnson, M. A. Moore, P. L. Werner, and S. M. Byer (1978). A study of fifty-six in-use catalyst vehicles: Emissions and fuel economy. _SAE Tech. Pap. 780645_.

Givens, L. (1980). Engineering highlights of the 1981 automobiles. _Automot. Eng._ 88(10):51-74.

Glassman, I. (1980). Phenomenological models of soot processes in combustion systems. AFOSR TR 79-1147.

Gliken, P. E., D. F. Mowbray, and P. Howes (1979). Some developments on fuel injection equipment for diesel powered cars. _Inst. Mech. Eng._ pp. 165-182.

Goff, E. V., J. R. Coombs, D. H. Fine, and T. M. Baines (1980). Nitrosamine emissions from diesel engine crankcases. _SAE Tech. Pap._ 801374.

Gordon, A. S., S. R. Smith, and J. R. McNesby (1959). Study of the chemistry of diffusion flames. _Symp. Int. Combust.Proc. 7th_, p. 317.

Graham, S. C. (1976). The collisional growth of soot particles at high temperatures. _Symp. Int. Combust. Proc. 16th_, p. 663.

Green, G. L., and D. Wallace (1980). Correlation studies of an in-line, full flow opacimeter. _SAE Pap. No. 801373_.

Greeves, G., C. H. T. Wang, and G. A. Kyriazis (1980). Inlet port design and fuel injection rate requirements for direct injection diesel engines. FISITA 80.4.1.4. _VDI Ber._ Nr. 370.

Groblieki, D. J., and C. R. Begeman (1979). Particle size variation in diesel car exhaust. _SAE Pap. No. 790421_.

Grosshandler, W. L. (1980). Washington State University. Personal communication with R. D. Matthews.

Grover, E. C., and R. D. H. Perry (1978). An experimental passenger car diesel engine. _SAE Tech. Pap. 790443_.

Gruden, D. (1972). Exhaust emission and exhaust odor from four-stroke spark ignition engines. _Autmobiltech. Z._ 74:180-188. (In German.)

Hames, R. J., R. J. Slone, J. M. Perez, and J. H. Johnson (1980). Cooperative evaluation of techniques for measuring diesel exhaust odor using the diesel odor analysis system (DOAS). _SAE Tech. Pap._ 800422.

Hardenberg, H. (1972). Investigations into the formation and reduction of fuel smoke and exhaust gas odor. _Motortech. Z._ 33(6):248-254.

Hare, C. T. (1979). Characterization of gaseous and particulate
    emissions from light-duty diesels operated on various fuels. Rep.
    No. EPA-460/3-79-008.   Southwest Res. Inst., San Antonio, Tex.
Hare, C. T., and T. M. Baines (1977). Characterization of diesel
    crankcase emissions. SAE Tech. Pap. 770719.
Hare, C. T., and T. M. Baines (1979). Characterization of particulate
    and gaseous emissions from two diesel automobiles as functions of
    fuel and driving cycle. SAE Tech. Pap. 790424.
Hare, C. T., and R. B. Bradow (1979). Characterization of heavy-duty
    diesel gaseous and particulate emissions and effects of fuel
    composition. The measure and control of diesel particulate
    emissions. SAE Publication PT-79/17.
Hare, C. T., and K. J. Springer (1971). Public response to diesel
    engine exhaust odors. Final Report to EPA, Contract No. CPA-70-44.
Hare, C. T., K. J. Springer, J. H. Somers, and T. A. Huls (1974).
    Public opinion of diesel odor. SAE Tech. Pap. 740214.
Harris, C. M. (Ed.) (1979). Handbook of Noise Control, 2nd ed.
    McGraw-Hill, New York.
Haskell, W. W., and G. E. Legate (1972). Exhaust hydrocarbon emissions
    from gasoline engines--Surface phenomena. SAE Tech. Pap. 720255.
Helms, H. E. (1980). Advanced gas turbine powertrain system
    development. NASA Contract DEN 3-168. In Fifth International
    Symposium on Automotive Propulsion Systems. Conf-800419, vol. 1.
Hergenrother, K. M. (1979). Comparison of fuel economy and emissions
    for diesel and gasoline powered taxicabs. Rep. No.
    UMTA-MA-06-0066-79-1. U.S. Dep. of Transp. NTIS No. PB 298609.
Herschfeld, F. (1978). Problems and progress in developing the
    automotive gas turbine. Mech. Eng. May:40-50.
Hikosaka, N., and K. Takeuchi (1980). Isuzu New 4FB1 1.8L High Speed
    Diesel Engine. Fifth International Symposium on Automotive
    Propulsion Systems. Manuscript.
Hilliard, J. C., and R. W. Wheeler (1977). Catalyzed oxidation of
    nitric oxide to nitrogen dioxide. Combust. Flame 29:15.
Hilliard, J. C., and R. W. Wheeler (1979). Nitrogen dioxide in engine
    exhaust. SAE Tech. Pap. 790691.
Hills, F. J., and C. G. Schleyerbach (1977). Diesel fuel properties
    and engine performance. SAE Tech. Pap. 770316.
Hiroyasu, H., M. Arai, and K. Nakanishi (1980). Soot formation and
    oxidation in diesel engines. SAE Tech. Pap. 800252.
Hoehn, F. W., and M. W. Dowdy (1974). Feasibility demonstration of a
    roade vehicle with hydrogen-enriched fuel. In Proceedings of the
    9th Intersociety Energy Conversion Conference, San Francisco,
    California, August 26-30.
Hoffman, D., and E. L. Wynder (1968) Chemical analysis of carcinogenic
    bioassays of organic particulate pollutants. In Air Pollution,
    vol. II, 2nd ed., pp. 187-247. Academic Press New York.
Hoffman, K. C., J. J. Reilly, F. J. Salzano, C. H. Waide, R. H.
    Wiswall, and W. E. Winsche (1976). Metal hydride storage for
    mobile and stationary applications. SAE Tech. Pap. 760569.
Holtzberg, M. (1979). The plastic engine. Formula (February):34-36.
Homann, K. H. (1967). Carbon formation in premixed flames. Combust.
    Flame 11:265.

Homann, K. H., and H. G. Wagner (1967). Some new aspects of the mechanism of carbon formation in premixed flames. _Symp. Int. Combust. Proc. 11th._

Hord, J. (1976). Is hydrogen safe? _Nat. Bur. Stand. U.S. Tech. Note 690._

Hornig, D. F. (1968). _Noise--Sound Without Value._ Committee on Environmental Quality, Federal Council for Science and Technology. U.S. Government Printing Office, Washington, D.C.

Horwitz, J. A. (1980). Personal communication with R. D. Matthews.

Houseman, J. and D. J. Cerini (1974). On-board hydrogen generator for a partial hydrogen injection internal combustion engine. _SAE Tech. Pap. 740600._

Houseman, J., and F. W. Hoehn (1974). A two-charge engine concept: Hydrogen enrichment. _SAE Pap. 741169_, presented at SAE International Stratified Charge Engine Conference, Troy, Michigan, October 30-November 1,

Howard, J. B., B. L. Wersborg, and G. C. Williams (1973). Coagulation of carbon particles in premixed flames. _Faraday Symp. Chem. Soc._ 7:109.

Howes, P. (1980). The new CAV microinjecter. _SAE Tech. Pap. 800509._

Hsieh, F. T., N. P. Cernansky, and C. W. Savery (1980). Fuel effects on diesel odor in a spray burner. _SAE Tech. Pap. 800425._

Hsieh, F. T., R. S. Cohen, N. P. Cernansky, and C. W. Savery (1979a). Diesel odor modification in a spray burner by reactant dilution. Pap. No. CSS/CI 79-11, presented at the 1979 Central States Section/Combustion Institute Meeting, Columbus, Ind.

Hsieh, F. T., M. N. Savliwala, R. S. Cohen, C. W. Savery, and N. P. Cernansky (1979b). Diesel odor studies in a laboratory spray burner. _Symp. Int. Combust. Proc. 17th_, pp. 485-499.

Huebner, G. J., Jr. (1979). Energy and pollution vs alternative piston and gas turbine powerplants. _SAE Tech. Pap. 790020._

Huisingh, J., R. Bradow, R. Jungers, L. Claxton, R. Zweidinger, S. Teljada, J. Bungarner, F. Duffield, M. Waters, V. F. Simmon, C. Hare, C Rodriguez, and L. Snow. Application of a mutagenicity bioassay to monitoring light-duty diesel particle emissions. Paper presented at Symposium on Application of Short-term Bioassays in the Fractionazation and Analysis of Complex Environmental Mixtures, Williamsburg, Va., February 1978.

Ingaus, A. M. (1926). Aiding digestion in gasoline motors. _Sci. Am._ 134:42.

Ingram, C. (1977). Diesel exhaust odor. Paper presented at CRC-APRAC Diesel Exhaust Measurement Symposium, Chicago, April.

Ingram, C. (1978). Diesel exhaust odour of small, high speed, direct injection engines. _SAE Tech. Pap. 780114._

Isuzu Corp. (1980). Presentation to Technology Panel, Diesel Impacts Study Committee, National Academy of Sciences.

Jackson, M. W. (1978). Effects of catalytic emission control on exhaust hydrocarbon composition and reactivity. _SAE Tech. Pap. 780624._

Jambekar, A. B., and J. H. Johnson (1978). An emission and fuel usage computer model for trucks and buses. _SAE Tech. Pap. 780630._

Jambekar, A. B., and J. H. Johnson (1980). The effect of truck dieselization on the population, mileage, and fuel useage. Mich. Technol. Univ. Final Report to EPA on P.O. A-0079-NAEX, September 15, 1980.

Jet Propulsion Laboratory, California Institute of Technology (1975). Communication to Robert Sawyer, Technology Panel, Diesel Impacts Study Committee, National Academy of Sciences.

Johnson-Matthey, Ltd. (1980). Presentation to Diesel Impacts Study Committee, National Academy of Sciences.

Johnson, G. L., and R. C. Anderson (1961). An electron microscope study of carbon formation in the pyrolysis of hydrocarbons. In Proceedings of the Fifth Carbon Conference, p. 395. New York, Pergamon Press.

Johnson, L. R., R. J. Kevala, J. C. Tomlinson, M. R. Monarch, and M. J. Bernard III (1979). Environmental Development Plan for Light Duty Diesel Vehicles. Argonne National Laboratory, U.S. Department of Energy.

Johnston, L. E. (1980). Advanced Fuel Systems. General Motors Engineering Staff. Manuscript.

Jones, J. H., W. L. Kingsbury, H. H. Lyon, P. R. Mutty, and K. W. Thurston (1978). Development of a 5.7-litre V8 automotive diesel engine. SAE Tech. Pap. 780412.

Kadota, T., N. A. Henein, and D. U. Lee (1980). Effect of fuel properties on the time resolved soot particulates in a simulated diesel spray. In Fifth International Symposium on Automotive Propulsion Systems. Conf-800419, vol. 1.

Keirns, M. H., and E. L. Holt (1978). Hydrogen cyanide emissions from three-way catalyst prototypes under malfunctioning conditions. SAE Tech. Pap. 780201.

Kendall, D. A., P. L. Levins, and G. Leonardos (1974). Diesel exhaust odor analysis by sensory techniques. SAE Tech. Pap. 740215.

Kerrebrock, J. L., and C. E. Kolb (1978). An assessment of the potential impact of combustion research on internal combustion engine emissions and fuel consumption. Aerodyne Res. Rep. ARI-RR-131.

Khan, I. M., and C. H. T. Wang (1971). Factors affecting emissions of smoke and gaseous pollutants from direct injection diesel engines. In Air Pollution Control in Transport Engine, p. 293. Institution of Mechanical Engineers, London.

Khan, I. M., S. Greeves, and D. M. Probert (1971). Prediction of soot and nitric oxide concentrations in diesel engine exhaust. In Air Pollution Control in Transport Engine, p. 205. Institution of Mechanical Engineers, London.

Khan, I. M., C. H. T. Wang, and B. E. Langride (1971b). Coagulation and combustion of soot particles in diesel engines. Combust. Flame 17: 409.

Khan, M. S., and P. J. Langdon (1979). The role of the computer in turbocharger design development and testing. SAE Tech. Pap. 790278.

Khatri, N. J., J. H. Johnson, and D. G. Leddy (1978). The characterization of the hydrocarbon and sulfate fractions of diesel particulate matter. SAE Tech. Pap. 780111.

Killough, P., J. Watson, and R. Bradow (1979). Filter type, high volume particulate sampler for automotive diesel emission studies. Draft Paper Based on EPA Contract No. 68-02-2566.

Kimberley, J. A., and R. A. DiDomenico (1977). UFIS--A new diesel injection system. SAE Tech. Pap. 770084.

Kirk, R. S., and P. W. Davis (1980). A view of the future potential of electric and hybrid vehicles. Paper presented at the 7th Energy Technology Conference, Washington, D.C., March 24-26.

Kittelson, D. B., and D. F. Dolan (1979). Diesel exhaust aerosols. Paper presented to the Division of Environmental Chemistry, American Chemical Society, Honolulu, Hawaii, April.

Kryter, K. O. (1970). The Effects of Noise on Man. Academic Press, New York.

Lasheras, J. C., I. M. Kennedy, and F. L. Dryer (1980). An initial study of the effects of alcohol and water addition on the burning of isolated distillate fuel droplets. Paper presented at Technical Meeting, Western States Section, The Combustion Institute, Irvine, Calif.

Lassiter, D. V., and T. H. Milby (1979). An initial assessment of the literature on the measurement, control, transport, transformation and health effects of unregulated diesel engine emissions; part D, Carcinogenic Aspects. Final Report from South Coast Technology Inc., to DOT/NHSTA. Contract No. DOT-HS-7-01790. Rep. No. DOT HS-804010.

Lave, L. B. (1980). Conflicting Objectives in Regulating the Automobile. April.

Lawrence, D. K., D. A. Plautz, B. D. Keller, and T. O. Wagner (1980). Automotive fuels--refinery energy and economics. SAE Tech. Pap. 800225.

Lawson, A., and A. J. Last (1979a). Development of an on-board mechanical fuel emulsifier for utilization of diesel/methanol and methanol/gasoline fuel emulsions in transportation. Paper presented at Alcohol Fuels Technology, Third International Symposium, Asilomar, Calif., May.

Lawson, A., and A. J. Last (1979b). Modified fuels for diesel engines by application of unstabilized emulsions. SAE Tech. Pap. 790925.

Lee, K. B., M. W. Thring, and J. M. Beer (1962). On the rate of combustion of soot in a laminar flame. Combust. Flame 6:137.

Lesley, S., and C. C. J. French (1976). Diesel exhaust odor. SAE Tech. Pap. 760554.

Levins, P. L. (1969-1971). Chemical identification of the odor components in diesel engine exhaust. CRC Proj. Rep. CAPE 7-68. NTIS PB 194061 (July 1969); PB 194144 (June 1970); and PB 205273 (June 1971). Arthur D. Little.

Levins, P. L. (1973). Analysis of the odorous compounds in diesel exhaust. CRC and EPA Contract 68-02-0087, June 1972. NTIS PB 228756/AS, September 1973.

Levins, P. L. (1979). Review of diesel odor and toxic vapor emissions. Report under Contract DOT-TSC-1242. Arthur D. Little.

Levins, P. L., D. A. Kendall, A. B. Caragay, G. Leonardos, and J. E.
Oberholtzer (1974). Chemical analysis of diesel exhaust odor
species. SAE Tech. Pap. 740216.

Lewis, J. M., and W. T. Tierney (1980). United Parcel Service applies
Texaco stratified charge engine technology to power parcel delivery
vans--Progress report. SAE Tech. Pap. 801429.

Lindsley, E. F. (1980). Traction drive transmission. Pop. Sci.
(March):83-86.

Lindvall, T., and S. E. Mortstedt (1970). Investigation of the odor
intensity of internal combustion engine exhaust, 13 pp. Rep.
BIL-52. AB Atomenergi, Stockholm. (In Swedish.)

Lipkea, W. H., J. H. Johnson, and C. T. Vuk (1978). The physical and
chemical character of diesel particulate emissions--Measurement
techniques and fundamental considerations. SAE Tech. Pap. 780108.

Lipscomb, D. M., and A. C. Taylor (Eds) (1978). Noise Control:
Handbook of Principles and Practices. Van Nostrand Reinhold, New
York.

Lucas, CAV, Ltd. (1980). Presentation to Diesel Impacts Study
Committee, Technology Panel, National Academy of Sciences.

Lynch, F. (1980). Turbocharging the Hydrogen-Fueled Internal
Combustion Engine. 3rd World Hydrogen Energy Conference Report.
Manuscript.

MacDonald, J. S. (1976). Evaluation of the hydrogen-supplemented fuel
concept with an experimental multicylinder engine. SAE Tech. Pap.
760101.

MacLeod, P. (1972). Procede et dispositif d'olfactometrie
differentielle. French Patent 2.210-298.

Magnussen, B. F. (1970). The rate of combustion of soot in turbulent
flows. In Symp. Int. Combust. Proc. 13th, p. 869.

Maine, State Department of Transportation, Materials and Research
Division (1979). Energy usage and other comparisons between diesel
and gasoline pickup trucks. Rep. No. DOT-TSC-OST-77.

Mantell, C. H. (1968). Carbon and Graphite Handbook. Interscience,
New York.

Marshall, A. E. (1980). The automobile as a component of community
noise. APCA Pap. No. 80-47.5.

Marshall, W. G., and D. E. Seizinger (1978). Diesel odor measurement:
Comparison of two methods. Rep. No. BETC/TPR-78/5. Bartlesville
Energy Technol. Center, U.S. Energy Res. and Devel. Admin.

Matthews, R. D. (1980a). Toxicology and atmospheric chemistry of the
nitrogenous products of combustion: The oxides of nitrogen. J.
Combust. Toxicol. 7:99-118.

Matthews, R. D. (1980b). Toxicology and atmospheric chemistry of the
nitrogenous products of combustion: The amines. J. Combust.
Toxicol. 7:23-41.

Matthews, R. D. (1980c). Estimated permissible levels, ambient
concentrations, and adverse effects of the nitrogenous products of
combustion: The cyanides, nitro-olefins, and nitroparaffins. J.
Combust. Toxicol. 7:157-192.

Matthews, R. D., and R. F. Sawyer (1979). Combustion sources of
unregulated gas phase nitrogenous species. Rep. LBL-10280, UC-11.
Lawrence Berkeley Lab.

Mayer, W. J., D. C. Lechman and D. L. Hilden (1980). The contribution of engine oil to diesel exhaust particulate emissions. <u>SAE Tech. Pap.</u> 800256.

McCann, J., N. E. Spingarn, J. R. Kobi, and B. N. Ames (1975). Detection of carcinogens as mutagens: Bacterial tester strains with R factors plasmids. <u>Proc. Natl. Acad. Sci. U.S.A.</u> 72:979-83.

McKee, D. E., F. C. Ferris, and R. E. Goeboro (1978). Unregulated emissions from a PROCO engine powered vehicle. <u>SAE Tech. Pap.</u> 780592.

Middlemiss, I. D. (1978). Characteristics of the Perkins 'squish lip' direct injection combustion system. <u>SAE Tech. Pap. 780113.</u>

Miller, H. (1980). DOT/TSC. Personal communication with R. D. Matthews.

Milliken, R. C. (1962). Non-equilibrium soot formation in premixed flames. <u>J. Phys. Chem.</u> 66:784.

Misch, H. L. (1979). Memo of Comments at a Public Hearing before EPA Standards Development and Support Branch, March 19-20.

Miwa, K., M. Ikegami, and R. Nakano (1980). Combustion and pollutant formation in an indirect injection diesel engine. <u>SAE Tech. Pap.</u> 800026.

Mobil Oil Corporation (1980). Communication to David Christison, Consultant to Technology Panel, Diesel Impacts Study Committee, National Academy of Sciences.

Monaghan, M. L. (1979). Two ways to boost a light duty diesel. <u>SAE Tech. Pap. 790038</u>.

Monaghan, M. L., C. C. J. French, and R. G. Freese (1974). A study of the diesel as a light-duty power plant. Rep. EPA-460/3-74-011.

Morrison, D., and B. J. Challen (1979). A survey of passenger car noise levels. <u>SAE Tech. Pap. 790442</u>.

Morrison, D., B. J. Challen, and T. Trella (1980). Passenger car noise control measures and their effects on fuel economy, weight and cost. <u>SAE Tech. Pap. 800439</u>.

Motoyoshi, E., T. Yamada, and M. Mori (1976). The combustion and exhaust emission characteristics and starting ability of Y.P.C. combustion system. <u>SAE Tech. Pap. 760215</u>.

Moulin, M., M. Regneault, M. Perez, M. LeCreurer, and M. LeBorne (1979). The supercharged diesel engine of the Peugeot 604. <u>SAE Tech. Pap. 790762</u>.

Moulton, D. G. (1980). Physiologic and morphologic effects of odorants. APCA Pap. No. 80-23.1.

Murphy, M. J., L. J. Hillenbrand, and D. A. Trayser (1979). Assessment of diesel particulate control: Direct and catalytic oxidation. Rep. 600/7-79-2326.

Murray, R. G. (1980). Performance and emission characteristics of a semi-adiabatic engine. <u>ASME Tech. Pap. 80-DGP-44</u>.

Murrell, J. D., J. A. Foster, and D. M. Bristor (1980). Passenger car and light truck fuel economy trends through 1980. <u>SAE Tech. Pap. 800853</u>.

Naegeli, D. W., and C. A. Moses (1980). Effect of fuel molecular structure on soot formation in gas turbine engines. <u>ASME Pap. No. 80-GT-62</u>. Am. Soc. of Mech. Eng., New York.

Nagle, J., and R. F. Strickland-Constable (1962). Oxidation of carbon between 1000-2000 C. In Proceedings of the Fifth Carbon Conference, vol. 1, p. 154. New York: Pergamon Press.

National Bureau of Standards (1977). Noise emission measurements for regulatory purposes. In NBS Handbook 122.

National Research Council, Board on Toxicology and Environmental Health Hazards (1979). Odors from Stationary and Mobile Sources. National Academy of Sciences, Washington, D.C.

National Research Council (1980). Refining synthetic liquids from coal and shale. Report of the Panel on R & D Needs in Refining of Coal and Shale Liquids. National Academy of Sciences, Washington, D.C.

Nightingale, D. R. (1975). A fundamental investigation into the problem of NO formation in diesel engines. SAE Tech. Pap. 750848.

Nissan Motor Company (1979). Presentation to DISC Technology Panel, November 15-16, 1979.

Nissan Motor Company (1980). Report to the Diesel Impacts Study Committee. Presentation to DISC Panel, June 19.

Noguchi, M., S. Sanda, and N. Nakamura (1976). Development of the Toyota lean burn engine. SAE Tech. Pap. 760757.

Norrish, R. G., and G. W. Taylor (1956). The oxidation of benzene. Proc. R. Soc. London, Ser. A 234:160.

Obert, E. F. (1973). Internal Combustion Engines and Air Pollution, 3rd ed. Harper and Row, New York.

Oblander, K., M. Fortnagel, H. J. Feucht, and U. Conrad (1978). The turbocharged five-cylinder diesel engine for the Mercedes-Benz 300SD. SAE Pap. 780633.

O'Donnell, A., and A. Dravnieks (1970). Chemical species in engine exhaust and their contributions to exhaust odors. Final Report to Coordinating Research Council, Inc., and National Air Pollution Control Administration, PHS, HEW. IIT Res. Inst., Chicago.

Oetting, H. (1979). Impact of emission standards on fuel economy and consumer attributes. SAE Tech. Pap. 790230.

Oetting, H., and S. Papez (1979). Reducing diesel knock by means of exhaust gas recirculation. SAE Tech. Pap. 790268.

Otto, K., M. H. Sieg, M. Zimbo, and L. Bartosiewicz (1980). The oxidation of soot deposits from diesel engines. SAE Tech. Pap. 800336.

Pachernegg, S. J. (1975). Efficient and clean diesel combustion. SAE Tech. Pap. 750787.

Parker, R. F. (1976). Future fuel injection system requirements of diesel engines for mobile power. SAE Tech. Pap. 760125.

Pefley, R. K. (1979). An alcohol fuel alternative. Mech. Eng. November:52-53,

Perez, J. M. (1979). Measurement of unregulated emissions--Some heavy-duty diesel engine results. Caterpillar Tractor Company. Paper presented at the International Symposium on Health Effects of Diesel Engine Emissions, Cincinnati, Ohio.

Petroleum Industry Research Foundation, Inc. (1980). Oil in the U.S. Energy Perspective--A Forecast to 1990.

Petrow, E. D., and N. P. Cernansky (1979). A comparison of the odorous emissions from a direct injection and an indirect injection diesel engine. Pap. No. CSS/CI 79-10, presented at 1979 Central States Section/Combustion Institute Meeting, Columbus, Ind.

Petrow, E. D., M. N. Savliwala, F. T. Hsieh, N. P. Cernansky, and R. S. Cohen (1978). An investigation of diesel odor in an air aspirated spray burner and a CFR diesel engine. SAE Tech. Pap. 780632.

Peugeot (1979). Technology of diesel engines--Fuels and emission controls. Presentation to DISC Technology Panel, November 15-16.

Pitts, J. N. Jr., K. A. Van Cauwenberghe, and A. M. Winer (1979). An initial assessment of the literature on the measurement, control, transport, transformation and health effects of unregulated diesel engine emissions, part C, Atmospheric Transport, Transformation and Microbiological Assay. Final Report from South Coast Technology, Inc., to DOT/NHSTA. Contract No. DOT-HS-7-01790, Rep. No. DOT HS-804010.

Plee, S. L., and J. S. McDonald (1980). Some mechanisms affecting the mass of diesel exhaust particulates collected following a dilution process. SAE Tech. Pap. 800186.

Priede, T. (1975). The problems of noise of engines in different vehicle groups. SAE Tech. Pap. 750795.

Priede, T. (1979). Problems and developments in automotive engine noise research. SAE Tech. Pap. 790205.

Priede, T. (1980). In search of origins of engine noise--A historical review. SAE Tech. Pap. 800534.

Pupp, C., R. Lao, J. Murray, and R. Pottie (1974). Equilibrium vapor concentrations of some polycyclic aromatic hydrocarbons, As 40^6, and SeO^2 and the collection efficiencies of these air pollutants. Atmos. Environ. 8:915.

Rackley, R. A. (1980). Advanced gas turbine powertrain system development project. NASA Contract DEN 3-167. In Fifth International Symposium on Automotive Propulsion Systems. Conf-800419, vol. 1.

Ray, S. K., and R. Long (1964). Polycyclic aromatic hydrocarbons from diffusion flames and diesel combustion. Combust. Flame 8:139.

Reading, A. R., and G. Greeves (1980). Measurement of diesel exhaust odorants and effect of engine variables. SAE Tech. Pap. 800424.

Renner, R., and H. M. Siegel (1979). Fuel economy of alternative automotive engines--Learning curves and projections. SAE Tech. Pap. 790022.

Ricardo & Co., Ltd. (1980). Presentation to Diesel Impacts Study Committee, National Academy of Sciences.

Rife, J. M. (1980). Massachusetts Institute of Technology, Sloan Laboratory, Private communication.

Robert Bosch (1980). Presentation to Diesel Impacts Study Committee, Cambridge. National Academy of Sciences, Washington, D.C.

Roessler, W. U. (1979). Diesel engine research and development status. Presented to NAS DISC, November.

Roessler, W. U., J. E. Clark, R. R. Covey, and G. J. Mascetti (1980). Light-Duty Diesel Engine Development Status and Research Needs. DOE Contract No. DE-AC03-78C552184. The Aerospace Corporation, El Segundo, Calif.

Roessler, W. U., _et al_. (1978). Diesel engine research and development status and Needs. Prepared by the Aerospace Corporation for DOE. Rep. No. SAN-2184-1.

Rounds, F. G., and H. W. Pearsall (1957). Diesel exhaust odor--Its evaluation and relation to exhaust gas composition. _SAE Trans._ 65:608-627.

Rupe, J. (1973). Hydrogen enrichment of hydrogen fuels as a means for obtaining low emissions and high thermal efficiency in a spark ignition engine. Paper presented at the Combustion Institute Western States Section Fall Meeting, October.

Russell, M. F. (1973). Automotive diesel engine noise and its control. _SAE Tech. Pap. 730243_.

Russell, M. F. (1977). Recent CAV research into noise, emissions, and fuel economy of diesel engines. _SAE Tech. Pap. 770257_.

Russell, M. F. (1979). Automotive diesel engine noise analysis, diagnosis and control. _Lucas Eng. Rev._ 7(4):82-103.

Sapienza, R., T. Butcher, C. Krishna, and J. Gaffney (1980). Soot reduction in diesel engines by catalytic effects. In _Proceedings: Fourth Workshop on Catalytic Combustion_, pp. 365-383, EPA-600/9-80-035.

Savitz, M. (1980). Exerpts from testimony before the House Subcommittee on Energy Research and Production. _Electric and Hybrid Vehicle Quarterly Report_, February 1980, DOE/CS-0026-9.

Schalla, R. L., and G. E. MacDonald (1955). Mechanism of smoke formation in diffusion flames. _Symp. Int. Combust. Proc. 5th_, p. 316.

Science News and Journal of Physical Chemistry (1980). Benzene. 118 (July 12):2.

Seizinger, D. E. (1975). Oxygenates in exhaust gases subjected to catalytic or thermal conversion. Rep. BERC/RI-75/1. Bartlesville Energy Res. Center, U.S. Energy Res. and Devel. Admin.

Seizinger, D. E., and B. Dimitriades (1972). Oxygenates in exhaust from simple hydrocarbon fuels. _J. Air Pollut. Control Assoc._ 22(1):47.

Seizinger, D. E., and B. Dimitriades (1973). Oxygenates in automotive exhausts: Effect of an oxidation catalyst. Rep. RI 7837. Bartlesville Energy Res. Center, U.S. Bureau of Mines.

Shahed, S. M., P. F. Flynn, and W. T. Lyn (1978). A model for the formation of emissions in a direct injection diesel engine. GM Symposium on Combustion Modeling in Reciprocating Engines.

Shiomoto, G. H., R. F. Sawyer, and B. D. Kelly (1978). Characterization of the lean misfire limit. _SAE Tech. Pap. 780235_.

Sigsby, J. (1980). Private telephone communication. EPA/RTP, July 2, 1980.

Simko, A., M. A. Choma, and L. L. Repko (1972). Exhaust emission control by the Ford Programmed Combustion Process--PROCO. _SAE Tech. Pap. 720052_.

Smaby, S. A., and J. H. Johnson (1979). An initial assessment of the literature on the measurement, control, transport, transformation and health effects of unregulated diesel engine emissions; part B, <u>Control Technology</u>. Final Report from South Coast Technology, Inc., to DOT/NHSTA. Contract No. DOT-HS-7-01790. Rep. No. DOT HS-804010.

Smaby, S. A., D. G. Leddy, and J. H. Johnson (1979). An initial assessment of the literature on the measurement, control, transport, transformation and health effects of unregulated diesel engine emissions; part A, <u>Measurement and Characterization of Emissions</u>. Final Report from South Coast Technology, Inc., to DOT/NHTSA. Contract No. DOT-HS-7-01790. Rep. No. DOT HS-804010.

Smith, L., M. A. Parness, E. R. Fanick, and H. E. Dietzman (1980). Analytical procedures for characterizing unregulated emissions from vehicles using middle-distillate fuels. EPA Contract No. 68-02-2497. EPA Rep. No. 600/2-80-068.

Society of Automotive Engineers (1975). Diesel engine noise. <u>Spec. Publ. SP-397</u>.

Society of Automotive Engineers (1979). <u>Proceedings of the Diesel Engine Noise Conference</u>, P-80.

Society of Automotive Engineers (1980). Standards and recommended practices for measurement of sound levels. In <u>SAE Handbook</u> J986b, J1030, J1169, J1074.

Sovran, G. (1980). GMRL. Personal communication with R. D. Matthews.

Springer, K. J. (1967-1969). <u>An Investigation of Diesel Powered Vehicle Odor and Smoke</u>, Part I (September 1967), Part II (February 1968), Part III (October 1969). Southwest Res. Inst., San Antonio, Tex.

Springer, K. J. (1974a). Combustion Odors--A Case Study. In <u>Human Responses to Environmental Odors</u>, pp. 227-262. Academic Press, New York.

Springer, K. J. (1974b). Emissions from diesel and stratified charge powered cars. EPA Rep. No. 460/3-75-001A. Southwest Res. Inst., San Antonio, Tex.

Springer, K. J. (1977a). Investigation of diesel-powered vehicle emissions, part VII. EPA Rep. No. 460/3-76-034. Southwest Res. Inst., San Antonio, Tex.

Springer, K. J. (1977b). The diesel odor by trained panel. Paper presented at CRC-APAC Diesel Exhaust Emission Measurement Symposium, Chicago, Ill. Southwest Res. Inst., San Antonio, Tex.

Springer, K. J. (1979). Characterization of sulfates, odor, smoke, POM and particulates from light- and heavy-duty engines, part IX. EPA Rep. No. 460/3-79-007. Southwest Res. Inst., San Antonio, Tex.

Springer, K. J., and T. M. Baines (1977). Emissions from diesel versions of production passenger cars. <u>SAE Tech. Pap. 770818</u>.

Springer, K. J., and C. T. Hare (1970). A field survey to determine public opinion of diesel engine exhaust odor. Final Report. HEW. Contract PH 22-68 36. Southwest Res. Inst., San Antonio, Tex.

Springer, K. J., and R. C. Stahman (1975). Emissions and economy of four diesel cars. <u>SAE Tech. Pap. 750332</u>.

Springer, K. J., and R. C. Stahman (1977). Unregulated emissions from diesels used in trucks and buses. <u>SAE Tech. Pap. 770258</u>.

Standard Oil Company of America (1980). Communication to David Christison, Consultant to Technology Panel, Diesel Impacts Study Committee, National Academy of Sciences.

Stebar, R. F., and F. B. Parks (1974). Emission control with lean operation using hydrogen supplemented fuel. *SAE Tech. Pap. 740187*.

Stein, L. E. (1978). On the high temperature chemical equilibria of polycyclic aromatic hydrocarbons. *J. Phys. Chem.* 82:566.

Stone, R. J., R. W. Scheffel, S. M. Shahed, and B. Petersen (1980). Nitrosamine emissions from diesels. *SAE Tech. Pap. 801375*.

Strait, J., J. J. Boedicker, and K. C. Johansen (1979). Diesel oil and ethanol mixtures for diesel-powered farm tractors. *SAE Tech. Pap. 790958*.

Stuart, B. (1980). Letter to R. F. Sawyer, April 9, 1980.

Such, C. H. (1978). Diesel odor consortium project final report. Rep. DP78-1142. Ricardo Consult. Eng.

Sullivan, R. F. (1979). Transportation fuels from shale oil. DOE Summary Rep. 791082. Highway Vehicle Systems Contractors' Coordination Meeting, October 23-25.

Surber, F. T. (1980). Hybrid vehicle potential assessment. Excerpt and summary in DOE (1980b), in press.

Tanzawa, T., and W. C. Gardner, Jr. (1978). Mechanism of acetylene thermal decomposition. *Symp. Int. Combust. Proc. 17th*, p. 563.

Tenniswood, D. M., and H. A. Graetzel (1967). Minimum road load for electric cars. *SAE Tech. Pap. 670177*.

Tesner, P. A., E. I. Tsygankova, V. P. Guilazetdinov, V. P. Zuyer, and G. V. Loshakova (1971). The formation of soot from aromatic hydrocarbons in diffusion flames of hydrocarbon-hydrogen mixtures. *Combust. Flame* 17:279.

Thien, G. E. (1979). A review of basic design principles for low-noise diesel engines. *SAE Tech. Pap. 790506*.

Thomas, A. (1962). Carbon formation in flames. *Combust. Flame* 6:46.

Toepler, O. Bernauer, and H. Buchner (1980). Use of hydrides in motor vehicles. *Fifth International Symposium on Automotive Propulsion Systems*. Conf-800419, vol. 2.

Toyota Motor Company (1979). Presentation to Diesel Impacts Study Committee, Technology Panel, National Academy of Sciences.

Toyota Motor Company (1980). Presentation to Diesel Impacts Study Committee, Technology Panel, National Academy of Sciences, June.

Tsunemoto, H., and H. Ishitani (1980). The role of oxygen in intake and exhaust on NO emission, smoke and BMEP of a diesel engine with EGR system. *SAE Tech. Pap. 800030*.

Turk, A. (1967). *Selection and Training of Judges for Sensory Evaluation of the Intensity and Character of Diesel Exhaust Odors*. PHS Contract No. PH 27-66-96. U.S. Department of Health, Education and Welfare, Public Health Service, Cincinnati, Ohio.

Tuttle, J. H. (1980a). Controlling engine load by means of late intake-valve closing. *SAE Tech. Pap. 800794*.

Tuttle, J. H. (1980b). GMRL. Personal communication with R. D. Matthews.

Urban, C. M. (1980a). Regulated and unregulated exhaust emissions from malfunctioning three-way catalyst gasoline automobiles. Rep. EPA-460/3-80-004. Southwest Res. Inst., San Antonio, Tex.

Urban, C. M. (1980b). Regulated and unregulated exhaust emissions from a malfunctioning three-way catalyst gasoline automobile. Rep. EPA-460/3-80-005. Southwest Res. Inst., San Antonio, Tex.

Urban, C. M. (1980c). Regulated and unregulated exhaust emissions from malfunctioning noncatalyst and oxidation catalyst gasoline automobiles. Rep. EPA-460/3-80-003. Southwest Res. Inst., San Antonio, Tex.

Urban, C. M., and R. J. Garbe (1979). Regulated and unregulated exhaust emissions from malfunctioning automobiles. SAE Tech. Pap. 790696.

Urban, C. M., and R. J. Garbe (1980). Exhaust emissions from malfunctioning three-way catalyst-equipped automobiles. SAE Tech. Pap. 800511.

Uyehara, O. A. (1980). Effect of burning zone A/F, fuel H/C on soot formation and thermal efficiency. SAE Tech. Pap. 800093.

Vaughn, S. N., J. F. Merklin, and T. W. Lester (1980). Isotopic studies of the chemical mechanisms of soot formation. Paper presented at Spring Technical Meeting, Central States Section, The Combustion Institute, Baton Rouge, La.

Voissard, D. (1979). Measuring of unregulated pollutants, on Peugeot diesel vehicles. Presented to Diesel Impacts Study Committee, Technology Panel, National Academy of Sciences.

Volkswagen Corporation (1980). Presentation to Diesel Impacts Study Committee, Technology Panel, National Academy of Sciences.

Vuk, C. T., M. A. Jones, and J. H. Johnson (1976). The measurement and analysis of the physical character of diesel particulate emissions. SAE Tech. Pap. 760131.

Wade, W. R. (1980). Light-duty diesel NO_x-HC-particulate trade-off studies. SAE Tech. Pap. 800335.

Wagner, H. G. G. (1978). Soot formation in combustion. Symp. Int. Combust. Proc. 17th, p. 3.

Wagner, T. O. (1977). Economics of manufacturing automotive diesel fuel. SAE Tech. Pap. 770758.

Wang, T.-S., R. A. Matula, and R. C. Farmer (1980). Soot formation from toluene. Pap. CSS/CI-80-03, presented at Spring 1980 Technical Meeting, Central States Section, The Combustion Institute, Baton Rouge, La.

Watson, J. F. (1977). Can battery cars co-exist with other vehicles? Electr. Rev. London 201:19, 27-29.

Webster, J. C., and G. L. Cluff (1974). Speech interference by noise. Geophys. Res. Lett. 3(12):751.

Wei, E. T., Y. Y. Wang, and S. M. Rappaport (1980). Diesel emissions and the Ames test: A commentary. Air Pollut. Control Assoc. 30.

Wersborg, B. L., J. B. Howard, and G. C. Williams (1973). Physical mechanisms of carbon formation in flames. Symp. Int. Combust. Proc. 14th, p. 929.

Wersborg, B. L., A. C. Yeung, and H. B. Howard (1974). Concentration and mass distribution on charged species in sooting flames. _Symp. Int. Combust. Proc. 15th_, p. 1439.

Wessel, W. G., and E. U. Joachim (1979). A strategy for optimization of diesel fuel injection system. _SAE Tech. Pap. 790036_.

White, R. A. (1980). University of Illinois. Personal communication with R. D. Matthews.

Whitehouse, N. D., E. Clough, and S. O. Uhunmwangho (1980). Development of some gaseous products during diesel engine combustion. _SAE Tech. Pap. 800028_.

Wiedemann, B., and P. Hofbauer (1978). Data base for light-weight automotive diesel power plants. _SAE Tech. Pap. 780634_.

Wiedemann, B., and R. Schmidt (1979a). Data base for light-weight automotive diesel power plants. Final Report Volume I: Executive summary. DOT Rep. No. DOT-TSC-NHTSA-79-62.I.

Wiedemann, B., and R. Schmidt (1979b). Data base for light-weight automotive diesel power Plants. Final Report Vol. II: Discussion and results. DOT Rep. No. DOT-TSC-NHTSA-79-62.II.

Williams, R. L., and D. P. Chock (1979). Characterization of diesel particulate exposure. GMR Rep. 3177, ENV-71.

Wojciki, S., T. J. Rychter, R. Saragih, and T. Lezanski (1980). Catalytic activation of a change in the prechamber of a SI lean-burn engine. Paper presented at the Eighteenth Symposium (International) on Combustion, University of Waterloo, Canada, August.

Wolber, W. G. (1980). Automotive engine control sensors, '80. _SAE Tech. Pap. 800121_.

Wright, F. J. (1969). The formation of carbon under well-stirred conditions. _Symp. Int. Combust. Proc. 12th_, p. 867.

Wright, F. J. (1970). Carbon formation under well-stirred conditions, part II. _Combust. Flame_ 15:217.

Yu, R. C., V. W. Wong, and S. M. Shahed (1980). Sources of hydrocarbon emissions from direct injection diesel engines. _SAE Tech. Pap. 800048_.

Zimmerman, K. D. (1979). Perspectives for the diesel engine and injection equipment. Extracts from a presentation made available by Robert Bosch, GMBH.

Zweidinger, R., B. S. Tejada, D. Dropkin, J. Huisingh, and L. Claxton (1978a). Characterization of extractable organics in diesel exhaust particulates. In _Symposium on Diesel Particulate Emissions and Measurement_. U.S. Environmental Protection Agency, Research Triangel Park, N.C.

Zweidinger, R., B. S. Tejada, D. Dropkin, and F. Black (1978b). _Collection and Characterization of Diesel Exhaust Particulates Bound Organics_. U.S. Environmental Protection Agency, Research Triangel Park, N.C.

GLOSSARY OF TERMS

Additive: A product added to a material that will improve a certain
    characteristic of the material, such as an emulsifier added to
    hydrocarbon fuels to improve the water compatibility of the fuel.
Air/Fuel Ratio: Proportionate mixture of air and fuel by weight.
    Kilograms of air per kilograms of fuel.
Aromatic Hydrocarbons: Hydrocarbons with a benzenoid ring form that
    have naturally high octane numbers and correspondingly low cetane
    numbers.
Bdc: Bottom dead center. The position of the piston where it is
    farthest from the engine head, yielding lowest compression.
Barrel: The opening in a carburetor through which air-fuel mixture
    enters the manifold of spark ignition engines.
BMEP: Brake mean effective pressure. The average pressure within a
    cylinder during the power stroke necessary to produce the
    horsepower determined on a dynamometer.
Calorific content: Chemical energy available in fuel and released upon
    combustion.
Catalytic Converter: A chamber in the exhaust system of vehicles
    containing a catalyst system which aids in oxidizing the carbon
    monoxide and unburned hydrocarbons in the exhaust gases or in
    reducing nitrogen oxides in the exhaust gases to innocuous products
    (carbon dioxide, N_2, O_2, and water).
Cetane Index: An empirical method for finding the cetane number of a
    diesel fuel, based upon the density of the fuel and its mid-boiling
    point.
Cetane Number: The equivalent percentage by volume of cetane (=
    n-hexadecane, cetane number 100) in a mixture with alpha-
    methylnaphthyene (cetane number 0); indicates the relative
    ignitability of the fuel upon injection into the engine cylinder.
Chassis Dynamometer: See Testing Methods.
Check Valve: A valve that will allow flow in one direction but will
    prevent flow in the opposite direction.
Compression Ignition: A term applied to the diesel engine operating
    principle. Heat to initiate combustion (ignition) is produced by
    rapid compression of air and fuel mixture in the cylinder.
Compression Ratio: Ratio between the volume above the piston when it
    is at bottom dead center to the volume above the piston when it is
    at top dead center.

Crossover: An opening through the intake manifold below the carburetor of spark ignition engines that connects the two exhaust manifolds.

Dead Center: Refers to the maximum upper or lower piston position when all movement in one direction stops before it reverses direction.

Distillate: A distilled low-volatility petroleum fuel used in diesel and turbine engines.

Dynamometer: A device that is used to measure loads, engine torque, and driving forces.

Engine Load: The composite vehicle resistance applied to the engine crankshaft.

Engine Speed: Crankshaft revolutions per minute.

Four-Stroke Cycle: An engine cycle consisting of a downward intake stroke, an upward compression stroke, a downward power stroke, and an upward exhaust stroke. The cycle then repeats on the next four strokes.

Heavy-Duty: Applications more severe than passenger car service; usually applied to buses or trucks in excess of 6,500 pounds gross weight that operate daily for long-duty cycles.

Horsepower: A value calculated from engine torque and rotating speed. One horsepower equals 33,000 foot-pounds per minute, measured at the testing brake.

Hydrocarbons: Substances the molecules of which are made solely from hydrogen and carbon atoms, with multiple carbon atoms linked to one another and with hydrogen atoms attached to the carbon atoms.

Manifold: Branched-pipe passage device that connects openings from each cylinder to a common opening.

Manifold Runners: A single passage in a manifold from one cylinder to the major manifold opening.

Mean Effective Pressure: The calculated average pressure in the combustion chamber during the power stroke.

Peak Pressure: The highest pressure in the combustion chamber, which occurs during the early part of the power stroke after fuel injection.

Performance: Effective operation at the maximum designed specification.

Poppet Valve: A valve that rises perpendicularly from its seat.

Preflame Reaction: Chemical reactions that occur in air-fuel gases before ignition by crossing the flame front.

Preignition: Ignition of the combustion charge before spark ignition.

Quench: Cooling to temperatures at which the flame will go out.

Sac Volume: A relatively small geometric volume included in the port area of a diesel fuel injector.

Spark Ignited or Spark Ignition: Engines whose combustion is started by an electrical discharge (spark).

Spent Gases: Hot gases remaining in the combustion chamber after combustion is complete.

Squish Area: An area in the combustion chamber where the piston approaches very close to the head.

Steady State: Constant operating conditions with no variation in fuel supply or load.

Stoichiometric: A chemically balanced air-fuel mixture, such that the
hydrocarbon is burned to produce only water and carbon dioxide, and
the oxygen is totally consumed.
Stroke: Piston movement in one direction, from one extreme position to
the other.
Supercharger: Device which forces the charge into the combustion
chamber under greater than atmospheric pressure.
Surface Ignition: Ignition caused within the combustion chamber at
some surface hot spot other than the electrical arc (spark) between
the sparkplug electrodes.
Tdc: Top dead center. The highest a piston can move within the
cylinder See also Bdc, Dead Center.
Testing Methods: The following referenced methods are key test
procedures, standardized and accepted for measuring fuel economy,
power performance or emissions from vehicles manufactured for
routine use on U.S. highways or within specific U.S.
jurisdictions. Although results of some tests may be useful for
estimating both fuel economy and emission levels, separate
consideration of these two features will simplify the treatment in
this glossary.
- Emission Test Methods:
-- Evaporative Emissions - Two methods are used to measure the
atmospheric discharge of fuel components: the carbon-cannister
test and the SHED test. In the carbon-cannister test method,
absorbent carbon in two preweighed cannisters is reweighted
after appropriate "runs" during which the cannisters have been
connected to fuel tank vents and carburetor external vent. In
the Sealed Housing for Evaporative Determination (SHED) method,
the entire vehicle is placed in a sealed enclosure during the
test run. The hydrocarbon concentration at the end of the test
is taken as the prima facie basis for computing the evaporative
emission loss.
-- Tail Pipe Emissions
-- Chassis Dynamometer - A general test procedure for vehicles,
making use of a test-bay facility at constant temperature/
humidity. Gaseous and particulate emissions are collected, and
fuel economy is measured during a semiautomated fast cycle,
providing variable vehicle inertia weight from 1,750 pounds to
above 3,000 pounds. Measurements of fuel economy can be made
in accord with the Federal Test Procedure or other prescribed
and defined methods.
-- Federal Test Procedure (FTP) - A multistage (multimodal) test
procedure for new car certification by the Environmental
Protection Agency. Carefully recording the run conditions,
operators take exhaust samples and subject these dynamometer
samples to routine chemical and ionization tests for regulated
emissions after submitting the gross exhaust to empirically
established dilution, filtration, and aging. Prescribed
dispensing from a constant-volume sampler (CVS) is also among
the pretreatment conditions imposed upon the exhaust sample.
The residence volume of a conventional CVS is 10 m^3/min, but

flow capacity of the dilution tunnel naturally exceeds this value.

-- Other test procedures used to measure fuel economy are also considered appropriate for emission measurement.

• Fuel Economy Test Methods: Various procedures have been devised and approved for testing fuel economy. Depending upon the approving authority and the correlation sought, the following procedures may also be used for emission data.

-- Highway Federal Economy Test (HFET) - A standardized multimodal procedure that permits accurate measurement of a vehicle's fuel consumption under empirically established driving conditions that resemble highway use.

-- LA-4 - The Urban Los Angeles Test Procedure; a driving cycle over a 12-mile route centered on the downtown Los Angeles area, simulating weekday morning peak driving conditions.

-- UDDS - The EPA Urban Dynamometer Driving Schedule ("Urban Cycle"); a 7.5 -mile track to be covered in programmed driving in 1,372 seconds. Coupled with the FET, the UDDS comprises the official emission and fuel economy test cycle.

Thermal Efficiency: The equivalent heat energy of the engine brake horsepower produced divided by the total heat energy content of the fuel consumed.

Timing: Events present to occur in relation to the angle of crankshaft rotation.

Torque: A power measurement that expresses the twisting force on a shaft in foot-pounds (or ft-lb).

Turbine: A finned disc whose axle is made to rotate by the force of a high-velocity stream of gases or liquid.

Valve Seat: The machined surface of the head on which the valve rests to close a port.

Vapor Pressure: Partial pressure of the volatile portion of an evaporating fluid.

Viscosity: A fluid's characteristic molecular resistance to flow.

Volatility: The tendency of a substance to evaporate at normal atmospheric pressure.

Volumetric Efficiency: The volume of air actually consumed by an engine, compared to the cubic displacement of the engine.

APPENDIX A

1978 AND 1979 MODEL YEAR DIESEL CERTIFICATION VEHICLE IDENTIFICATION AND SPECIFICATIONS

Manufac-turer	Model	Trans-mission	Rated Power, hp	N/V Ratio[a]	A/C Simu-lation[b]	Vehicle[c] Identification
			1978 Model Year			
GM	Delta 88	Auto	120	30.6	Yes	GM9
	Custom Wagon	Auto	120	33.9	Yes	GM10
MB	240D	Auto	62	53.6	Yes	MB5
	300D	Auto	77	50.7	Yes	MB6
	300 SD	Auto	110	44.8	Yes	MB7
	240D	Manual	62	51.4	Yes	MB8
Peugeot	504	Auto	71	52.3	Yes	P6
	504	Auto	71	52.5	No	P7
	504 Wagon	Auto	71	56.0	No	P8
	504 Wagon	Auto	71	56.0	Yes	P9
	504	Manual	71	51.4	No	P10
	504	Manual	71	51.4	Yes	P11
	504 Wagon	Manual	71	56.0	No	P12
	504 Wagon	Manual	71	56.0	Yes	P13
VW	Rabbit	Manual	48	44.0	No	VW4
	Rabbit	Manual	48	57.6	No	VW5
			1979 Model Year			
GM	Cutlass	Auto	96	32.7	Yes	GM1, 2
	Cutlass Wagon	Auto	120	31.1	Yes	GM3
	Ninety Eight	Auto	120	30.6	Yes	GM4
	Delta 88	Auto	120	31.3	Yes	GM5, 6
	Deville	Auto	120	28.3	Yes	GM7
	Cutlass	Manual	96	34.9	Yes	GM8
MB	240D	Auto	62	53.7	Yes	MB1
	300D	Auto	77	50.7	Yes	MB2
	300SD	Auto	110	44.8	Yes	MB3
	240D	Manual	62	51.4	Yes	MB4
Peugeot	504	Auto	71	52.5	Yes	P1
	504 Wagon	Auto	71	56.0	No	P2
	504 Wagon	Auto	71	56.0	Yes	P3
	504	Manual	71	51.4	No	P4
	504 Wagon	Manual	71	55.9	Yes	P5
VW	Rabbit	Manual	48	57.6	No	VW1
	Rabbit	Manual	48	45.1	No	VW2
	Dasher Wagon	Manual	48	56.9	No	VW3

[a]Ratio of engine speed to road speed, in top gear.
[b]"Yes" indicates vehicle was tested with air conditioning load simulated.
[c]Vehicle identification used in figures.

SOURCE: Roessler et al., 1980.

APPENDIX B:

FUNDAMENTALS OF PARTICULATE FORMATION
IN FLAMES WITH APPLICATION TO DIESEL ENGINE
PARTICULATE EMISSIONS

Despite the number of studies of particulate formation in flames, relatively little quantitative information on the mechanism of particulate formation and the governing rate processes is available. The available information is likely to be in semiempirical form, severely limiting its application to systems other than those from which the measurements were taken.

Experimental evidence from a variety of combustion systems (premixed and diffusion flames, perfectly stirred reactors, etc.) using a variety of fuels indicates that chemical kinetics is the dominant rate process governing the emission of particulates. Even well mixed systems emit particulates when the carbon to oxygen ratio in the fuel-oxidizer mixture exceeds 0.5. As described by Wagner (1978), from a thermodynamic point of view, particulate emission should begin at a carbon/oxygen ratio of unity, corresponding to the condition $\underline{m} = 2\underline{y}$:

$$C_{\underline{m}}H_{\underline{n}} + \underline{y}\ O_2 = 2\underline{y}\ CO + (\underline{n}/2)H_2 + (\underline{m}-2\underline{y})\ C_{\underline{s}}.$$

The fact that particulate, water, and carbon dioxide are observed as combustion products when the ratio of carbon to oxygen is greater than 0.5 indicates that some of the oxygen is tied up in relatively stable combustion products (carbon dioxide and water) or intermediates, and is unavailable for reaction with solid carbon in the time period character- istic of combustion systems. Wagner (1978) explains that this is because one of the important oxidizing species in flames, the hydroxl radical, is destroyed relatively quickly through the reactions

$$H_2 + OH \longrightarrow H_2O + H$$
$$CO + OH \longrightarrow CO_2 + H\ .$$

Because the reverse reactions of both require large activation energies, and thus occur more slowly at flame temperatures than the corresponding forward reactions, they effectively tie up oxygen in an unreactive form. Fenimore and Jones (1969) have extended this concept to develop a criterion for particulate emissions from flames.

The carbon/oxygen ratio will always exceed 0.5 in some region of a diffusion flame; hence, particulate will always be emitted from that region. Whether particulate will be observed as a final combustion

314

product depends on the speed at which it is destroyed in the flame region where the ratio of carbon to oxygen is less than 0.5. Particulate oxidation in that region is favored by low emissions of particulates from the region of formation and growth, small particle diameter, large concentrations of the predominant oxidizing species, and high temperatures.

The diesel engine is unsuitable for use in the study of the fundamental processes controlling combustion-generated particulate. Experimental difficulties in following the course of particulate formation and oxidation within the diesel cylinder are enormous because of the high temperatures and pressures, and the presence of extremely reactive intermediate species. Even if such data could be obtained, interpretation of those data would be complicated because of the unsteady environment and the role of turbulent mixing. Thus, the most useful studies of fundamental particulate formation processes have been performed in relatively simple environments. One-dimensional premixed flames (Homann, 1967; Bonne _et al._, 1965; Wersborg _et al._, 1973, 1974; Delfan _et al._, 1979; Howard _et al._, 1973; Bittner and Howard, 1980), diffusion flames (Schalla and MacDonald, 1955; Gordon _et al._, 1959; Tesner _et al._, 1971; Chakraborty and Long, 1968), and stirred reactors (Wright, 1969) have produced useful results. Recent experiments in such unsteady devices as shock tubes (Graham, 1976; Wang _et al._, 1980) and constant-volume bombs (Flower and Dyer, 1980) have also resulted in significant contributions.

As evidenced by the volume of related literature, the fundamental mechanism of particulate formation in flames has been of interest to combustion researchers for many years. Only recently, however, has the process begun to be understood. Considerable progress was made during the 1970s, largely as a result of the application of sensitive, non-obtrusive diagnostic techniques. These techniques allow measurement of the relative abundance of high-molecular-weight, unstable hydrocarbon radicals and ions, which are now thought to be the direct precursors of particulates and of embryonic particulates with diameters as small as 10 Å. Optical techniques and molecular beam sampling combined with mass spectrometric analysis have proved to be the most successful methods (Wagner, 1978; Homann, 1967; Bonne _et al._, 1965; Wersborg _et al._, 1973, 1974; Delfan _et al._, 1979; Howard _et al._, 1973; Bittner and Howard, 1980; Graham, 1976; Wang _et al._, 1980; Flower and Dyer, 1980). Although most of these studies were performed on premixed flames at pressures well below those present in diesel engines, in many respects the systems exhibit similar characteristics. These similarities include: the size and microscopic structure of the particles produced; the qualitative effects of changes in the local equivalence ratio, temperature, and pressure; and the effect of such additives as halogenated or alkali-metal-containing species. This has led to the presumption that the basic mechanism of particulate formation is similar in both systems (Broome and Khan, 1971). Nevertheless, a recent study of particulate formation in rich, premixed propane-air mixtures burned at higher pressures (0.45 to 1.0 MPa) in a constant volume bomb showed that in many respects the system exhibited markedly different behavior (Flower and Dyer, 1980). This is disturbing because these conditions might well be expected to provide a bridge between investigations conducted at constant pressure in laboratory flames and less detailed studies carried out in engines. Inconsistencies between the results of the Flower and Dyer study and laboratory flame studies will be explored as they arise.

The currently accepted mechanism of particulate formation from the vapor phase in flames is described in Figure B.1. Each of the processes is thought to occur in diesel engines.

Diesel particulate takes the form of roughly spherical aggregates of individual spherical elements; particles from rich, premixed flames are generally chainlike aggregates. The latter are formed late in the overall process, with residence times on the order of 100 ms (Bonne et al., 1965). The relatively short residence times (1 to 10 ms) in the diesel engine (Broome and Khan, 1971), together with the faster initial growth and coagulation rates caused by the higher pressures and mixing rates encountered, are apparently responsible for the difference in particulate appearance. Particulate formation by polymerization and carbonization within individual liquid droplets has also been observed in laboratory devices (Fenimore and Jones, 1968). With properly adjusted fuel injection equipment the liquid phase mechanism contribution to total particulate emissions of diesel engines (except perhaps at very light load) is expected to be negligible.

Concurrently with the formation process outlined in Figure B.1, destructive processes will also occur in flame regions where oxidizing species are present. As was pointed out by Fenimore and Jones (1968), these are important even in fuel-rich premixed systems. Destructive processes should play an even larger role in diesel engines, which operate at less than unit overall equivalence ratio.

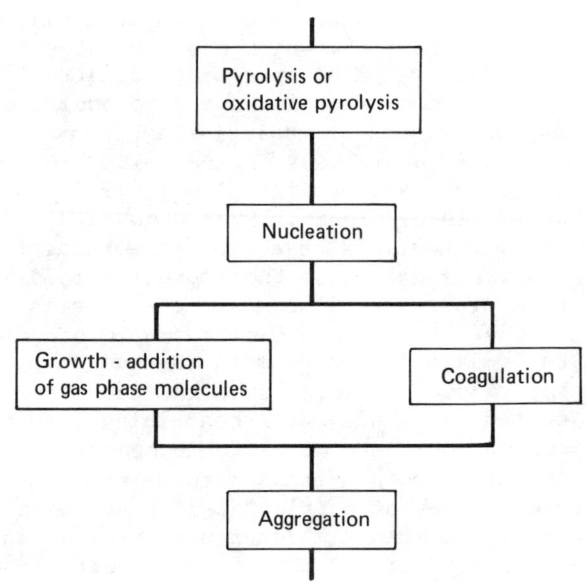

FIGURE B.1 The mechanism of particulate formation in combustion systems.

    As was noted in a recent review, most modeling and experimental
studies of particulate formation in diesel engines have been focused on
direct injection diesels (Amann _et al._, 1980). This focus is
unfortunate because there are profound differences in the combustion
characteristics of the direct injection and indirect injection engines.
The committee is primarily concerned with the particulate emissions
charac teristic of light-duty diesels, which are almost entirely of the
indirect injection design.

FUNDAMENTALS OF PARTICULATE FORMATION

The formation of diesel particulate starts with a fuel molecule,
typically of 12 to 22 carbon atoms, with a hydrogen/carbon ratio of
about 2. The particulates formed from these molecules typically
contain approximately 10^5 carbon atoms and have a hydrogen/carbon
ratio of about 0.1. The carbon atoms are packed in hexagonal face-
centered arrays, commonly referred to as platelets (Lipkea _et al._,
1978). As is illustrated in Figure B.2, the mean layer spacing is 3.55
Å, which is only slightly larger than that of graphite. Platelets
are arranged in layers to form crystallites. There are typically 2 to
5 platelets per crystallite and on the order of 10^3 crystallites per
spherical particle (Howard _et al._, 1973; Mantell, 1968). The crystallites are arranged in turbostatic fashion, with planes more or less

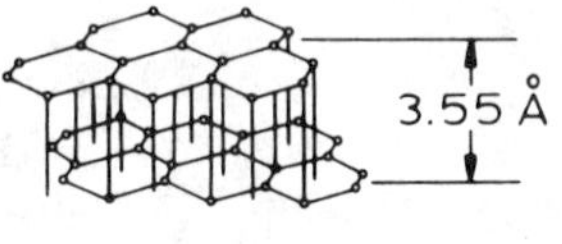

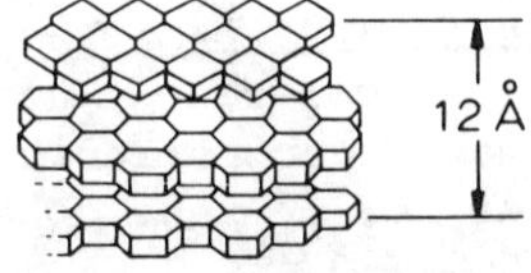

FIGURE B.2 Substructure
of the carbon particle.

parallel to the particle surface. This turbostatic structure is also characteristic of pryrolytic graphite and is thought to be responsible for its unusually high resistance to oxidation (Appleton, 1973). During long residence times spherical particles may agglomerate to form branched or unbranched chains.

Pyrolysis

Most, if not all, fuel hydrocarbon molecules decompose and rearrange atomically before particulate formation begins. Such processes, usually leading to less highly saturated, lower-molecular-weight hydrocarbons than the parent species, are generally termed pyrolytic, especially when they occur in the absence of oxygen. Pyrolysis reactions are generally endothermic, and as a result, their reaction rates are often highly temperature dependent. Pyrolysis typically occurs through a free-radical mechanism; thus even small concentrations of singlet oxygen, of doublet oxygen, or of the hydroxl free radical might be expected to accelerate the reaction process significantly by providing an additional path for radical formation through branching reactions.

Johnson and Anderson (1961) found that the addition of 3 percent doublet oxygen to a pyrolyzing mixture of 20 percent acetylene in an inert carrier gas results in a small reduction in particle size at temperatures above 750°C; however, no effect on the rate was reported. Ray and Long (1964), in a review of polycyclic aromatic formation from diffusion flames and diesels, note that the influence of doublet oxygen and the hydroxyl radical on the pyrolysis mechanism is not known. However, a more recent study by Fujii and Asaba (1973) reported that the addition of a small amount of oxygen ($O_2/C_6H_6 = 1/10$) to a pyrolyzing mixture of benzene in an inert gas results in an order-of-magnitude increase in the rate. Additional evidence of the importance of oxidative pyrolysis in combustion systems is provided by Bittner and Howard's (1980) study of rich benzene flame chemistry. Detailed flame structure was studied in a low pressure (2.67 kPa), laminar, premixed benzene-oxygen-argon flame, using a nozzle beam skimmer-sampling system followed by mass spectrometric analysis. Bittner and Howard interpret the rapid decay of benzene concentration and subsequent sequential appearance of C_6H_6O and C_5H_6 as an indication that the initial attack on the aromatic ring is predominantly by atomic oxygen; this results in the rapid formation and subsequent decomposition of phenol:

$$\text{(benzene)} + O \longrightarrow \text{(phenol)} - OH \begin{cases} \longrightarrow \text{(cyclopentadiene)} + CO \\ \longrightarrow \text{open chain } C_5H_6 + CO \end{cases}$$

An alternate route involving hydrogen atom abstraction by hydrogen or OH, followed by rapid oxidation by oxygen or OH of the resulting phenyl radical

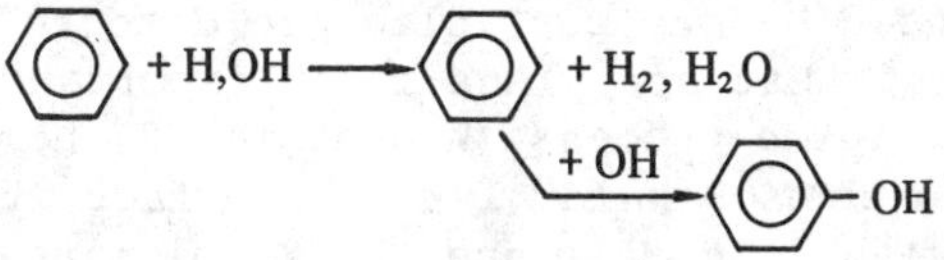

319

may also be important.

    Alternate routes for the destruction of the phenyl radical, such as
reaction with molecular oxygen or unimolecular decomposition, may be
eliminated on kinetic grounds. Some formation of C_6H_7 is observed
in the early part of the reaction zone; however, calculations indicate
that both the stabilization of this radical by the addition of the
hydrogen atom and its decomposition by unimolecular scission were too
slow to account for the rapid C_6H_6 consumption and low C_6H_7
concentrations observed

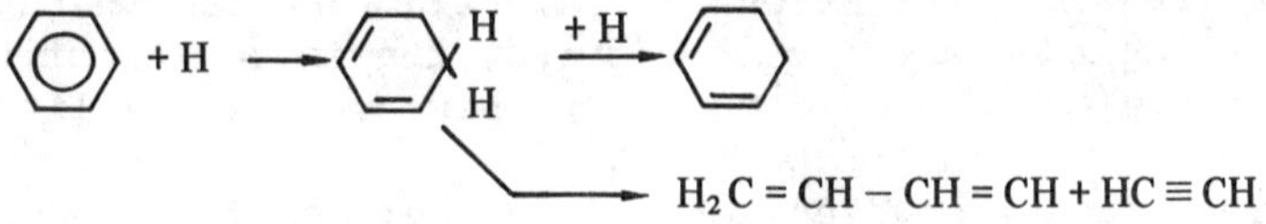

This illustrates the importance of the initial attack on the fuel
molecule by the oxidizing species (if available).

    Bittner and Howard (1977) have also recently reviewed studies of
benzene pyrolysis in the absence of oxygen. The principal products are
hydrogen, acetylene, diacetylene, and biphenyl; biphenyl predominates
at temperatures below 1800°K, and acetylene and its polymeric forms
predominate at higher temperatures. At high temperatures, such as
those encountered in the fuel-rich regions of combustion devices,
extensive ring fracturing is to be expected before particulate
formation. This is observed, for example, in benzene-hydrogen-air
diffusion flames, where the concentration of benzene is negligible
compared with the concentration of either methane or acetylene at the
onset of particulate formation (Tesner et al., 1971).

    The hexagonal arrays of carbon atoms that have been found to be
part of the structure of the particles have led some researchers to
propose that combustion generated particulate is initially composed of
heavy aromatic hydrocarbons formed more or less directly from the
aromatic fuel molecule (Gordon et al., 1959). With the exception of
the nonoxidative pyrolysis of aromatic molecules that occurs at low
temperatures, where large concentrations of phenyl radicals could be
obtained from biphenyl, it is apparent that this is not the case.
Aromatic hydrocarbon radicals do play a role in the formation of
particulate flames (they probably have an important role in nucleation),
but most of the mass of the particle is derived from less complex
aliphatic species.

    For a wide variety of aliphatic and aromatic fuels, acetylene is a
dominant product of high temperature and oxidative pyrolysis (Homann,
1967; Chakraborty and Long, 1968; Ray and Long, 1964; Fujii and Asaba,
1973; Bittner and Howard, 1977). Short of solid carbon, this is the
most highly saturated, lowest-molecular-weight hydrocarbon possible.

Experimental evidence indicates that acetylene is the last stable species to be produced before particles are detected, which has led to the widespread belief that acetylene is an important intermediate in the particulate formation process (Wagner, 1978; Homann, 1967; Bonne et al., 1965; Schalla and MacDonald, 1955). In fact, there is considerable evidence that acetylene radicals and their polymers are responsible for particulate formation in low pressure, acetylene-oxygen, premixed flames (Homann, 1967; Bonne et al., 1965).

As was pointed out by Glassman (1980), the tendency of fuels to pyrolyze primarily to acetylene in the presence of relatively small amounts of oxidative species (oxidative pyrolysis) or at high temperatures (above 1800°K), as opposed to the tendency of some fuel molecules, notably aromatics, to produce other products at lower temperatures and in the absence of oxygen, leads to important implications with regard to the effect of fuel structure on particulate formation. In particular, when the higher temperatures and relatively large concentrations of oxidizing species in the preheat zone of a premixed flame are considered, the molecular structure of the fuel molecule would be expected to have relatively little effect on its particulate forming properties.
In diffusion flames, however, where the fuel molecule may well have a long residence time in lower-temperature regions (which are essentially free of oxidizing species), some sensitivity to fuel structure is to be expected.

Nucleation

The most succinct definition of the nucleation process in particulate formation is given by Glassman (1980): "Nucleation involves the formation of an embryonic species that grows faster than it decomposes or otherwise reacts." Thus, hydrocarbon precursors to the particulate nuclei must be sufficiently stable to resist disproportionation at the high temperatures encountered in combustion systems, yet sufficiently reactive to account for the rate at which particles are formed (typically with characteristic time of less than 1 ms). These conflicting requirements led Thomas (1962) and Glassman (1980) to postulate that the direct precursors of the nucleation process have a polar conjugated structure. Resonance stabilization, made possible by the conjugated structure, provides the needed stability with respect to decomposition, especially for radicals and/or ions derived from these species. The polar nature of these resonance forms makes rapid reaction with other radicals and ions more probable. A more rapid rate of nucleation might be expected for the fuel molecules, such as aromatics, from which these polar-conjugated species can be more or less directly formed.

Detailed measurements of flame structure in low-pressure, particle-producing premixed acetylene-oxygen-argon flames have provided circumstantial evidence that the polymerization of acetylene plays an important role in the eventual nucleation of particulate (Homann, 1967; Bonne et al., 1965). Polyacetylene ($C_n H_2$, n = 4 to 10) formation takes place in the reaction zone, where concentrations of hydroxyl radicals and hydrogen atoms are relatively large. Although poly-acetylenes are formed in oxygen-free systems, they form at a faster rate in premixed flames than is observed for pyrolysis under similar conditions (Bonne _et al._, 1965).

It is postulated that the mechanism of polyacetylene formation is of the radical chain type. Homann (1967) suggests the following sequence:

$$C_2H + C_2H_2 \longrightarrow C_4H_3 \xrightarrow[+C_2H_2]{} C_6H_3 \xrightarrow[+C_2H_2]{} C_8H_3 \longrightarrow \text{etc.}$$

$$C_4H_3 \longrightarrow C_4H_2 + H \qquad C_6H_3 \longrightarrow C_6H_2 + H \qquad C_8H_3 \longrightarrow C_3H_2 + H$$

$$C_6H_3 \longrightarrow + H_2 \qquad C_8H_3 \longrightarrow + H_2$$

As the chain length increases, radical attack at positions other than the chain ends becomes more probable. Through ring closure, the branched chain hydrocarbons that are formed are expected to eventually lead to cyclic or polycyclic structures. These heavy hydrocarbons (molecular weights greater than 150) are more saturated than aromatics and apparently still have considerable radical character. They are capable of further rapid reaction among themselves and polyacetylenes and have been identified as particulate precursors by Homann (1967).

In a recent review, Glassman (1980) summarized certain problems with a strictly polyacetylene radical mechanism for nucleation. He notes the following:

- Some difficulty is encountered in explaining the rapid formation of the particulate nuclei when reasonable estimates of the free radical reaction rates are used (Tanzawa and Gardner, 1978).
- There is some difficulty in proposing a reasonable mechanism by which the hydrocarbon chains produced are rearranged to form aromatics (Cullis, 1976).
- There are thermodynamic constraints that may limit the growth of aromatics from hydrocarbon radicals (Stein, 1978).

Acetylene, which does not have a conjugated structure, derives its stability from the strength of its bonds. It can, however, react to form such polar-conjugated molecules as butadiene, vinylacetylene, and diacetylene (Glassman, 1980). Diacetylene and vinylacetylene have been observed in both particulate-producing flames (Homann, 1967) and pyro-lyzing benzene mixtures (Vaughn _et al._, 1980). Small concentrations of a C_4H_6 species, presumably butadiene, have also been observed when sophisticated sampling techniques are used (Homann, 1967). Introduction of these species, the radicals and ions of which can be represented by very reactive polar resonance structures, helps to overcome the kinetic objections to the acetylenic nucleation mechanism outlined earlier. Taking butadiene, for example, we may write the resonance structures

$$H_2C = CH - CH = CH_2 \longrightarrow H_2C^+ - CH = CH - CH_2.$$

Resonance stabilization increases for the positive ion:

$$H_2C = CH - CH = CH_2 + H^+ \longrightarrow H_3C - CH - CH = CH_2$$
$$+ HC - CH = CH = CH_2.$$

Furthermore, these molecules can undergo fast Diels-Alder reactions to form aromatic structures:

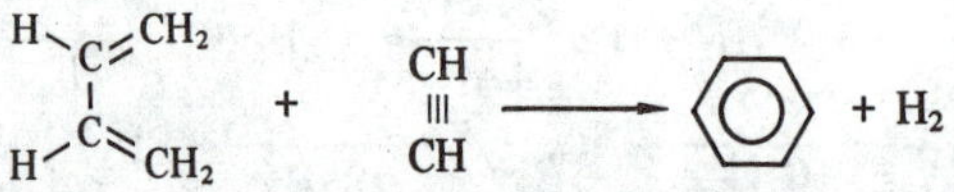

thus helping to overcome the mechanistic problems detailed earlier.

Additional evidence of the important role played by polar-conjugated species in particulate nucleation is provided by Bittner and Howard's (1980) study. In the premixed benzene-oxygen-argon flame that they employed, phenyl and benzene radicals are formed more or less directly from the fuel molecule. The benzene radical is presumably formed by π-complex stabilized electrophilic substitution of a methyl radical, followed by hydrogen atom abstraction from the methyl group. The structure of these aromatic radicals is polar resonant and, therefore, likely points to the initiation of the nucleation process. On the basis of their measurements of flame structure and thermodynamic considerations, Bittner and Howard suggest that nucleation begins when vinylacetylene or acetylene is added to the phenyl radical to form naphthalene. Their mechanism for the addition of vinylacetylene is depicted as follows:

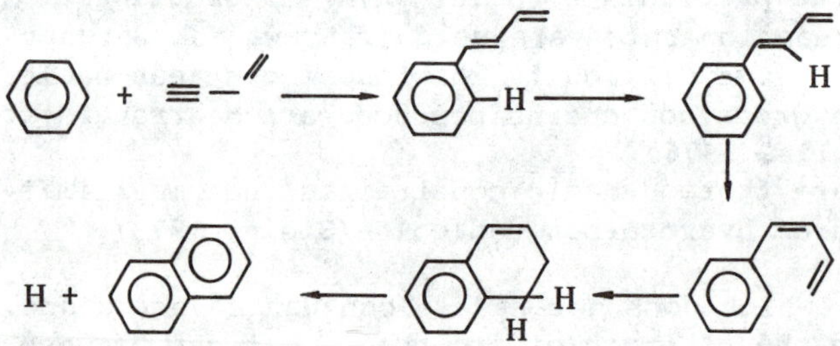

By a similar mechanism, the addition of two acetylene molecules can lead to naphthalene, with the possibility of phenylacetylene and styrene (both of which have been observed in the flame) as side products. Alternately, the addition of both vinylacetylene and acetylene to the benzene radical could lead to a methylnaphthalene radical. When acetylene is added, indene is a side product observed in the flame. Following these mechanisms, larger polycyclic aromatics, from naphthalene, are then thought to be formed by π-complex stabilized substition of a methyl radical at the one position. Hydrogen atom abstraction and acetylene addition are then thought to form phenalene:

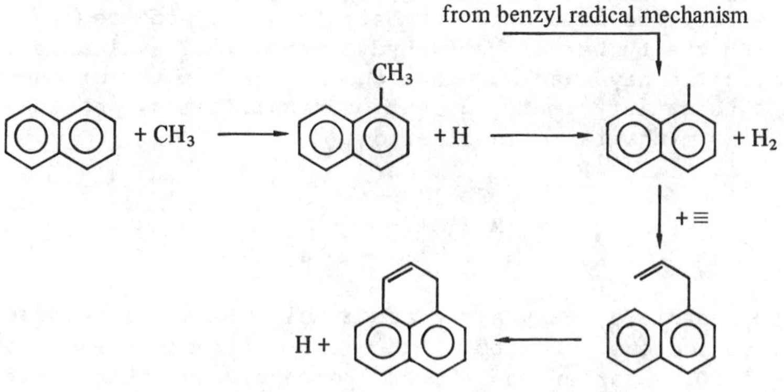

from benzyl radical mechanism

Repetition of the same process leads to pyrene:

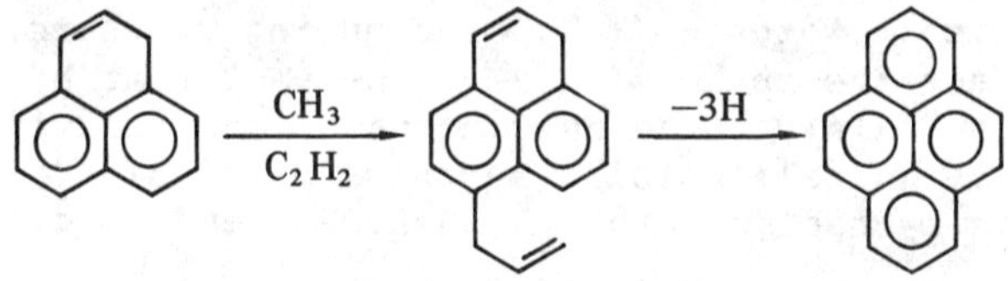

Because, as rings are added, the adduct formed by methyl substitution
is more strongly stabilized by the resulting λ-complex, and subsequent addition of CH_3 and C_2H_2 to pyrene should be more thermodynamically favorable and thus tend to eliminate the previously posed
objection to Homann's mechanism on thermodynamic grounds.

Recent studies (Wersborg $\underline{et}$ $\underline{al}$., 1973; Wersborg $\underline{et}$ $\underline{al}$., 1974;
Delfan $\underline{et}$ $\underline{al}$., 1979; Howard $\underline{et}$ $\underline{al}$., 1973) have provided much evidence
of the role of positive ions in the nucleation process. Wersborg and
co-workers (1974) observed that the predominant positive charge carriers
in particle producing acetylene-oxygen flames are heavy hydrocarbons of
molecular weight of 300 or more. A striking correlation noted between
the peak concentration of these molecules and the onset of nucleation
is illustrated in Figure B.3. Delfan and co-workers (1979), in an
investigation employing a similar system, note that for mixtures that
do not produce particulates, the predominant charge carrier has a mass
of 39 atomic mass units (amu) and that the carrier reaches its maximum
concentration in the reaction zone. In Figure B.4, the curve for this
carrier is labeled I_1+. Characteristics of the curve mass may be
attributed to C_3H_3+, which is produced by chemi-ionization. For
particulate producing mixtures, the predominant charge carriers have
masses of 700 amu or more, denoted I_2+ on the curve. These
carriers reach their peak somewhat later. As was reported by Wersborg
$\underline{et}$ $\underline{al}$. (1974), the peak concentration of these species and the onset of
particulate formation correspond. The mass of these heavy ions at the
onset of nucleation is between 5×10^3 and 6×10^3 amu, corresponding to spherical particles with diamters of 10 to 22 Å (assumed
to be 1.8 g/cm).

Wersborg and co-workers (1974) have shown that under the conditions of their study, the cumulative number of nuclei produced is a factor of 20 less than the number of heavy hydrocarbon ions available at peak concentration. They conclude that the number density of these ions is sufficient to satisfy ionic nucleation requirements, and they propose the following mechanism of nucleation:

$$M + M \longrightarrow s \tag{1}$$
$$M + M+ \longrightarrow s+ \tag{2}$$
$$M^+ + M^+ \longrightarrow s^{++} \tag{3}$$

where M represents a heavy hydrocarbon molecule and s represents the newly formed nucleus. Due to electrostatic effects, reactions 2 and 3 are thought to be faster and slower, respectively, than reaction 1. Thus, if all heavy hydrocarbons were charged, the rate of particulate nucleation would be considerably reduced. The authors reason that this condition can be approached by increasing temperature, reducing the equivalence ratio (towards unity), or reducing pressure. Additives should also affect the ratio of M^+/M. The fact that highly electrophilic additives (mainly halogens) increase, and additives with low electronegativity (alkali metals) suppress particulate formation, lends credence to this mechanism (Wagner, 1978; Ray and Long, 1964).

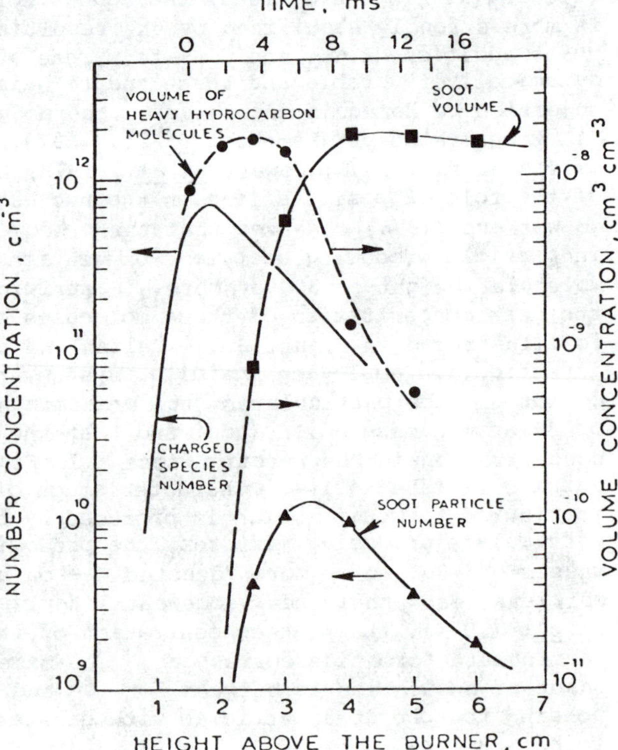

FIGURE B.3 Concentrations of charged species, heavy hydrocarbons, and particulates in low pressure, one-dimensional, premixed acetylene-oxygen flame. Cold gas velocity is 50 cm/s, $\phi = 3$. SOURCE: Wersburg, <u>et al.</u>, 1974.

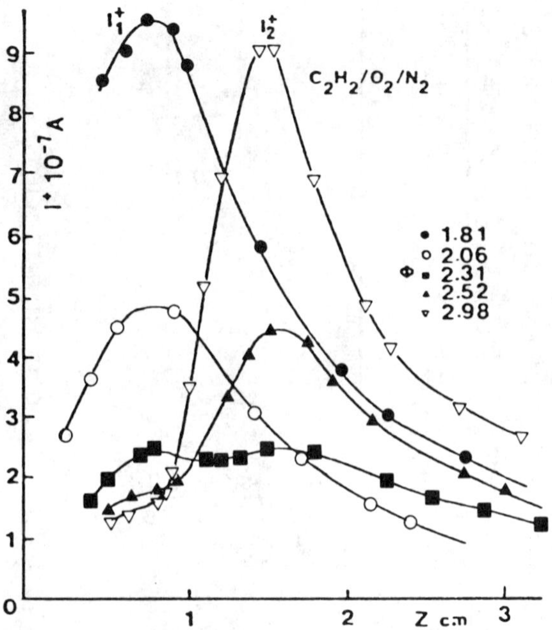

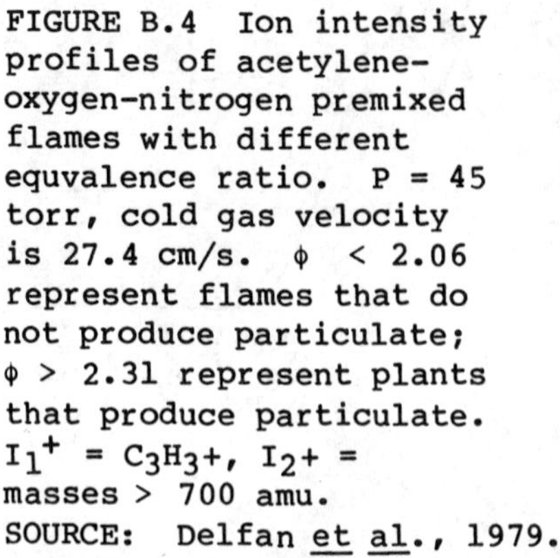

FIGURE B.4 Ion intensity profiles of acetylene-oxygen-nitrogen premixed flames with different equvalence ratio. P = 45 torr, cold gas velocity is 27.4 cm/s. ϕ < 2.06 represent flames that do not produce particulate; ϕ > 2.31 represent plants that produce particulate. I_1^+ = C_3H_3+, I_2+ = masses > 700 amu.
SOURCE: Delfan et al., 1979.

The rate of nucleation can be derived from the dynamics of the particle number and size distribution. Howard and co-workers (1973) undertook a study of this type using a premixed, flat, acetylene-oxygen flame at low pressures. Their results are shown in Figure B.5. Note that in this system, nucleation and surface growth occurred over the same time interval.

Figure B.6 depicts the critical steps in the formation and oxidation of combustion generated particulate. Beginning with an unconjugated aliphatic fuel molecule, acetylene is produced by thermal and oxidative pyrolysis. If significant amounts of oxygen or the hydroxl radical are present, some acetylene will be oxidized to relatively inert products. A fraction of the remainder will react to form such polar conjugated aliphatics as butadiene, vinylacetylene, and diacetylene. The pyrolysis reactions that form acetylene, and subsequently butadiene, etc., typically have large activation energies; however, the concentrations of oxygen and the hydroxl radical in the flame are also an extremely strong function of temperature. Glassman (1980) has suggested that the particulate producing tendency of flames is largely determined by the route by which acetylene is consumed. If the rate of acetylene oxidation is assumed to be more temperature dependent (due to changes in oxygen and the hydroxl radical) than the rate of the overall process that results in the formation of polar-conjugated species from the fuel molecule (as was originally proposed by Milliken (1962)), the qualitative particulate-producing tendencies of many aliphatics can be predicted in both premixed and diffusion flames.

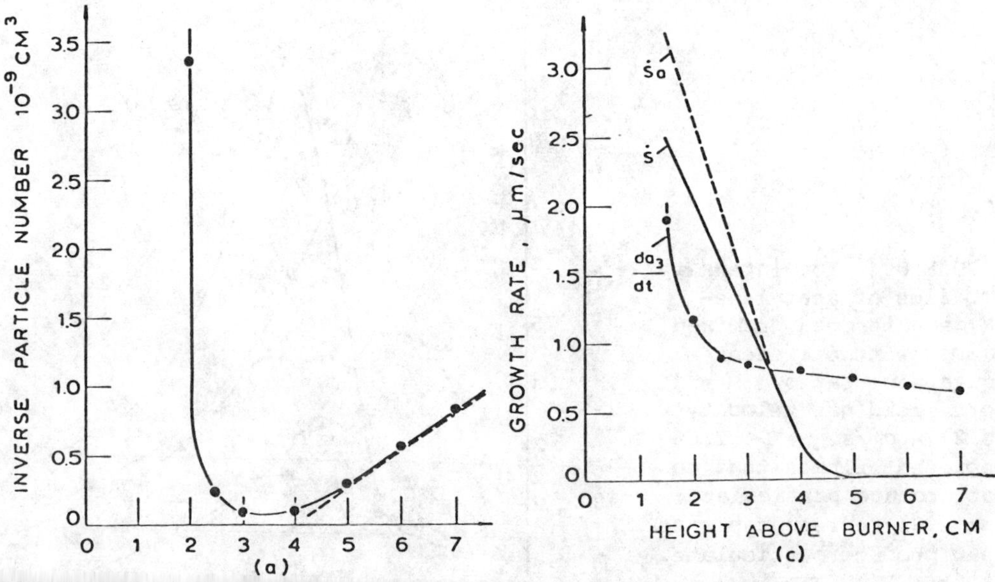
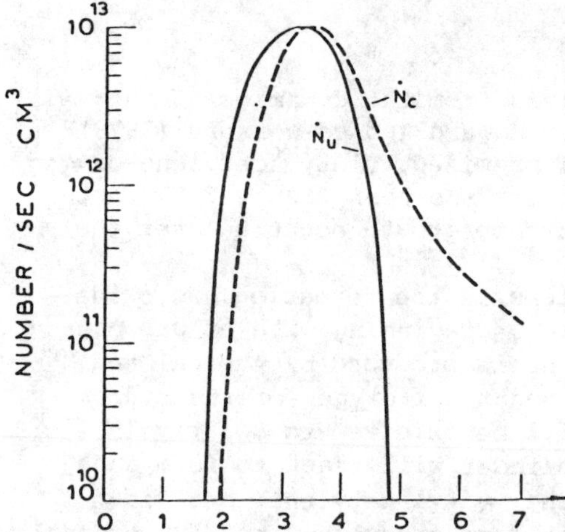

FIGURE B.5 (a) Inverse particle number profile. Aggregation is observed in excess of 5 cm from the burner; (b) particle nucleation ($\dot{N}_u$) and coagulation ($\dot{N}_c$) rate profiles; (c) particule surface growth rate profile. All results from P = 20 torr acetylene-oxygen flames (ϕ = 3) with cold gas velocity of 50 cm/s. SOURCE: Wersborg et al., 1974.

The very strong enhancement of particulate formation by sulfur trioxide provides further evidence in support of this idea (Homann, 1967). Sulfur trioxide catalyzes the recombination of oxygen atoms through the rapid reactions

$$SO_3 + O > SO_2 + O_2$$
$$O + SO_2 + M > SO_3 + M$$

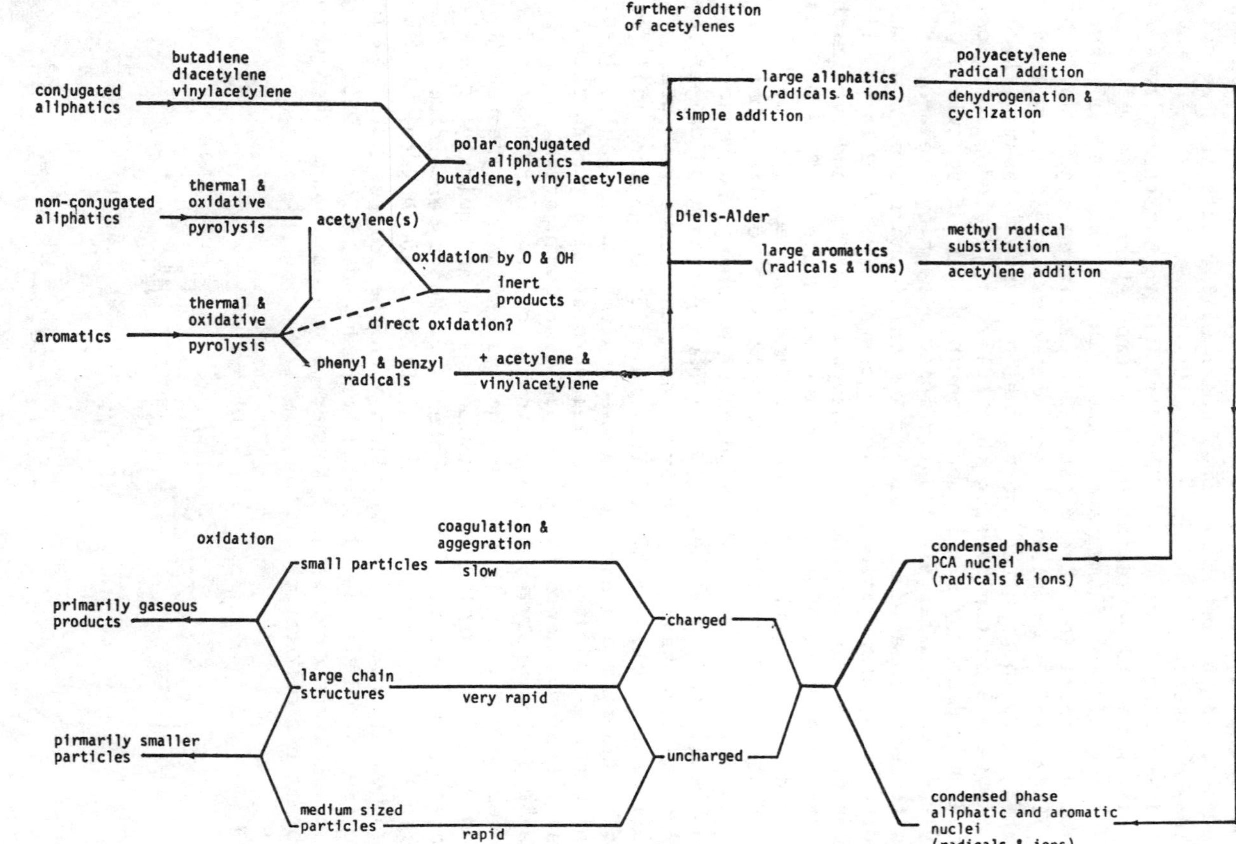

FIGURE B.6 Overview of the particulate formation process.

Fast radical shuffle reactions involving hydrogen, oxygen, and the hydroxl radical result in rapid relaxation of the superequilibrium overshoots in hydrogen and the hydroxl radical concentrations often produced in the reaction zone of a flame. Thus, the addition of sulfur to a flame is expected to lower concentrations of both oxygen and the hydroxl radical in the reaction zone, which will result in a reduction in acetylene oxidation.

At a given equivalence ratio, the particulate nucleation rate for a premixed system should decrease when the temperature increases. However, with diffusion flames an increase in temperature will increase the particulate formation rate because few oxidizing species will be present. This tends to explain the opposite particle-forming tendencies of aliphatic fuels in premixed and diffusion flames (Glassman, 1980).

From Figure B.6 it can be seen that polar-conjugated aliphatics, once formed, may react by simple addition (probably of acetylene or polyacetylene) to form large aliphatic molecules or by Diels-Alder reactions to form aromatic molecules. A significant fraction of these heavy hydrocarbons have radical or ionic characters. The large aliphatics may grow by further addition, dehydrogenation, and cyclization to form condensed phase aromatic and aliphatic nuclei. This process is outlined by Homann (1967). The heavy aromatics react with methyl radicals and acetylene to form condensed phase polycyclic aromatics (Bittner and Howard, 1980). The studies of Wersborg and co-workers (1973) and Delfan and co-workers (1979) show that the radical or ionic character of the reactants is preserved to a considerable degree.

The aliphatic fuels capable of forming polar conjugates will join their nuclei and form a branch of the mechanism. As a result, less oxidation occurs, and particulate nuclei form in larger quantities than would result from aliphatic fuels incapable of conjugation.

Aromatic fuels pyrolyze, through ring fracture, to form acetylenes or, at lower temperatures, phenyl and benzyl radicals. The phenyl and benzyl radicals add acetylene or vinylacetylene and quickly join the particulate forming branch of the mechanism. Under certain conditions their formation may be rate controlling (Bittner and Howard, 1980). Acetylenes usually predominate and then join the process outlined for aliphatic fuels. Glassman (1980) and Norrish and Taylor (1956) suggest that low temperature oxidation of aromatics can be accomplished through the use of maleic or oxalic acid:

$$C_6H_6 \xrightarrow{2\,OH} C_6H_4(OH)_2 \rightleftharpoons[\ ]{O_2} \text{(quinone)} \rightarrow \text{maleic acid} + C_2H_2$$

$$C_2H_2 + H_2 + 2CO_2 \longleftarrow$$

As was previously discussed, the formation of carbon dioxide removes oxygen (essentially irreversibly in the relevant time period), which results in a richer fuel mixture and leads to an increase in the amount

of unburned fuel and increased particulate formation. Bittner and
Howard (1980), however, found no evidence for this process in a near-
particulate-producing benzene premixed flame.

Growth

Growth is defined as the addition of mass to a particle (after
nucleation) through reaction with gas phase molecules. As with the
development of particulate nuclei, the reacting gas phase hydrocarbons
seem to be principally acetylenes, and the larger polymers add faster
than the smaller ones (Homann, 1967; Bonne et al., 1965). Small polyacetylenes predominantly undergo further polymerization, presumably by
the same mechanism outlined in the preceeding section on nucleation.
As a result of preferential addition of the larger polymers, the
hydrogen/carbon ratio of the particulates (initially ~0.4 at 15 Å
diameter) decreases toward its fully developed value. This indicates
that most of the polyacetylene added must be of very high molecular
weight or that dehydrogenation also takes place.

These reactions are extremely rapid, and much of the radical and/or
ionic character of the original nuclei is apparently preserved. Strong
electron spin-resonance signals from young particles have been observed
during the initial stage of growth (Wagner, 1978). As evidenced by the
sharply reduced growth rates of older particles, the particles
apparently lose much of their affinity for polyacetylenes as growth
proceeds, even in flame regions where considerable acetylene and
polyacetylenes are still present (Wagner, 1978; Bonne et al., 1965).
Electron spin-resonance signals from the older particulates are lower
than signals from the young particles, indicating loss of radical
character. The fraction of charged particles also declines, from a
number fraction of ~38 percent at a mean diameter of 50 Å to
~13 percent at 100 Å (Wersborg et al., 1973). Wagner (1978)
attributes the decreased growth rates for older particles to increased
temperatures late in the reaction zone of the premixed flames in which
these results were obtained. Increased temperature results in increased
thermal dissociation, which competes with the polymerization process.
For example, larger radicals also may be deactivated by eliminating a
hydrogen atom.

This loss of reactivity is illustrated in Figure B.5(c) for a low-
pressure, premixed acetylene-oxygen flame. In these systems, growth is
practically complete by the time the mean particle diameter reaches
~100 Å (Homann, 1967; Howard et al., 1973).

Coagulation

Small particles, each growing by addition from the gas phase, have also
been shown to coalesce. Because of the low number density of the
particulates, as compared with gas phase hydrocarbons during the growth
period, increases in the mean particle size diameter have been shown to
predominately result from growth (Howard et al., 1973). Thus, even

though coalescence occurs, the resulting particles quickly reacquire
their original spherical shape. The process by which small spherical
particles coalesce to form a larger spherical particle is referred to
as coagulation. Coagulation, as defined, only affects the relatively
small particles characterized by high rates of growth. As was
mentioned in the previous section, this occurs up to a diameter of 100
Å in low-pressure, premixed systems.

When the growth rate slows because gas phase hydrocarbons are
depleted or the particle has become less active, coalescence results in
chains of spherical particles. This process, which is different from
coagulation, is called aggregation. It is discussed briefly in the
next section.

Particle size in flames gradually changes from a Gaussian (for
young particulates) to a log normal distribution during coagulation
(Howard et al., 1973). The predicted deviation from the log normal
distribution is attributed to the continuous addition of particles at
the small end of nucleation. As the nucleation rate becomes much less
than the coagulation rate, the distribution assumes a log normal form.

Because it is esentially a physical process, the rate of coagulation can be calculated with relatively little difficulty. Assumptions
characteristic of such calculations are:

1. Particulates are in the free molecular regime.
2. Each collision of two particles results in coagulation.
3. All particles are spherical.

The first assumption is valid for particles that are smaller than the
mean free path. This assumption is valid in low pressure flames during
the entire coagulation process (15 Å < $\underline{d}$ < 100 Å) (Homann,
1967; Howard et al., 1973). In diesel engines, where the mean free
path varies between 40 and 70 Å during the period of combustion, the
first assumption is invalid; however, the second and third assumptions
seem to remain valid for coagulating particles (Wagner, 1978; Homann,
1967; Wersborg et al., 1973; Delfan et al., 1979).

With these assumptions, a relatively simple mathematical description
of the coagulation process can be developed, even when accounting for
polydisperse systems interacting through electronic and dispersion
forces. An alternate treatment for continuum systems is of similar
complexity. Each has been fully developed in studies by Friedlander
(1977). In a study of particulate coagulation in shock-heated
hydrocarbon-argon mixtures, Graham (1976) has shown that the previously
listed assumptions apply, resulting in a coagulation rate (expressed in
terms of the rate of particle number decrease) of the form

$$\frac{-dN}{dt} = \frac{5}{6} \, \underline{k}_{tn} \, \phi \underline{N}^{11/16} \, ,$$

where

$$\underline{k}_{tn} = \frac{5}{12} \left(\frac{3}{4\eta}\right)^{1/6} \left(\frac{6kT}{\rho}\right)^{1/2} G'\alpha \ .$$

Here ϕ is the soot volume fraction, $\underline{k}$ is Boltzmann's constant, ρ is the particle density, α is a factor related to the polydisperse nature of the system, and $\underline{G}'$ is a factor accounting for the increase in the collision cross section over the hard sphere value that results from electronic dispersion forces (note that the equation $\underline{k}_{tn}$ is misprinted in Wagner (1978)). Graham's results indicate that $\underline{G}'$ ~2 and $\underline{d}$ = 6.55.

A less accurate treatment of coagulation at the rarefied limit is given by Wersborg and co-workers (1973), who express the coagulation rate as

$$\frac{-dn}{dt} = \frac{1}{2} \underline{K}n^2 \ ,$$

where

$$\underline{K} = 16\underline{a}_2\underline{y}\left(\frac{\pi kT}{m}\right)^{1/2} \ .$$

Here $\underline{n}$ is the particle number density, $\underline{a}$ is the radius of the particle, $\underline{m}$ is its mass, and $\underline{y}$ corresponds to the factor $\underline{G}'$ in the previous treatment. This simplified treatment does not account for the poly-disperse nature of the evolving system. By assuming that $\underline{K}$ is constant in the coagulation region of their system, Wersborg and co-workers derived a value of $\underline{K}$ ~ 10^{-7} cm^3/s from the measured dynamics of the particle size distribution. This assumption, which is equivalent to the assumption that the product of $\underline{T} \cdot \underline{d}$ remains constant through-out the coagulation region, is mainly justified by its utility. The value of $\underline{y}$ is 30, as opposed to $\underline{G}'$ ~2 in the previous case. Theoretical estimates of the effect of electronic image forces on the collision cross section place its magnitude of ≤ 3 for particles with only one charge (Howard $\underline{et}$ $\underline{al}$., 1973).

Turbulent or laminar shear layer mixing may also affect the coagulation rate. For turbulent environments where the average eddy size is much greater than the diameter of the coagulating particles, the change in particle number density with time is given by (Friedlander, 1977)

$$\frac{-dn}{dt} = 1.24\phi \left(\frac{\varepsilon_d}{\nu}\right)^{1/2} \underline{n} \ .$$

Here ε_d denotes the turbulent energy dissipation rate, ν the kinematic viscosity of the fluid phase, and ϕ the volume fraction of suspended phase. The rate of Brownian coagulation is dependent on the square of the number density; however, shear dominated coagulation is directly proportional to $\underline{n}$.

The treatments of coagulation presented in this section are based on a system of constant volume. Modifications must be made if they are to be applied to engines.

Aggregation

After particle growth ceases, continued coalescence of the spherical particles results in the formation of chains. The fact that these structures are not found in diesel particulates indicates that the time scale of the diesel combustion process may be too short for this process to be important (Broome and Khan, 1971).

Wersborg and co-workers (1973) observed that 20 percent of the particles in the region where aggregation occurs are charged. They report that these charges are predominately found on particle chains and attribute the clusters' chainlike shape to the charge. Because no growth or nucleation occurs during the process, the aggregation rate should be proportional to the square of the particle number density (n_2). Thus, the number density at relatively long times should be inversely proportional to time. This condition is clearly observed in Figure B.5(a) at distances greater than 5 mm from the burner face. Wersborg and co-workers calculated the aggregation rate from the slope of this curve. Although not strictly applicable, the equations used are the same as those used to calculate the coagulation rate (see previous section). Results are displayed in Figure B.5(b).

Oxidation

Oxidation of the particles, their nuclei, and their precursors will occur in the regions of the flame where oxidizing species are present. Boisdron and Brock (1972) interpreted the moments of the particulate size distribution in premixed flames measured by Bonne and co-workers (1965), to show that oxidation occurs simultaneously with growth and coagulation, but to a relatively small extent. Fenimore and Jones (1969) explain the onset of particulate emissions from their composite premixed-diffusion flames in terms of competition between agglomeration and oxidation by the hydroxl radical. The dominant process is assumed to be controlled by competition for the hydroxl radical by carbon monoxide and agglomerating particulates. They derive a criterion for particulate emissions by assuming that a hydroxl radical leads to removal of a carbon atom once in every 10 collisions.

Most of the information available on the heterogeneous mechanism of particulate oxidation is ambiguous. There is disagreement regarding the predominant oxidizing species involved. Fenimore and Jones (1968, 1969) attribute particulate oxidation to singlet oxygen and the hydroxl radical; others attribute it directly to doublet oxygen (Wagner, 1978; Appleton, 1973; Khan $\underline{et}$ $\underline{al}$., 1971; Magnussen, 1970; Lee $\underline{et}$ $\underline{al}$., 1962).

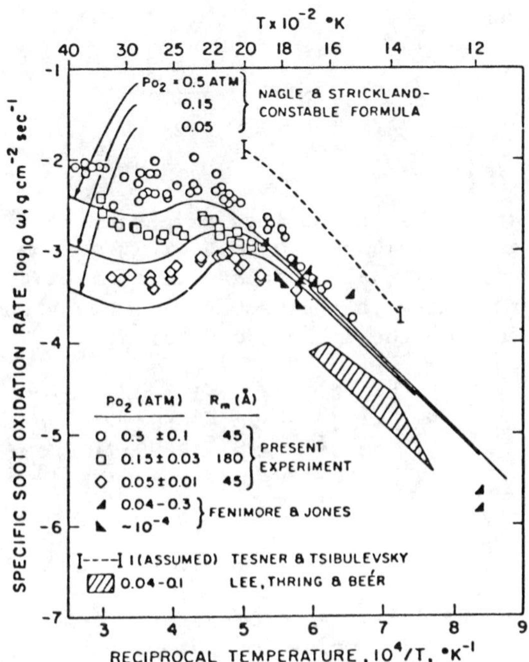

FIGURE B.7 Specific soot
oxidation rate as a function of
reciprocal temperature for
various O_2 partial pressures.
SOURCE: Appleton, 1973.

Appleton (1973) showed that the semiempirical expression for the
rate oxidation of pyrolytic graphite first proposed by Nagle and
Strickland-Constable (1962) also correlates satisfactorily with
particulate oxidation data in most cases (Fenimore and Jones, 1969; Lee
et al., 1962). Specifically, it appears that the surface reaction rate
becomes independent of the partial pressure of doublet oxygen at low
temperatures and/or high partial pressures of doublet oxygen because of
interactions that change the nature of the surface. This is illustrated
in Figure B.7. Note the drastic decrease in pressure dependence as the
temperature passes through 2000°K.

The change in surface characteristics resulting from changes in
temperature and the partial pressure of oxygen may be accounted for by
assuming the existence of two types of surface sites--(a) and (b)--
which differ in their rate of oxidation (Blyholder et al., 1958). The
fraction of the surface covered by type (a) sites, denoted x, is given
by the rate of oxidation of type (b) sites to type (a), balanced by the
rate of thermal annealing of type (a) sites to type (b). The overall
specific surface reaction rate is then given by

$$\frac{\omega}{Wc} = x \frac{k_a P_{O_2}}{1 + k_z P_{O_2}} + k_b P_{O_2} (1 - x),$$

where

$$\underline{x} = \frac{[1 + \underline{k}t]^{-1}}{\underline{k_b}P_{O_2}} \; .$$

Here ω is the surface oxidation rate (g-atoms/cm$_2$-s); $\underline{W_C}$ is the atomic weight of carbon (12 g/g-atom); and $\underline{k_a}$, $\underline{k_b}$, $\underline{k_t}$, and $\underline{k_z}$ are empirical constants related to the rate coefficients for various reactions in the mechanism. Nagle and Strickland-Constable (1962) provide the following data:

$$\underline{k_a} = 10^{6.30} \exp(-15100/\underline{T}) \; \text{g/cm}^2\text{-s-Pa},$$
$$\underline{k_b} = 10^{2.65} \exp(-7650/\underline{T}) \; \text{g/cm}^2\text{-s-Pa},$$
$$\underline{k_t} = 10^{5.18} \exp(-48820/\underline{T}) \; \text{g/cm}^2\text{-s},$$
$$\underline{k_z} = 10^{6.33} \exp(+2063/\underline{T}) \; \text{Pa}^{-1}.$$

APPLICATIONS TO DIESEL ENGINE PARTICULATE EMISSIONS

Engine design and operating parameters such as injection timing and pressure, compression ratio, and inlet port design influence particulate emissions by affecting both the size and the time-temperature history of particulate forming and oxidizing regions of the combustion chamber. As is illustrated in Figure B.8, design and operating parameters affect the parameters that have been identified in laboratory studies (only in an extremely indirect manner) as important in particulate formation/oxidation. This figure shows that relatively few of the directly controllable parameters (underlined in the figure) affect the more fundamental quantities (listed in the fourth column from the right). The large degree of synergism found among the variables listed is also to be noted.

The influence of most of the directly controllable variables on diesel smoke emissions is described by Khan and co-workers (Broome and Khan, 1971; Khan and Wang, 1971), who have also modeled the influence of many of those variables in three direct injection engines of unspecified design (Khan et al., 1971). These variables influence the particulate formation/oxidation process by affecting fuel-air mixing and time-temperature history. The relation between injection timing and ignition delay has been shown to be particularly important. For fuel injected prior to ignition, atomization and evaporation result in a gaseous mixture similar to a rich, premixed flame. Fuel injected after ignition, however, burns before much mixing can occur and results in a diffusion flame. With respect to injection timing, this leads to smoke emission characteristics that are opposite for the direct injection and indirect injection engines; these effects are illustrated in Figures B.9 and B.10. For the direct injection engine (Figure B.9), injection timing advanced beyond 5° before top dead center (btdc) increases the ignition delay period and thereby decreases smoke emissions, which results in a more homogeneous mixture before ignition.

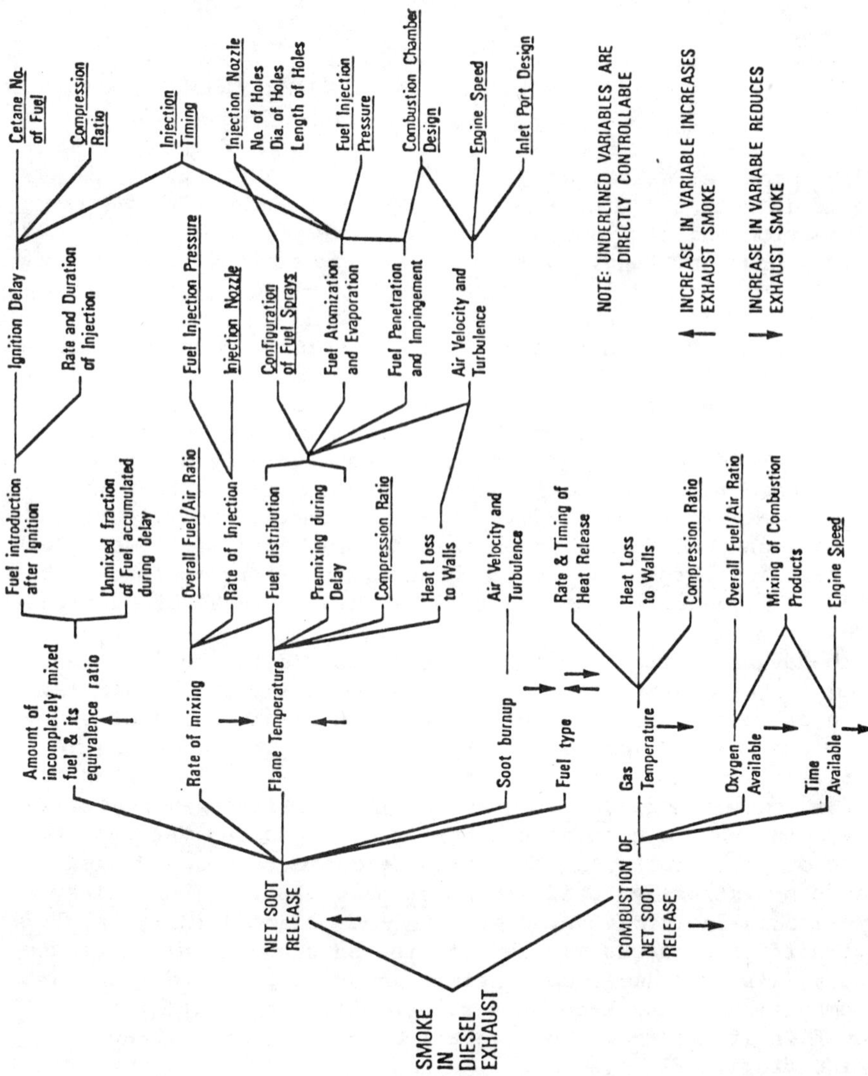

FIGURE B.8 Relationship of diesel engine design and operation parameters to particulate emissions. SOURCE: Broome and Khan, 1971.

336

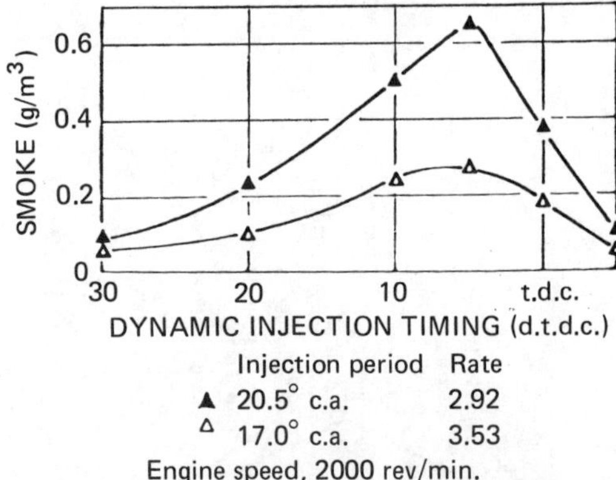

FIGURE B.9 Smoke emissions as a function of injection timing in a direct injection engine. SOURCE: Adapted from Khan and Wang, 1971.

For the indirect injection engine, smoke emissions increase as injection timing is advanced beyond 5° btdc (Figure B.10). Ignition delay is less important here, because the fuel injected can only mix with the air in the first chamber (designated the prechamber or swirl chamber, depending on engine design). This occurs very quickly due to the high swirl rates used (up to 6 times those of conventional direct injection engines) (Broome and Khan, 1971). As a result, the equivalence ratio at ignition increases as injection is retarded because a lower fraction of the total air charge will be contained in the first chamber. The increased local equivalence ratio leads to higher rates of particulate formation.

This behavior is the basis for an essential difference between combustion in the direct injection and indirect injection engines. The combustion process is illustrated for each engine in terms of simplified conceptual models in Figure B.11. In the direct injection engine (Figure B.11(a)), evaporation of fuel droplets takes place near the moment of injection, producing an extremely fuel-rich vapor jet. This generally occurs under conditions supercritical to the various chemical components of the droplet, so that transition to the vapor phase is expected to be extremely rapid (Amann et al., 1980). Thus, flame envelopes around individual droplets are not observed (Chigier, 1976). Turbulent diffusion causes mixing with the surrounding air near the edge of the fuel jet, and a combustible vapor is produced. At that point, combustion takes place at a rate controlled by the rate of turbulent fuel-air mixing. The overall process can be viewed as a turbulent diffusion flame.

Higher swirl rate and prechamber confinement result in a different process in the indirect injection engine (Figure B.11(b)). Because of the typically higher swirl rates in the prechamber and the action of the jet exhausting from the prechamber, more backmixing occurs in the indirect injection engine. Of the laboratory combustors discussed in the previous section, the perfectly stirred reactor most closely approximates the intracylinder environment (Rife, 1980). Both the

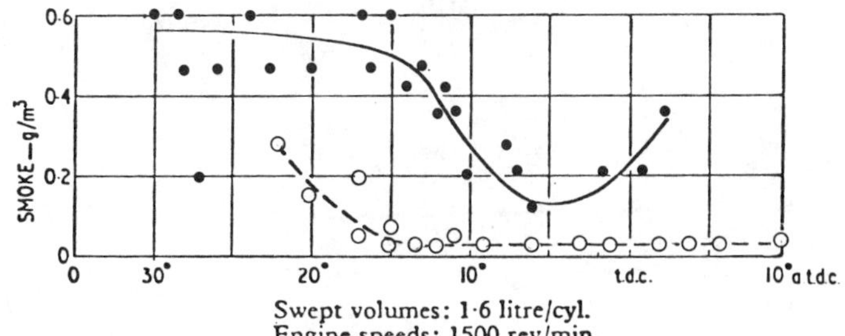

FIGURE B.10   Smoke emissions as a function of
injection timing in an indirect injection engine.
SOURCE: Adapted from Broome and Khan, 1971.

particulate producing characteristics (Glassman, 1980) and the heat
release rates of perfectly stirred reactors are similar to those of
premixed flames; Broome and Khan (1971) found that the heat release
rates in indirect injection engines are more characteristic of premixed
than of diffusion controlled combustion. Thus, combustion in an
indirect injection engine more closely resembles combustion in a
premixed flame.

As was previously explained, the fundamentals of particulate
evolution in a direct injection engine may best be modeled in terms of
a diffusion flame. Application of the fundamental processes of Figure
B.1 to the conceptual model of Figure B.11(a) results in the following
scenario: After evaporation from the droplet surface, a fuel molecule
undergoes pyrolysis in the region between the surface and the heat
release zone. In this region, the concentration of oxidizing species is
very small (Chakraborty and Long, 1968); nevertheless, it may play a
significant role because of the mechanism by which pyrolysis reactions
occur. Nucleation of particles could occur in the pyrolysis zone,
depending on the residence time and temperature histories of the fuel
molecules. These histories primarily depend on the thermodynamic
properties of the fuel and the rate of mixing. Near the heat release
zone, the expected concentration of oxidizing species should be
sufficient for the operative mechanisms of particulate nucleation and
growth to be similar to those studied in rich premixed flames. The
rates of these processes are expected to decay to near zero at the
fuel-lean side of the heat release zone because most heavy, unstable
organic molecules on which they are thought to depend will oxidize.
Particle coagulation should terminate at this point. Further particle
growth will be exclusively by aggregation, with the volume fraction of
particulate decreasing through oxidation. The oxidation rate is
expected to decrease quickly as the gas temperature decays to ambient
cylinder values.

In the indirect injection engine (Figure B.11(b)), the prechamber
is described as a very rich, premixed region where particulate is
formed by the same fundamental processes that occur in a premixed
flame. These are oxidative pyrolysis, nucleation, growth, and

(a)

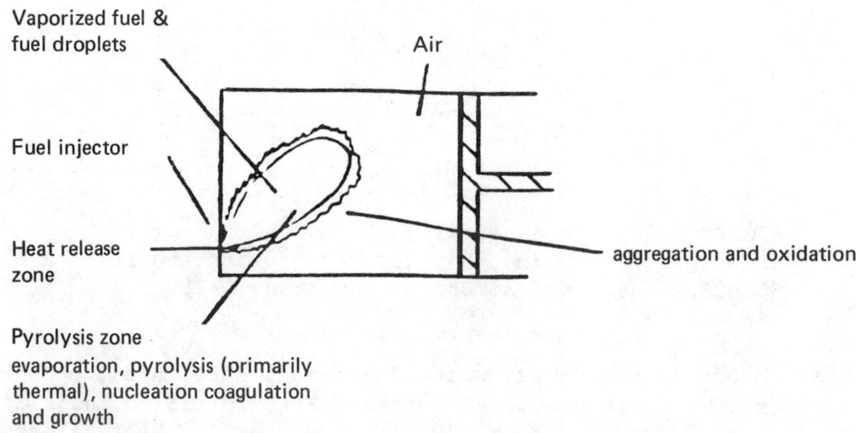

(b)

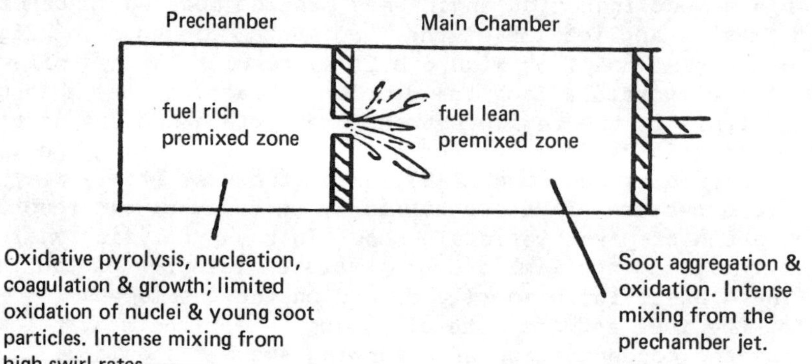

FIGURE B.11 Schematic representation of direct injection engine combustion process (a) and indirect injection engine combustion process (b).

coagulation. There may also be limited oxidation of particulate nuclei and very young particles.

The main chamber is also described as a premixed zone, however, in this chamber the mixture is fuel-lean. In the main chamber the premixed processes are, as in the fuel-lean zone of the direct injection engine model, particle aggregation and oxidation. Particle growth and further nucleation are negligible.

IS PARTICULATE FORMATION INHERENT IN DIESEL ENGINES?

Particulate is observed in a wide variety of combustion systems whenever the local ratio of carbon to oxygen exceeds 0.5. It is apparent that elimination of particulate formation in a diesel engine would require that the local carbon/oxygen ratio at the time of combustion be less than 0.5 throughout the cylinder. To achieve sufficient homogeneity, some combination of high mixing rate and long ignition delay would be required. With known mixing technology, the ignition delay times characteristic of current diesel fuels are too short to accomplish the required homogeneity. Even if this could be achieved, the increased peak cylinder pressures needed (up to 2.5 times the current values) would cause serious structural problems (Amann et al., 1980). The conclusion of all researchers in this area (Amann et al., 1980; Aoyagi et al., 1980; Hiroyasu et al., 1980; Dimick, 1980) is that particulate formation is inherent in the operation of compression ignition engines. This conclusion seems certain to be valid for the indirect injection engine, where lack of communication between the chambers increases the equivalence ratio of the first chamber even in the presence of perfect mixing.

METHODS FOR REDUCTION OF PARTICULATE EMISSIONS IN DIESELS

If particulate formation is inherent in the diesel combustion process, one goal of engine design should be to oxidize as much of the particulate as possible within the cylinder. For the indirect injection engine the goal is to (1) form minimum particulate in the first chamber and (2) oxidize as much particulate as possible to gaseous products in the main chamber.

Previous studies have produced conflicting results regarding the best point of attack for meeting these goals. Khan and co-workers (1971b) conclude that particulate formation is the dominant process and that oxidation plays only a minor role. Hiroyasu and co-workers (1980), however, reach the opposite conclusion. Both are based on interpretation of data from direct injection engines.

Suppressing Nucleation

To facilitate complete oxidation, it is desirable to minimize the mass of particulate produced and, perhaps more importantly, to maximize the surface/volume ratio of the particles formed. One means of accomplishing this is to suppress nucleation through fuel modification. According to Figure B.6, suppression of nucleation can be accomplished by removing conjugated molecules from the fuel, especially polycyclic and substituted aromatics. This serves to suppress entry to the nucleation mechanism downstream of the reaction of acetylene to form either conjugated aliphatics or inert products of oxidation, which is thought to determine the degree to which particulate is produced in premixed (indirect injection) environments (Glassman, 1980). Studies

of premixed flames by Homann and Wagner (1967) support this nucleation
scheme; they found that substitution of benzene for acetylene at the
same carbon/oxygen ratio increased the amount of particulate formed by
increasing the number density (not the size) of particulate. Studies
by the General Motors engineering staff have demonstrated decreases in
particulate emissions of up to 70 percent in a dynamometer-mounted
indirect injection engine, and of up to 43 percent during the Federal
Test Procedures for a diesel-powered automobile, using fuel blends of
low aromatic content and low 90 percent point (Dimick, 1980). Fuel
modification is expected to be even more effective in the direct
injection engine because the path from acetylene oxidation to inert
products is largely removed. In this case (diffusion flame), the
nucleation rate is related directly to the rate of conjugate formation
by pyrolysis. This in turn tends to be strongly related to molecular
structure (Glassman, 1980).

Removing sulfur from diesel fuel should decrease the nucleation
rate. Fuel sulfur species are quickly converted to sulfur oxides (99
percent sulfur dioxide and ~1 percent sulfur trioxide) in the flame
reaction zone. As was previously discussed, these species catalyze the
recombination of oxygen and the hydroxl radical and thus reduce the
degree of acetylene oxidation and enhance the formation of
polar-conjugated particulate precursors (see Figure B.6). Sulfur
removal is expected to be most effective for indirect injection engines
and for fuels of low aromatic content.

Nucleation in premixed systems can also be suppressed by lowering
the density during combustion. This could be accomplished by increasing
the temperature of the reacting mixture, and thereby increasing the
fraction of acetylene oxidized, or by decreasing the pressure. In
fact, studies of rich (ϕ = 2.1), premixed propane oxidation under
constant volume conditions with initial pressures between 500 and 900
kPa indicate that no particulate is formed when the pressure/temperature
ratio is less than 500 Pa/K in the postflame gases (Flower and Dyer,
1980). These conditions are not possible in a compression ignition
engine. Furthermore, an increase in temperature is expected to increase
the nucleation rate in a direct injection engine, and there is some
evidence of sensitivity of the pressure dependence of the nucleation
rate to fuel structure (Friswell, 1979; Naegeli and Moses, 1980).

Faster mixing should also suppress nucleation, particularly in a
mixing controlled direct injection environment. Wright (1969) demonstrated reductions of particulate emission up to 90 percent of the
premixed flame levels by increased backmixing. However, it should
again be emphasized that some indirect injection engines may already
operate close to the well-stirred limit because of the high swirl rates
and the action of the divided chamber.

An interesting method for increasing charge homogeneity that may be
particularly effective in the direct injection engine is the use of
fuel emulsions. Under certain circumstances, droplets of petroleum
distillate/ethanol solutions and petroleum distillate/water emulsions
undergo explosive disruption because of homogeneous bubble nucleation
in the superheated droplet interior (Lasheras et al., 1980). General
Motors' tests (Dimick, 1980) show particulate emission reductions of 34

percent on the dynamometer, and of 7 percent during the Federal Test
Procedure (FTP) when an indirect injection engine is operated with
diesel fuel/water emulsions, although this may be due to the resulting
increase in prechamber concentration of oxidizing species.

Arresting Growth

If nucleation cannot be suppressed sufficiently, the size of the
resulting particles should be kept to a minimum to facilitate later
oxidation. It would be useful to stop particle growth from the gas
phase. As was previously stated, growth stops naturally (presumably
because of a loss of radical or ionic character) when the particles are
still fairly small, even in the presence of acetylene and radical, or
their ionic derivatives. For low-pressure premixed flames, growth
stops at ~100 Å; for other systems, including diesel engines, the
figure varies between 100 and 300 Å (Bittner and Howard, 1980;
Kittelson and Dolan, 1979). As additional knowledge about the
mechanism of growth and the reason why the process seems to stop at
roughly the same point in a wide variety of systems becomes available,
it may be possible to devise methods for arresting growth at a smaller
particle size. Aggregation would still occur, but the particulate
produced would likely have a chain structure, leaving a surface/volume
ratio that is considered higher than that of a sphere of equivalent
mass. Amann and co-workers (1980) have calculated nearly equal
residence times for complete combustion of unaggregated particles and
the volume-equivalent chain aggregate under diesel cylinder conditions.
A much longer residence time is required for the volume equivalent
sphere.

Minimizing Coagulation and Aggregation

Coagulation and aggregation may be minimized by using additives that
act to promote particle charging by lowering the ionization potential
of the particulate. Alkali metal compounds are quite effective
(Wagner, 1978; Ray and Long, 1964; Glassman, 1980). Calculations by
Ball and Howard (1971) indicate that coagulation is retarded between
particles of like charge. Holding the charge per particle constant,
the potential barrier (E_m) that must be overcome before the
attractive image forces dominate increases quickly as the particle
diameter is decreased. E_m is displayed as a function of charge per
particle for 200-Å particles in Table B.1. Coagulation should be
retarded significantly when E_m is greater than kT, so that for the
diesel cylinder environment a charge of at least $2e$ per particle would
be required. However, if the particulate is in equilibrium with a
fixed electron density in its surroundings, the average charge per
particle will increase with increasing diameter. This results in E_m
increasing as a function of particle diameter, so that further
coagulation between large particles should be eliminated (Ball and
Howard, 1971).

TABLE B.1 $\underline{E}_m$ as a Function of Change per Particle for 200-Å Particles

$\underline{n}$	$\underline{E}_m$, ergs	$\underline{n}$	$\underline{E}_m$, ergs
1	7.6×10^{-14}	5	2.2×10^{-12}
2	3.5×10^{-13}	10	4.2×10^{-12}
3	8.0×10^{-13}	15	2.1×10^{-11}
4	1.4×10^{-12}	20	3.9×10^{-11}

General Motors has demonstrated reductions in particulate emissions up to 30 percent for a dynamometer mounted indirect injection engine, and up to 17 percent during the FTP by using fuel additives (Dimick, 1980).

Because of the high swirl rates normally employed in the indirect injection engine, there is a possibility that coagulation and/or aggregation could be dominated by shear layer mixing. The likelihood of this increases as the process proceeds because of the stronger dependence of the Brownian coagulation rate on number density (see section on coagulation). In a study of particulate coagulation and oxidation in direct injection engines Khan $\underline{et}$ $\underline{al}$. (1971) derived a value of $\sim 10^{-4}$ cm^3/s for the coagulation constant under diesel cylinder conditions. After allowing for the change in particle size and temperature between this study and the study of Wersborg and co-workers (1973), Khan's value is still approximately 3 orders of magnitude higher than values from the laminar flame experiment. In view of the fact that the lower value results in coalescence at, or slightly above, the neutral particle collision rate and that the effect of transition from molecular to continuum regimes is in the wrong direction, Khan's value can only be attributed to intense mixing. If this is true, the wrong form is used for the coagulation rate. Furthermore, it is probable that Khan $\underline{et}$ $\underline{al}$. (1971) have greatly overestimated the particle oxidation rate and that this has led to too large a value for $\underline{K}$. Nevertheless, if shear layer mixing dominates the coagulation rate at some point in the process, reducing the mixing rate at that point would be desirable.

Enhanced Particulate Oxidation

Oxidation of particulates and other partially oxidized products of combustion is expected to take place primarily in the main chamber. As was previously stated, long residence times under conditions of high temperature and large partial pressure of oxygen will enhance particle oxidation. Examination of Figure B.7 indicates that for partial pressures of oxygen between 5 and 50 kPa the specific particulate oxidation rate is a strong function of the oxygen concentration only when the temperature is greater than 2000°K. Because entrainment of colder surrounding air into the hot jet exiting the prechamber reduces the jet temperature while increasing the oxygen mole fraction, it is likely that the mixing rate could be optimized for particulate oxidation. The expansion lowers the temperature throughout the main chamber; hence, the optimum entrainment rate is also expected to be a function of engine speed.

Amann and co-workers (1980) performed calculations in which a particulate-containing packet was introduced into an expanding volume under diesel cylinder conditions. The surrounding air was then allowed to mix with the packet, and the amount of particulate oxidized was monitored. The results clearly show the existence of an optimum mixing rate. As would be expected from Figure B.7, it is desirable to mix quickly at first, introducing hot oxygen from the nearly fully compressed surroundings. As the expansion proceeds, air must be entrained at a slower rate in order to maintain a temperature of 2000°K or more for as long as possible. For the initial conditions chosen, a mixing schedule that reduces the packet equivalence ratio to 0.9 and holds it there throughout the expansion is found to be most suitable for a packet introduced early in the expansion. For particulate introduced later, the optimum equivalence ratio increases slightly because of the lower temperature of the surrounding air.

Amann and co-workers did not take heat loss from the packet into account. Radiation, in particular, has been shown to result in significant heat losses in such systems (Flower and Dyer, 1980). Such losses are likely to shorten drastically the period during which the temperature is sufficient to oxidize particulate effectively, so that the optimum mixing rate will probably increase considerably.

Khan et al. (1971) modeled particulate emissions from a direct injection engine in terms of competition between coagulation-aggregation and surface oxidation and have determined values of the coagulation and oxidation rate coefficients parametrically. They found the oxidation rate coefficent on the basis of data in the temperature range between 1300° and 1500°K, and partial pressure of oxygen between 0.4 and 0.8 atmospheres--conditions that, according to Figure B.7, lie in the fall-off and pressure-independent regimes. Their assumption of first order kinetics probably results in a great overestimation of the rate, especially in the low temperature environment of the partially expanded volume. Apparently, this error results in the high coagulation coefficient noted earlier, illustrating the danger involved in this type of blind parametric study. Their conclusion that particulate oxidation is unimportant after the end of heat release is therefore discounted.

CONCLUSIONS

Although particulate formation is an inherent characteristic of the compression ignition engine, there appears to be no fundamental reason why particulate emissions cannot be greatly reduced or even eliminated.

A number of techniques show promise for particulate emission reduction through suppression of the particulate formation rate and/or enhancement of the oxidation rate. At the present time, it is impossible to tell whether one or more of these methods may ultimately affect complete oxidation in the engine cylinder.

The state of scientific knowledge is such that modeling efforts to determine the effectiveness of many of these techniques are feasible. This is especially true for the indirect injection engine, where the mixing process is more easily handled. Where these studies can be performed, the results would be most useful in directing experimental programs. Several such studies are already under way (Rife, 1980).

In closing, it should be noted that some of the particulate control strategies discussed in the text may adversely affect the emissions of other harmful species. For instance, techniques used to enhance particulate oxidation are likely to increase emissions of nitric oxides (Amann et al., 1980). Moreover, their implementation into a practical (i.e., drivable, maintainable, etc.) engine in many cases poses a considerable engineering challenge.

APPENDIX C:
THE AMES TEST

The Ames Salmonella/microsomal test is the most popular short-term
bioassay (Ames et al., 1975) for measuring the activity of the soluble
organic fraction of diesel particulate, and its subfractions. With
this test a quantitative dose-response curve showing the mutagenicity
of a sample compound can be obtained. In its principal form, the Ames
test uses a mutant strain of Salmonella typhimurium that is incapable
of producing histidine. (Mutagenicity is defined as the ability of a
test compound to revert--back mutate--this bacterium to its wild state
where it regains its ability to produce histine.)

Along with the histidine mutations, the TA genotypes of Salmonella
typhimurium contain additional mutations that increase the bacteria's
ability to detect mutagenicity in a broad class of compounds. Principal
variants include the "deep rough" (rfa) mutation, which eliminates the
lipopolysaccharide (LPS) barrier on the surface of Salmonella
typhimurium (Ames, 1971). This allows large-molecular-weight compounds
access to the target DNA. The uvrB mutation eliminates the gene for
DNA excision repair, thereby increasing the DNA's susceptibility to
irreparable damage (Ames et al., 1973). The integration of an R-factor
plasmid, to create the TA98 and TA100 strains, gives these bacteria an
increase in sensitivity by channeling the repair of chemically damaged
DNA to an error prone pathway (McCann et al., 1975). Table C.1 depicts
the most commonly used strains and their mutations.

TA1535 and TA100 are base-pair substitution mutation indicators
that are reverted by various alkylating agents. Both strains contain
the histidine mutation his G46; a missense mutation possessing a base
substitution that alters the messenger RNA for the gene coding the
first enzyme of histidine biosynthesis (Ames, 1971; Ames et al., 1973).

TA1537, a frame-shift mutation indicator, contains the histidine
mutation his G3076. The frame-shift in TA1537 is caused by the addition
of the base-pair -C- and -G- in the aminotransferase gene. TA1537 is
reverted by various polycyclic hydrocarbons and other compounds capable
of deleting a single base pair (Ames, 1971; Ames et al., 1973).

Another frame-shift mutation, the histidine mutation his D3052, is
present in the TA1538 and TA98 strains. This mutation contains two
additional base pairs -GC-. Frame-shift mutagens capable of a double
base-pair deletion, e.g., aromatic amines and benzpyrene, will cause
these strains to revert (Ames, 1971; Ames et al., 1973).

TABLE C.1 <u>Salmonella</u> <u>Typhimurium</u> Strains and Mutation

Genotype of Ames <u>Salmonella</u> <u>Typhimurium</u> Strains

Histidine Mutation			Additional Mutations		
his G46	his C3076	his D3052	LPS	Repair	R. Plasmid
TA1535	TA1537	TA1538	rfa	uvrB	−R
TA100		TA98	rfa	uvrB	+R

The Ames test procedure consists of mixing tester bacteria with a
sample compound in a medium containing a limited amount of histidine.
The histidine allows the mutant <u>Salmonella</u> <u>typhimurium</u> to undergo
several growth divisions, during which time the sample compound is
assimilated by the bacteria. If the compound is not a mutagen of the
class capable of being detected by the tester strain, growth stops
after all the exogenous histidine has been incorporated. However, if
the compound is mutagenic, the mutant bacteria are reverted to their
histidine independent wild type; growth continues and a colony is
formed. The number of colonies formed is proportional to the mutagenic
activity of the compound. The response is expressed in terms of
revertants per microgram of extract (total SOF, one of the seven SOF
subfractions, or other fractions).

APPENDIX D:
EFFECTS OF INERTIA WEIGHT AND ROAD LOAD
HORSEPOWER ON NO$_x$ AND PARTICULATE EMISSIONS

The available Environmental Protection Agency (EPA) and industry data
on light-duty cars and trucks are shown in Figures D.1 and D.2. These
data show that road load has only a minimal effect on particulate
emissions levels, except for the Opel at unusually high inertia weight
settings. The higher road-load settings of light-duty trucks might
account for a few percent increase in particulate levels (EPA, 1980c).

Figure D.2 shows that the inertia weight settings of a light-duty
vehicle do have a significant effect on particulate emission levels.
On the basis of these data and comments from Chrysler, EPA concluded
that for each 1,000-pound decrease in inertia weight, particulate
emission decreased by approximately 17 percent (EPA, 1980c).

EPA further deduced that the combined effect of inertia weight and
road-load settings appears to be approximately 20 percent. If all
other considerations were equal, EPA decided that the light-duty-truck
particulate standards would be 20 percent higher than the corresponding
light-duty-vehicle particulate standards (EPA, 1980c). Thus, the light-
duty-truck particulate standard was set at the same level as the light-
duty-vehicle particulate standard for 1982-1984 (0.6 g/mi) because the
light-duty-truck NO$_x$ standard is 2.3 g/mi, while the light-duty
vehicle must meet a NO$_x$ standard between 1.0 and 1.5 g/mi, depending
on EPA waivers. EPA set the 1985 particulate standard at 0.26 g/mi
because the NO$_x$ level requires a 75 percent reduction, which will be
near the stringency level of the light-duty vehicle 1.0 g/mi standard,
and because of the 20 percent inertia weight horsepower effect. In
addition, the trend of downsizing to smaller engines in light-duty
vehicles will probably not take place as rapidly for light-duty trucks
(EPA, 1980c).

Figure D.3 illustrates Peugeot-Citroen data showing the effect of
inertia weight on NO$_x$ and particulate emissions. These data show an
approximate 28 percent decrease in NO$_x$ emissions per 1,000-pound
inertia weight change and decreases in particulates varying from 26 to
56 percent per 1,000-pound inertia weight change.

Figures D.4 and D.5 show the effect of inertia weight for data
calculated by using engine maps. Figure D.4 contains data as a function
of engine displacement/inertia weight. These data show a 20 to 40
percent decrease in NO$_x$ per 1,000-pound inertia weight decrease and a
17 to 23 percent decrease in particulate per 1,000-pound inertia weight
decrease.

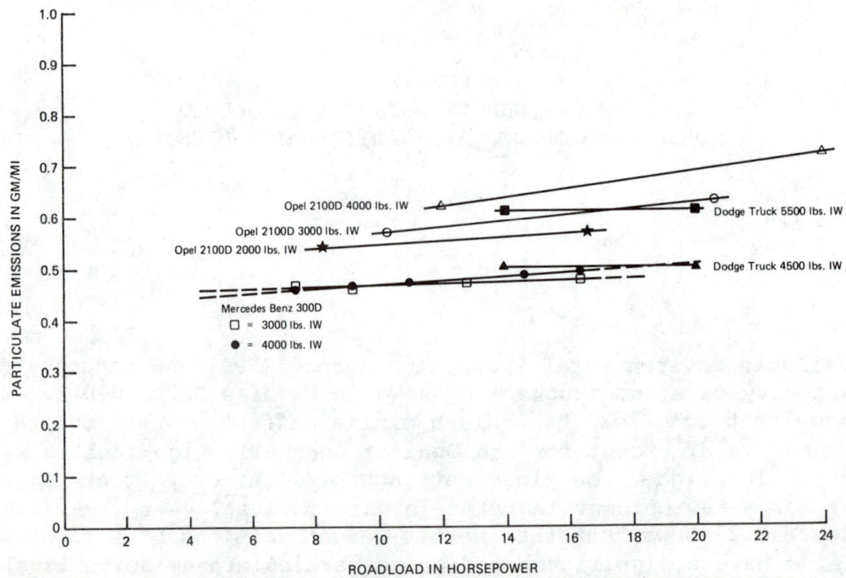

FIGURE D.1 Particulate emissions as a function of road-load federal test procedure driving cycle HP. SOURCE: EPA, 1980c.

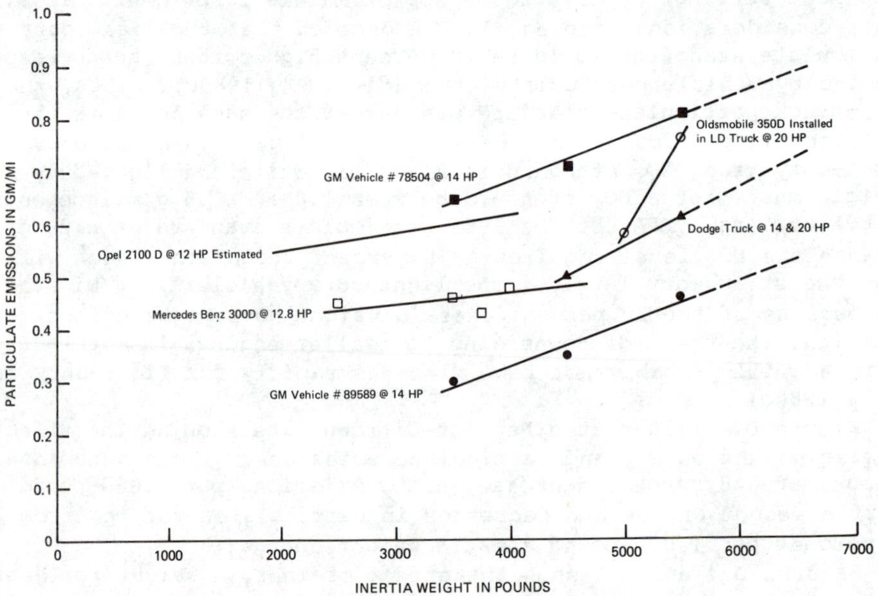

FIGURE D.2 Particulate emissions, as a function of inertia weight, for the federal test procedure driving cycle. SOURCE: EPA, 1980c.

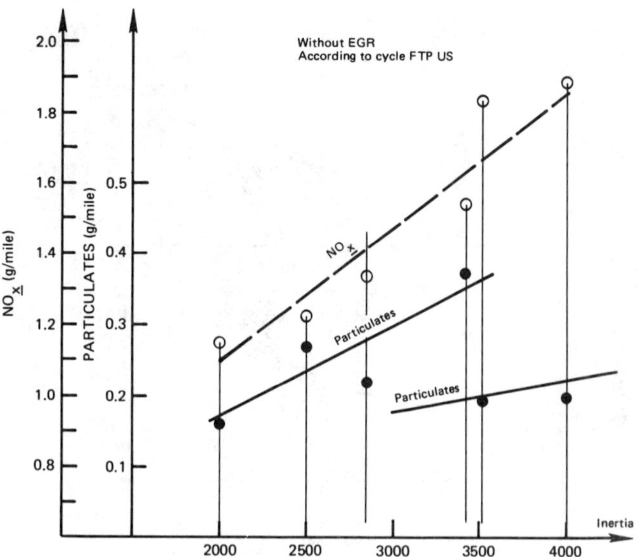

FIGURE D.3   Relationship of vehicle weight to emissions.   SOURCE: Peugot, 1979.

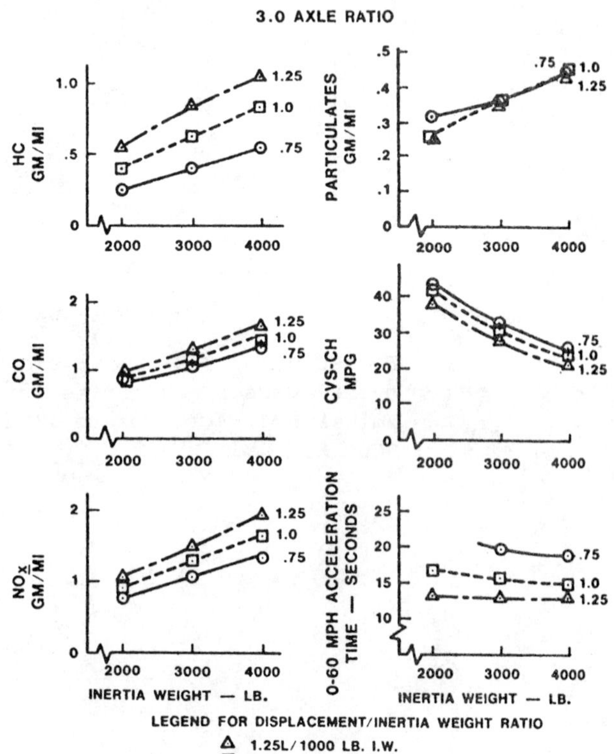

FIGURE D.4   Effect of vehicle inertia weight and engine displacement/ vehicle inertia weight ratio on emissions, fuel economy and performance of swirl chamber diesel vehicles (calculated from island maps).
SOURCE: Wade, 1980.

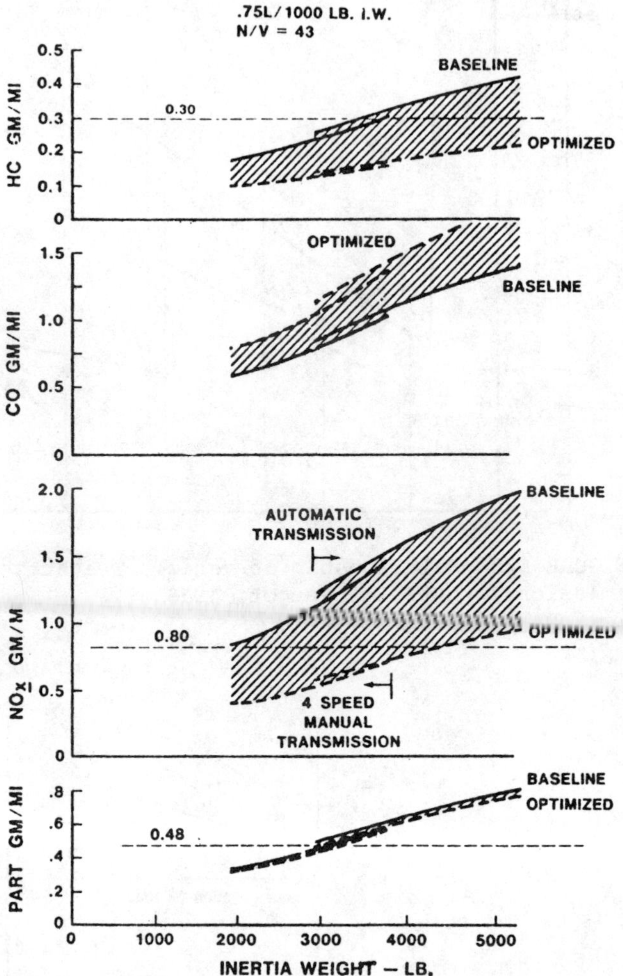

FIGURE D.5   CVS—CH emissions (in g/mi) for
diesel passenger cars as estimated from
emission indices versus vehicle inertia weight.
SOURCE: Wade, 1980.

APPENDIX E:
ANALYTICAL PROCEDURES FOR CHARACTERIZING UNREGULATED
EMISSIONS FROM VEHICLES USING MIDDLE-DISTILLATE FUELS*

Many techniques are available for measuring unregulated pollutants in automotive exhaust with varying degrees of sophistication, difficulty, and precision. Obviously not all of the researchers cited in this report used the same analytical procedures. The following is included as an example to describe state-of-the art techniques for quantifying unregulated emissions.

The Southwest Research Institute, San Antonio, Texas, recently published a report on the optimum analytical technology for providing qualitative and quantitative measurements of unregulated products from diluted exhaust. The analytical procedures were selected to measure the following compounds or groups of compounds:

* aldehydes and ketones;
* hydrogen cyanide and cyanogen;
* hydrogen sulfide;
* organic sulfides and carbonyl sulfide;
* ammonia;
* organic amines;
* nitrous oxide;
* sulfur dioxide;
* individual hydrocarbons;
* phenols;
* N-nitrosodimethylamine;
* benzo(a)pyrene; and
* soluble sulfate.

After a series of validation and qualification experiments, 10 analytical procedures were pinpointed as the optimum procedures currently available. These procedures are listed and briefly described below.

____

*The descriptions of the ten analytical procedures cited in this appendix were taken from Smith and co-workers (1980).

ALDEHYDES AND KETONES

The collection of aldehydes (formaldehyde, acetaldehyde, isobutyr-
aldehyde, and hexanaldehyde) and ketones (acetone and methylethyl-
ketone) is accomplished by bubbling CVS diluted exhaust through glass
impingers containing 2,4-dinitrophenylhydrazine (DNPH) in dilute hydro-
chloric acid. The aldehydes and ketones (also known as carbonyl
compounds) react with the DNPH to form their respective phenylhydrazone
derivatives. These derivatives are insoluble or only slightly soluble
in the DNPH/HCl solution and are removed by filtration followed by
pentane extractions. The filtered precipitate and the pentane extracts
are combined, and then the pentane is evaporated in the vacuum oven.
The remaining dried extract contains the phenylhydrazone derivatives.
The extract is dissolved in a quantitative volume of toluene containing
a known amount of anthracene as an internal standard. A portion of this
dissolved extract is injected into a gas chromatograph and analyzed by
using a flame ionization detector.

TOTAL CYANIDE (HYDROGEN CYANIDE PLUS CYANOGEN)

The collection of total cyanide is accomplished by bubbling CVS diluted
exhaust through glass impingers containing a 1.0 N potassium hydroxide
absorbing solution. This solution is maintained at ice bath temperature. An aliquot of the absorbing reagent is then treated with
KH_2PO_4 and chloramine-T. A portion of the resulting cyanogen
chloride is injected into a gas chromatograph equipped with an electron
capture detector (ECD). External CN-standards are used to quantify the
results.

INDIVIDUAL HYDROCARBONS

For measurement of selected individual hydrocarbons, methane (CH_4),
ethane (C_2H_6), ethylene (C_2H_4), acetylene (C_2H_2), propane
(C_3H_8), propylene (C_3H_6), benzene (C_6H_6), and toluene
(C_7H_8), a sample of CVS diluted exhaust is collected in a Tedlar
bag. This bagged sample is then analyzed for individual hydrocarbons
by using a gas chromatographic system containing four separate columns
and a flame ionization detector. The peak areas are compared to an
external calibration blend, and the individual hydrocarbon concentrations are obtained by using a Hewlett-Packard 3354 computer system.

ORGANIC AMINES

The collection of organic amines (monomethylamine, monoethylamine and
dimethylamine, trimethylamine, diethylamine, and triethylamine) is
accomplished by bubbling CVS diluted exhaust through glass impingers
containing dilute sulfuric acid. The amines are complexed by the acid
to form stable sulfate salts that remain in solution. A portion of

this solution is then injected into a gas chromatograph equipped with an ascarite loaded precolumn and a nitrogen phosphorus detector (NPD). External amine standards in dilute sulfuric acid are used to quantify the results.

SULFUR DIOXIDE

The concentration of sulfur dioxide in dilute exhaust is determined as sulfate by using an ion chromatograph. Sulfur dioxide is collected and converted to sulfate by bubbling dilute exhaust through two glass impingers containing a 3 percent hydrogen peroxide absorbing solution. The samples are analyzed on the ion chromatograph and compared to standards of known sulfate concentrations.

NITROUS OXIDE

For measurement of nitrous oxide, a sample of the CVS diluted exhaust is collected in a Tedlar bag. This bagged sample is then analyzed for nitrous oxide by using a gas chromatograph equipped with an electron capture detector. Calibration blends are used to quantify the results. Gas chromatograph peak areas are obtained by using a Hewlett-Packard 3354 system.

HYDROGEN SULFIDE

The collection of hydrogen sulfide is accomplished by bubbling CVS diluted exhaust through glass impingers containing a buffered zinc acetate solution that traps the sulfide ion as zinc sulfide. The absorbing solution is then treated with N,N-dimethylparaphenylene diamine sulfate and ferric ammonium sulfate. Cyclization occurs, forming the highly colored heterocyclic compound methylene blue (3,9-bisdimethyl-aminophenazothionium sulfate). The resulting solution is analyzed with a spectrophotometer at 667 nm in a 1- or 4-cm-pathlength cell, depending upon the concentration.

AMMONIA

Ammonia in CVS diluted automotive exhaust is measured in the protonated form, NH_4^+, after collection in dilute H_2SO_4. The acidification is carried out in a glass impinger maintained at ice bath temperature. A sample from the impinger is analyzed for ammonia in an ion chromatograph, and the concentration in the exhaust is calculated by comparison to an ammonium sulfate standard solution.

ORGANIC SULFIDES

The collection of carbonyl sulfide (COS) and the organic sulfides--
methyl sulfide (dimethylsulfide, $(CH_3)_2S$), ethyl sulfide (diethyl-
sulfide, $(C_2H_5)_2S$), and methyl disulfide (dimethyldisulfide,
$(CH_3)_2S_2$)--is accomplished by passing CVS diluted exhaust through
Tenax GC traps at -76°C. At this temperature the traps remove the
organic sulfides from the dilute exhaust. The organic sulfides are
thermally desorbed from the traps into a gas chromatograph sampling
system and injected into a gas chromatograph equipped with a flame
photometric detector for analysis. External organic sulfide standards
generated from permeation tubes are used to quantify the results.

PHENOLS

The collection of phenols (phenol; salicylaldehyde; m-cresol and
p-cresol; p-ethylphenol, 2-isopropylphenol, 2,3-xylenol, 3,5-xylenol,
and 2,4,6,-trimethylphenol; 2,3,5,-trimethylphenol; and 2,3,5,6,-
tetramethylphenol) is accomplished by bubbling CVS diluted exhaust
through two Greenburg-Smith impingers containing 200 ml of 1 N KOH. The
phenols react with the KOH and remain in solution. The contents of
each impinger are acidified and extracted with ethyl ether. The
samples are partially concentrated, combined, and then further
concentrated to about 1 ml. An internal standard is added, and the
volume is adjusted to 2 ml. The final sample is analyzed by using a
gas chromatograph, and concentrations of individual phenols are
determined by comparison to external and internal standards.

Tables F.1, F.2, and F.3 of this appendix contain the data upon which some of the summary tables in the section on gaseous emissions are based.

TABLE F.1 Unregulated Emissions From Noncactalyst Spark Ignition, Catalyst Spark Ignition, and Light-Duty Vehicles (Hydrocarbon and NO$_x$ in g/mi; all others in mg/mi[a])

Year	Vehicle	Total EGR	HC	NO$_x$	Carbonyls	Organic Sulfides	Amines	DMNA	NH$_3$	Nitriles (HCN)	H$_2$S	COS	N$_2$O	Formaldehyde	Benzene
								Noncatalyst							
	VW Rabbit	N	2.91	3.44	–	–	–	–	–	10.1	–	–	–	–	–
1974	Chev. Vega[b]		1.21	2.16	–	–	–	–	3.0	8.5	–	–	–	–	–
1977	Chev. Nova[b]	Y	6.84	1.50	–	–	–	–	3.7	12.0	–	–	–	–	–
1977	AMC Pacer	Y	1.175	1.95	21.5	0.30	0.04	–	1.9	3.8	0.06	0.18	4.8	15.6	64.4
1972	Chevrolet		–	–	71.5	–	–	–	–	–	–	–	–	30.0	–
	Average noncatalyst				46.5	0.30	0.04	–	7.2	8.6	0.06	0.18	4.8	25.8	64.4
								Oxidation							
	VW Rabbit	Y	0.59	2.18	–	–	–	–	–	12.5	–	–	–	–	–
1977	Chev. Nova 250I6	Y	0.76	1.73	–	–	–	–	3.0	2.5	–	–	–	–	–
1976	Olds Starfire 231I6		2.14	2.35	–	–	–	–	7.2	–	–	–	–	–	–
1978	Chev. Malibu 305V8[c]	Y	0.499	1.21	2.66	0.76	0.00[d]	0.12	4.9	1.0	0.0	0.34	66.0	1.8	28.7
1978	Chev. Malibu 305V8 (Cal.)	Y	0.306	0.69	2.38	0.39	0.02	0.18	8.6	0.16	0.0	0.19	8.1	1.9	8.1
1978	Ford Granada 302V8[c]	Y	0.435	1.96	5.49	0.18	0.05	0.05	3.1	0.37	0.00	0.06	33.8	4.7	12.9
1978	Ford Mustang 302V8[c]	Y	0.515	1.64	4.34	0.43	0.02	0.20	4.9	0.77	0.18	0.26	43.5	2.5	22.5
1978	Olds Cutlass	Y	0.499	1.27	12.48	0.00	0.00	–	2.7	0.97	0.00	0.11	17.7	3.8	16.4
1975	Plymouth Fury		–	–	11.00	–	–	–	–	–	–	–	–	5.0	–
1977	Olds Cutlass		0.338	1.37	23.00	–	–	–	–	–	–	–	–	4.0	9.0
1977	VW Rabbit		0.225	1.01	61.00	–	–	–	–	–	–	–	–	1.0	15.0
1975	Chev. Impala 350V8		1.03	1.79	–	–	–	–	0.2	–	–	–	–	–	–
1975	Chev. Impala 350V8		1.02	1.52	–	–	0.29	–	0.6	–	–	–	–	–	–
1976	Chev. Impala 350V8		0.75	3.38	–	–	–	–	1.2	–	–	–	–	–	–
1976	Chev. Caprice 350V8		0.60	3.12	–	–	–	–	5.6	–	–	–	–	–	–
1975	Olds Cutlass 350V8		0.71	0.97	–	–	0.89	–	2.4	–	–	–	–	–	–
1975	Pont. Bonneville 400V8		1.57	1.97	–	–	–	–	0.7	–	–	–	–	–	–
1975	Buick Electra		1.37	1.05	–	–	–	–	5.9	–	–	–	–	–	–
1977	Olds Delta 88		0.44	1.58	–	–	0.29	–	6.9	–	–	–	–	–	–
1978	Chev. Malibu 305V8[e]	Y	0.346	0.69	1.88	–	–	–	–	–	–	–	–	1.2	6.6

Year	Model															
1978	Ford Mustang^c,e	Y	0.781	1.46	3.69	-	-	-	-	-	-	-	-	-	2.3	25.6
	Average oxidation				12.79	0.35	0.20	0.09	3.9	2.6	0.04	0.19	33.8	2.8	15.9
								Dual								
1979	Mercury Marquis^e		0.242	0.79	2.95	-	-	-	-	-	-	-	-	1.18	6.3
1979	Mercury Marquis		0.209	1.34	4.21	0.34	0.00	-	8.1	0.5	0.00	0.21	38.8	3.90	7.5
1978	Ford Pinto		0.113	0.76	1.74	1.30	0.05	-	9.7	0.08	0.02	1.09	25.8	1.30	3.1
P	GM, 4.3-liter engine		0.460	0.55	7.00	-	-	-	0.00	1.0	-	-	-	-	-
P	AMC 230I6		-	-	-	-	-	-	-	4.44	-	-	-	-	-
	Average dual				3.98	0.82	0.03	-	5.9	1.51	0.01	0.65	32.3	2.13	5.6
								3-Way								
1974	Chev. Vega 140I4		0.40	0.43	5.0	-	-	-	21.5	10.0	-	-	-	-	-
1978	Pontiac Sunbird	Y	0.258	0.84	1.03	0.322	0.00	-	17.9	1.63	0.03	0.21	28.3	0.56	12.2
1978	Saab		0.161	0.18	0.64	0.515	0.016	-	97.9	1.77	0.19	0.47	17.7	0.15	31.8
1978	Saab		0.209	0.19	2.03	-	-	-	-	-	-	-	-	0.31	23.0
1977	Volvo		0.209	1.01	-	-	-	-	24.9	0.00	0.32	-	-	-	-
P	VW Rabbit	N	0.34	1.40	-	-	-	-	-	5.40	-	-	-	-	-
	Average 3-way				2.18	0.419	0.008	-	40.6	3.76	0.18	0.34	23.0	0.34	22.3
								Diesel								
P	GM, 5.7 liter engine	N	0.38	1.24	19.0	-	-	-	1.0	2.0	-	-	-	-	-
P	GM, 5.7 liter engine	Y	0.44	0.71	27.0	-	-	-	2.0	2.0	-	-	-	-	-
1973	Opel Record		0.39	1.32	61.0	-	-	-	-	-	-	-	-	29.0	-
1973	Peugeot 504D		1.96	1.00	150.0	-	-	-	-	-	-	-	-	74.0	-
1972	MBZ 220D		0.24	1.21	35.0	-	-	-	-	-	-	-	-	16.0	-
1973	Nissan^f		0.35	1.53	73.0	-	-	-	-	-	-	-	-	34.0	-
1976	MBZ 300D		-	-	22.0	-	-	-	-	-	-	-	-	-	-
1976	Peugeot 504D		-	-	60.0	-	-	-	-	-	-	-	-	-	-
1979	Olds Delta 88		0.395	1.59	10.6	-	-	-	-	-	-	-	-	7.5	13.8
1975	MBZ 240D		0.290	-	15.0	-	-	-	-	-	-	-	-	6.0	0.0
1975	MBZ 300D		0.161	-	25.0	-	-	-	-	-	-	-	-	6.0	4.0
1974	Peugeot 204D		1.111	-	51.0	-	-	-	-	-	-	-	-	18.0	0.0
1976	Olds Cutlass		0.757	1.13	132.0	-	-	-	-	-	-	-	-	25.0	19.0

TABLE F.1 (continued)

Year	Vehicle	Total EGR	HC	NO$_x$	Carbon-yls	Organic Sulfides	Amines	DMNA	NH$_3$	Nitriles (HCN)	H$_2$S	COS	N$_2$O	Formal-dehyde	Ben-zene
1977	VW Rabbit		0.370	0.87	64.0	–	–	–	–	–	–	–	–	26.0	8.0
1973	Nissan[f]		0.247	–	47.6	–	–	–	–	–	–	–	–	25.4	7.3
1975	Peugeot 504D		0.304	–	70.6	–	–	–	–	1.32	–	0.55	–	48.20	9.2
	VW Rabbit NA	N	0.16	1.20	34.4	–	–	–	–	–	–	–	–	–	–
	VW Rabbit NA	Y	0.40	0.47	34.0	–	–	–	–	–	–	–	–	–	–
	VW Rabbit TC	N	0.11	0.90	26.1	–	–	–	–	–	–	–	–	–	–
	VW Rabbit TC	Y	0.20	0.40	33.3	–	–	–	–	–	–	–	–	–	–
P	Fiat, 2.4-liter engine		0.31	1.60	58.0	–	–	–	–	–	–	–	–	15.0	–
	Peugeot 504D	N	–	–	23.0	–	–	0.0002	–	–	–	–	–	16.0	–
P	Peugeot 504D	Y	–	–	22.0	–	–	0.0016	–	–	–	–	–	16.0	–
P	Peugeot 504D Turbo	N	–	–	47.0	–	–	–	–	–	–	–	–	33.0	–
P	Peugeot 504D Turbo	Y	–	–	15.0	–	–	–	–	–	–	–	–	7.0	–
1976	MBZ 240D		0.19	1.26	23.0	–	–	–	–	–	–	–	–	8.0	–
1976	VW Rabbit		0.26	0.95	21.0	–	–	–	–	–	–	–	–	8.0	–
1973	Nissan[f]		0.247	1.37	51.5	–	–	–	–	–	–	–	–	27.2	–
	Average diesel		–	–	44.7	–	–	0.001	1.5	1.83	–	0.55	–	22.3	7.66

[a]Significant figures in this table may not reflect precision of analysis.
[b]Catalyst removed.
[c]Forty-nine states.
[d]0.00 indicates below detectable limits. Limits of detectability depends upon species and researcher.
[e]Bykowski, 1979. Same vehicle as tested by Urban and Garbe (1979) or Urban and Garbe (1980).
[f]Same vehicle.

SOURCE: Braddock and Bradow, 1975; Braddock and Gagele, 1977; Bykowski, 1979; Cadle et al., 1979; Cornetti, 1979a,b,c; Gabele et al., 1977a,b; GM, 1979a,b; Hare and Baines, 1979; Keirns and Holt, 1978; Oetting, 1979; Peugeot, 1979; Springer and Baines, 1977; Springer and Stahman, 1975, 1977; Urban, 1980a,b; Urban and Garbe, 1979, 1980; Wiedemann and Hofbauer, 1978.

TABLE F.2 Individual Carbonyl Emission Rates and Average Distribution[a]

Year	Vehicle	Form-aldehyde mg/mi	%Carb	Acet-aldehyde mg/mi	%Carb	Isobutyr-aldehyde mg/mi	%Carb	Croton-aldehyde mg/mi	%Carb	Hexane-aldehyde mg/mi	%Carb	Benz-aldehyde mg/mi	%Carb	Acetone[b] mg/mi	%Carb	Methylethyl Ketone mg/mi	%Carb	Total Carbonyl mg/mi	(%HC)	Total THC, mg/mi
									Noncatalyst											
1972	Chev	36.0		10.0		–		1.0		–		8.0		17.0		–		71.5		–
1977	AMC Pacer	15.6		3.8		0.0[c]		–		0.3		–		1.2		0.6		21.5		1175
	Average noncatalyst	25.8	55.5	6.9	14.8	0.0	0.0	0.5	1.1	0.2	0.4	4.0	8.6	1.5	3.2	0.3	0.6	46.5	3.8	
									Oxidation											
1977	Olds Cutalss	4.0		1.0		6.0		12.0		–		–		–		–		23.0		338
1977	VW Rabbit	1.0		–		4.0		52.0		–		4.0		–		–		61.0		225
1975	Plymouth Fury	5.0		2.0		–		1.0		–		1.0		2.0		–		11.0		–
1978	Malibu (49 states)	1.8		0.5		0.0		–		0.18		–		0.08		0.097		2.66		499
1978	Malibu (Cal.)	1.9		0.3		0.0		–		0.10		–		0.0		0.081		2.38		306
1978	Granada	4.7		0.2		0.0		–		0.24		–		0.08		0.27		5.49		435
1978	Ford Mustang II	2.5		1.0		0.05		–		0.29		–		0.24		0.21		4.34		515
1978	Chev. Malibu[d]	1.2		0.04		0.0		0.02		0.13		0.35		0.04		0.10		1.88		346
1978	Ford Mustang II[d]	2.3		0.0		0.0		0.0		0.27		0.76		0.10		0.26		3.69		781
1978	Olds Cutlass	3.8		1.3		1.9		–		0.38		–		4.8		0.57		12.48		515
	Average oxidation	2.8	21.9	0.6	4.7	1.1	8.6	6.5	50.8	0.16	1.3	0.61	4.8	0.73	5.7	0.16	1.3	12.8	2.8	
									3-Way											
1979	Pontiac Sunbird	0.56		0.10		0.0		–		0.18		–		0.5		0.14		1.03	0.40	258
1978	Saab	0.15		0.47		0.0		–		0.0		–		0.0		0.0		0.62	0.39	161
1978	Saab	0.31		0.31		0.0		–		0.35		0.66		0.14		0.26		2.03	0.97	209
	Average 3-way	0.34	27.6	0.29	23.6	0.0	0.0	–	0.0	0.18	14.6	0.22	17.9	0.06	4.9	0.13	10.6	1.23	0.59	
									Dual											
1979	Mercury Marquis[d]	1.18		0.03		0.16		0.06		0.0		1.26		0.15		0.11		2.95		242
1979	Mercury Marquis	3.88		0.18		0.13		–		0.0		–		0.02		0.0		4.21		209
1978	Ford Pinto	1.30		0.0		0.0		–		0.15		–		0.06		0.23		1.74		113
	Average dual	2.12	71.4	0.09	3.0	0.10	3.4	0.02	0.7	0.05	1.7	0.42	14.1	0.08	2.7	0.11	3.7	2.97	1.55	
									Diesel											
1973	Nissan[e]	27.2		9.4		–		7.2		–		2.0		5.7		–		51.5		247
P	Fiat	15.0		18.0		4.0		3.0		5.0		7.0		6.0		–		58.0		310
1979	Olds Delta 88	7.5		1.2		0.8		0.1		0.0		0.6		0.4		0.0		10.6		395

TABLE F.2   (continued)

Year	Vehicle	Form-aldehyde mg/mi	%Carb	Acet-aldehyde mg/mi	%Carb	Isobutyr-aldehyde mg/mi	%Carb	Croton-aldehyde mg/mi	%Carb	Hexane-aldehyde mg/mi	%Carb	Benz-aldehyde mg/mi	%Carb	Acetone[b] mg/mi	%Carb	Methylethyl Ketone mg/mi	%Carb	Total Carbonyl mg/mi	%HC	Total THC, mg/mi
1975	MBZ 240D	6.0		2.0		4.0		1.0		0.0		0.0		2.0		–		15.0		290
1975	MBZ 300D	6.0		2.0		0.0		2.0		0.0		0.0		15.0		–		25.0		160
1974	Peugeot 204D	18.0		7.0		14.0		7.0		0.0		0.0		5.0		–		51.0		1110
1976	Olds Cutlass	25.0		10.0		30.0		7.0		–		3.0		57.0		–		132.0		757
1977	VW Rabbit	26.0		8.0		26.0		–		–		–		4.0		–		64.0		370
1973	Nissan[e]	25.4		10.8		–		6.4		0.0		0.3		4.7		–		47.6		247
1975	Peugeot 504D	48.2		11.9		–		0.6		0.4		0.8		8.7		–		70.6		304
1976	MBZ 240D	8.0		1.0		1.0		6.0		2.0		0.0		5.0		–		23.0		190
1976	VW Rabbit	8.0		1.0		2.0		6.0		1.0		1.0		2.0		–		21.0		260
	Average diesel	18.4	38.8	6.9	14.6	6.8	14.3	3.9	8.2	0.7	1.5	1.2	2.5	9.6	20.3	0.0	0.0	47.4	13.8	

[a]Significant figures may not reflect precision of measurement.
[b]May include acrolein (acrylaldehyde) and proponal.
[c]0.0 indicates below detectable limits. Limits of detectability depend upon species and researcher.
[d]Bykowski, 1979. Same vehicles as tested by Urban and Garbe (1979, 1980).
[e]Same vehicle.

SOURCE: Braddock and Bradow, 1975; Braddock and Gabele, 1977; Bykowski, 1979; Cornett, 1979a,b,c; Hare and Baines, 1979; Springer and Baines, 1977; Springer and Stahman, 1977; Urban and Garbe, 1979, 1980.

TABLE F.3 Individual Hydrocarbon Emission Rates (in mg/mi) and Distributions

Year	Vehicle	Methane	Ethane	Acetylene	Propane	Benzene	NRHC	Toluene	Ethylene	Propylene	Total HC
						Noncatalyst					
1977	AMC Pacer	77.3	11.3	56.4	8.1	64.4		96.6	133.6	61.2	1175
	Average % by mass	6.58	0.96	4.80	0.69	5.48	18.51	8.22	11.37	5.2	
						Oxidation					
1977	Olds Cutlass	47.0	24.0	2.0	0.0	9.0		22.0	29.0	13.0	338
1977	VW Rabbit	53.0	10.0	4.0	0.0	15.0		20.0	24.0	6.0	225
1978	Malibu (49 states)	69.9	16.7	11.9	0.9	26.7		48.9	42.1	19.6	499
1978	Malibu (Cal.)	96.6	16.1	1.6	1.6	8.1		14.5	22.5	6.4	306
1978	Ford Granada	59.6	22.5	0.0[a]	1.6	12.9		33.8	19.3	8.1	435
1978	Mustang	72.4	12.9	0.5	3.2	22.5		46.7	38.6	19.3	515
1978	Olds Cutlass	77.3	24.2	8.9	2.7	16.4		27.4	43.4	20.7	499
1978	Chev. Malibu (Cal.)[b]	81.5	15.0	2.5	3.0	6.6		28.1	18.5	7.3	346
1978	Ford Mustang[b]	81.9	36.0	4.5	8.8	25.6		78.8	32.2	31.9	781
	Average rate, mg/mi	71.0	19.7	4.0	2.4	15.9		35.6	30.0	14.7	
	Average % by mass	16.2	4.5	0.91	0.54	3.63	25.77	8.12	6.84	3.35	
						Dual					
1978	Pinto	49.4	3.2	2.6	0.1	3.1		2.8	4.7	1.5	113
1978	Mercury Marquis	73.4	7.5	2.5	1.0	7.5		8.8	7.6	3.1	209
1978	Mercury Marquis[b]	78.4	5.4	5.1	2.3	6.3		23.2	5.0	3.1	242
	Average rate, mg/mi	67.1	5.4	3.4	1.1	5.6		11.6	5.8	2.6	
	Average % by mass	35.7	2.9	1.8	0.6	3.0	43.9	6.2	3.1	1.4	
						3-way					
1978	Pontiac Sunbird	50.7	6.5	10.1	2.1	12.2		18.6	20.1	9.8	258
1978	Saab	33.4	5.0	4.1	2.5	31.8		17.3	15.3	6.9	161
1978	Saab	32.9	5.1	3.1	0.5	23.0		20.1	15.2	10.6	209
	Average rate	39.0	5.5	5.8	1.7	22.3		18.7	16.9	9.1	
	Average % by mass	18.6	2.6	2.8	0.8	10.6	35.5	8.9	8.1	4.3	

TABLE F.3 (continued)

Year	Vehicle	Methane	Ethane	Acetylene	Propane	Benzene	NRHC	Toluene	Ethylene	Propylene	Total HC
						Diesel					
1979	Olds Delta 88	19.0	2.5	7.4	0.0	13.8		2.3	36.6	12.1	395
1975	MBZ 240D	9.0	–	2.0	–	0.0		–	29.0	0.0	290
1975	MBZ 300D	6.0	–	5.0	–	4.0		–	23.0	0.0	161
1974	Peugeot 204D	15.0	–	12.0	–	0.0		–	61.0	32.0	1111
1976	Olds Cutlass	20.0	7.0	8.0	2.0	19.0		28.0	79.0	28.0	757
1977	VW Rabbit	11.0	1.0	2.0	0.0	8.0		1.0	45.0	15.0	370
1973	Nissan	13.5	0.9	12.9	0.0	7.3		–	28.0	9.4	247
1975	Peugeot 504D	12.1	2.1	12.0	1.0	9.2		–	41.3	13.7	304
	Average rate, mg/mi	13.2	1.7	7.7	0.4	7.7		3.9	42.9	13.8	
	Average % by mass	2.9	0.4	1.7	0.1	1.7	6.8	0.9	9.4	3.0	

[a]0.0 indicates below detectable limits. Limits of detectability depend upon species and researcher.
[b]Bykowski, 1979. Same vehicles as tested by Urban and Garbe (1979, 1980).

SOURCE: Braddock and Gabele, 1977; Bykowski, 1979, 1980; Springer and Baines, 1977; Springer and Stahman, 1977; Urban, 1980a,b; Urban and Garbe, 1979, 1980.

APPENDIX G:

EFFECTS OF DILUTION RATIO AND MIXTURE TEMPERATURE

Figures G.1 and G.2 show some of the data that EPA used in its analysis of comments to the proposed 1981 and 1983 particulate standards. These data appear not to be of a statistical significance that would justify the 125°F maximum limit. A cycle average mixture temperature and dilution ratio would probably be more meaningful for comparison of the data because both the time varying mixture temperature and dilution ratio will continuously affect the filter face adsorption/desorption process.

Figures G.3 to G.4 show the effect of filter temperature, the soluble fraction, and the dilution ratio on the particulate mass (Plee and MacDonald, 1980). These data were obtained in tests with a 2.1-liter Opel engine operated under steady state conditions, using a dilution minitunnel to make the measurements. The changes indicated in the data were all attributed to the soluble portion of the particulate. The change in the soluble organic fraction was explained theoretically by an adsorption/desorption process, which involves adherence of hydrocarbon molecules to the surfaces of the particulate by either chemical and/or physical (van der Waals) forces. By assuming equilibrium between the rates of adsorption and desorption, the particulate data collected fit well into a Langmuir equation relating the fraction of surface occupied by the adsorbate to the partial pressure of the adsorbing hydrocarbon gas and the temperature (Plee and MacDonald, 1980). The dilution ratio of the tunnel ( ), the sample temperature (T_s) for adiabatic dilution, and constant specific heats are related by the following equation, which was derived from the energy equation (Amann et al., 1980)

$$T_s = \frac{T_\epsilon + (\Lambda - 1)\, T_a}{\Lambda},$$

where T_a is the absolute temperature of dilution air and T_ϵ is the absolute temperature of engine exhaust.

In Figure G.5, sample temperature is plotted against dilution ratio for two different exhaust gas temperatures. As is indicated in the figure, the temperature of the diluted steam falls rapidly during the

363

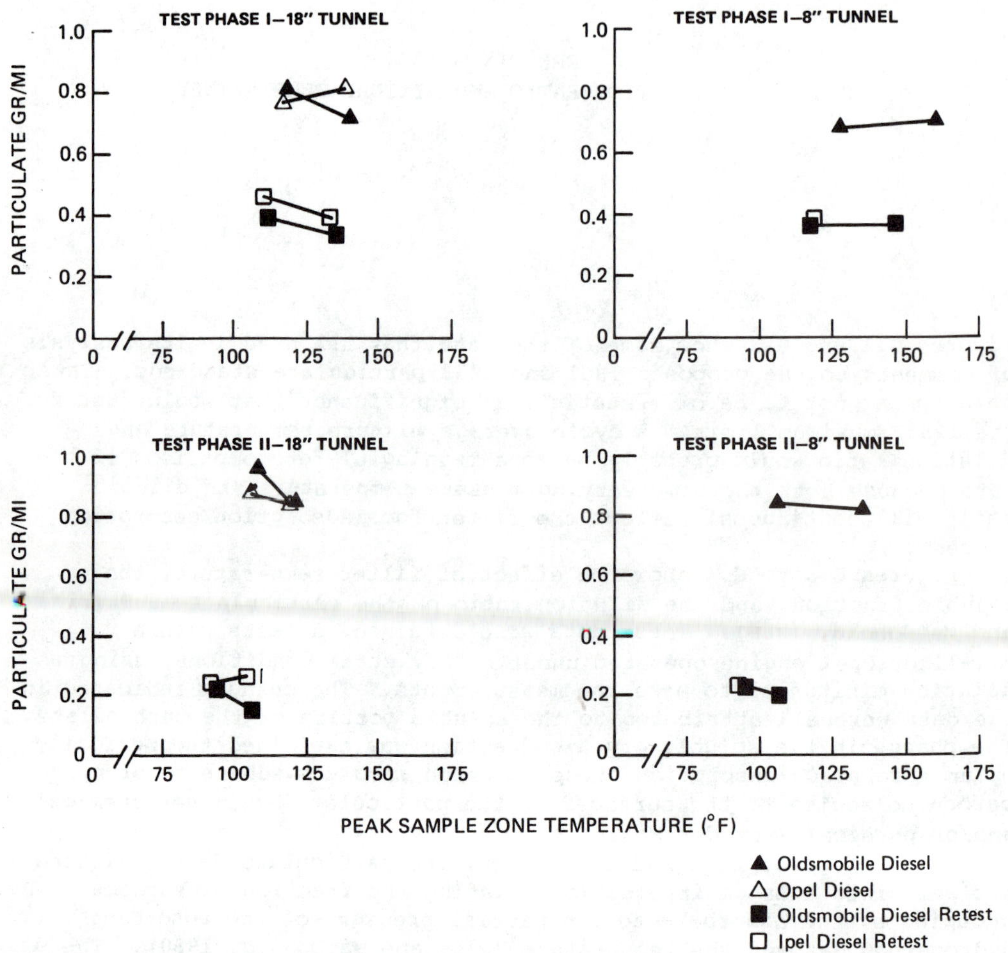

FIGURE G.1 Particulate as a function of temperature--summary of GM measurements. SOURCE: EPA, 1980c.

initial stages of dilution. The EPA maximum 125°F dictates minimum tunnel dilution ratios of 3.5 and 10 for these exhaust temperatures. The general coupling between sample temperature and dilution ratio in a CVS tunnel is explained by Equation 1 and Figure G.5. Using representative adsorption constants from Plee and MacDonald (1980), the effect of the dilution ratio, for two different exhaust temperatures and one engine operating condition, is shown in Figure G.6. The competition between dilution ratio and temperature in influencing the fraction of the particle surface occupied by adsorbed hydrocarbons determines the trends of Figure G.6. Beginning from a dilution ratio of 1, at low dilution ratios, the temperature effect dominates the adsorption process: the adsorbed fraction in Figure G.6 rises with

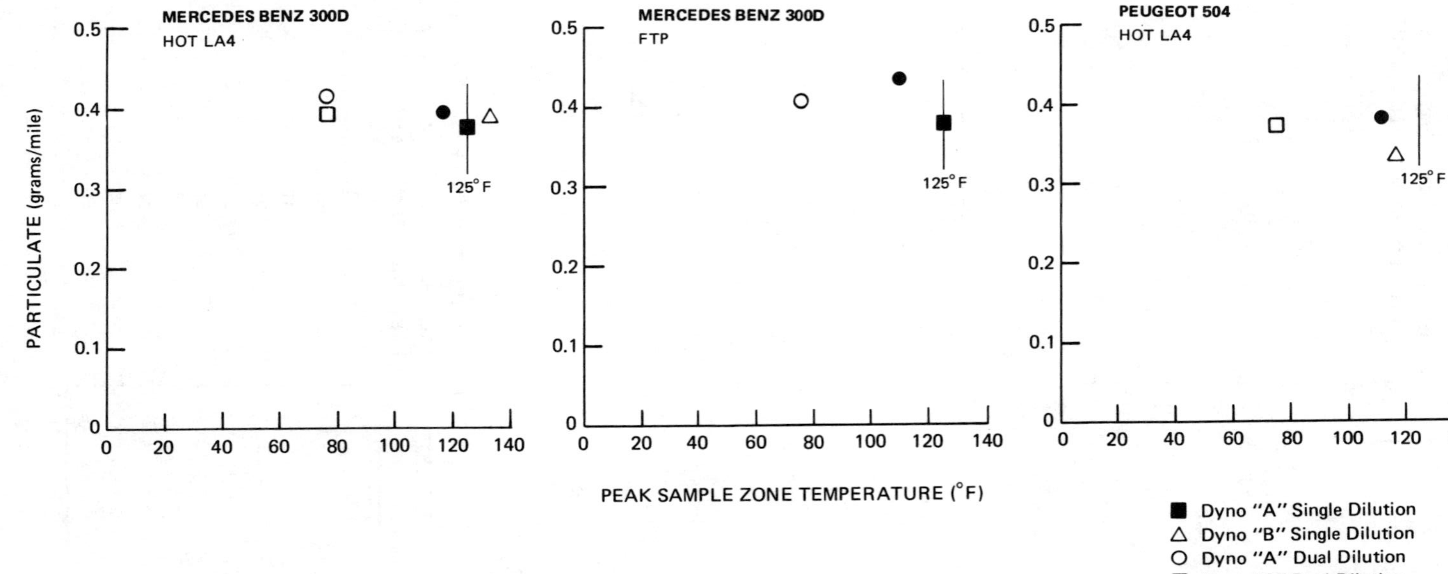

FIGURE G.2 Particulate as a function of temperature EPA measurements. SOURCE: EPA, 1980c.

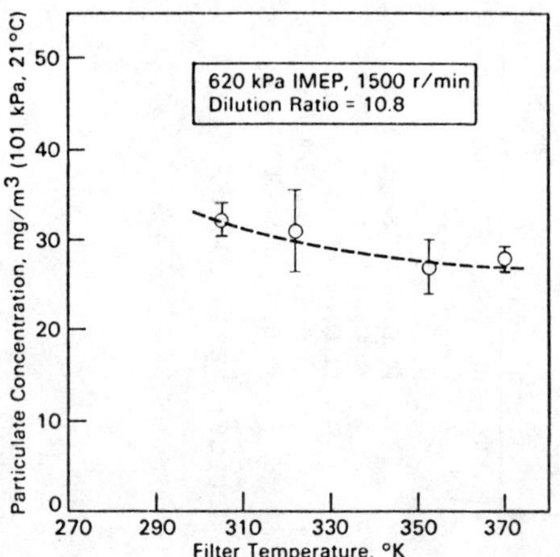

FIGURE G.3 Particulate concentration versus filter temperature. SOURCE: Plee and MacDonald, 1980.

FIGURE G.4 Soluble fraction versus filter temperature. SOURCE: Plee and MacDonald, 1980.

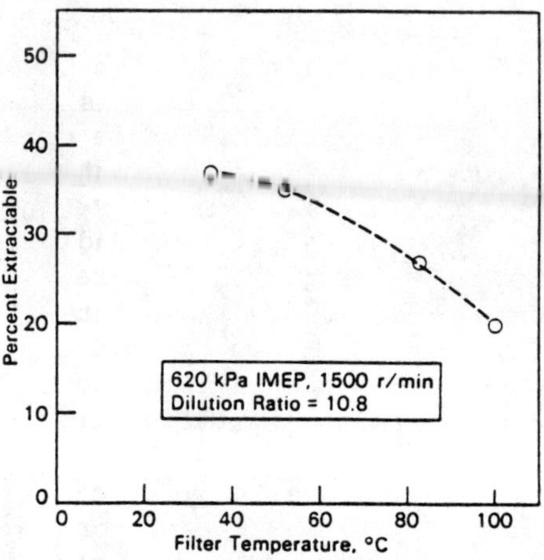

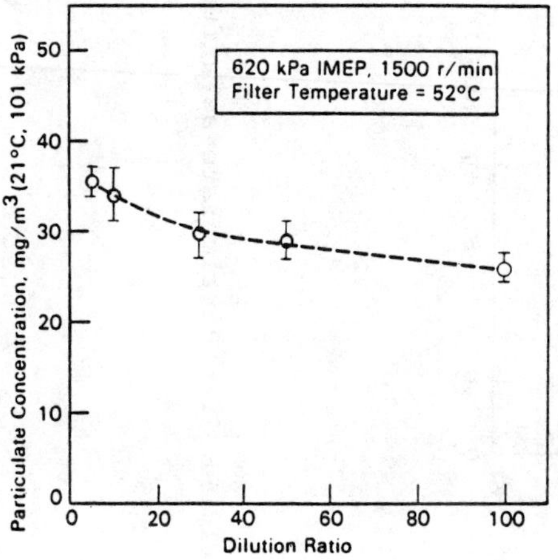

FIGURE G.5 Particulate concentration versus dilution ratio. SOURCE: Plee and MacDonald, 1980.

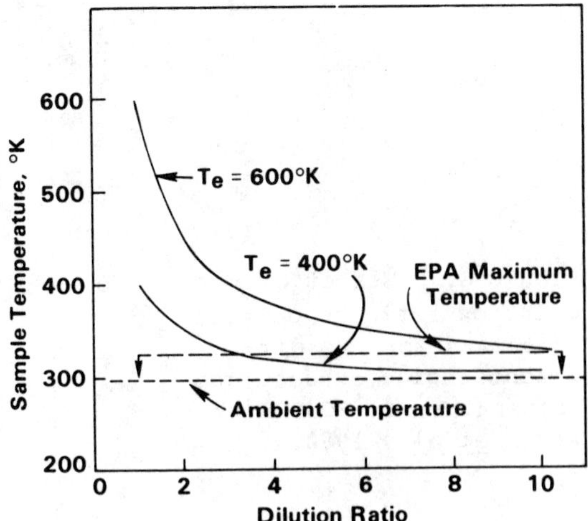

FIGURE G.6 Effect of dilution ratio on sample temperature for two different engine exhaust gas temperatures.
SOURCE: Amann et al., 1980.

increased dilution as shown in Figure G.5. At higher dilution ratios the effect of temperature is less, and the effect that dilution ratio has on the partial pressure of the gaseous hydrocarbons dominates. In addition, the adsorbed fraction falls. Similar experimental results to Figure G.8 were reported by Frisch and co-workers (1979), who studied the effects of fuel type and dilution ratio on the particulate emissions from a Caterpillar 3208 direct injection diesel engine under steady state conditions. These data are shown in Figure G.7. For all the fuels tested (heavy gas oil blend and diesel fuels No. 1 and No. 2; see Figure G.8), the particulate mass concentration increased as the dilution ratio was varied from 1 (no dilution) up to 25, and then it decreased in going to 50:1.

The EPA maximum 125°F filter temperature limit is cross-plotted in Figure G.6. Near the prescribed temperature limit, the total particulate concentration curves are reasonably flat. By only specifying a maximum temperature, large and small vehicles will not necessarily see the same dilution ratios for the FTP. It is recommended that an average cycle temperature be specified rather than a maximum temperature so that the dilution tunnel system will provide more comparable data for different-sized vehicles. The 125°F maximum temperature works because the tunnels operate near the broad peak of Figure G.6, although it is probably too restrictive on tunnel design as compared to the proposed approach of using an average mixture temperature that has a higher maximum limit to prevent desorption.

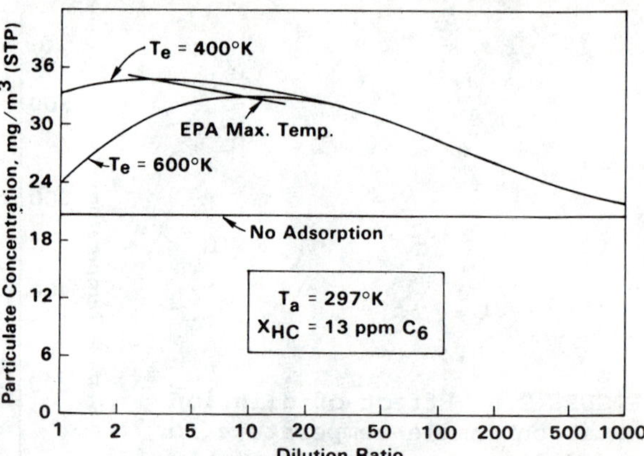

FIGURE G.7 Effect of dilution ratio on particulate mass for two different engine exhaust gas temperatures. SOURCE: Amann *et al.*, 1980.

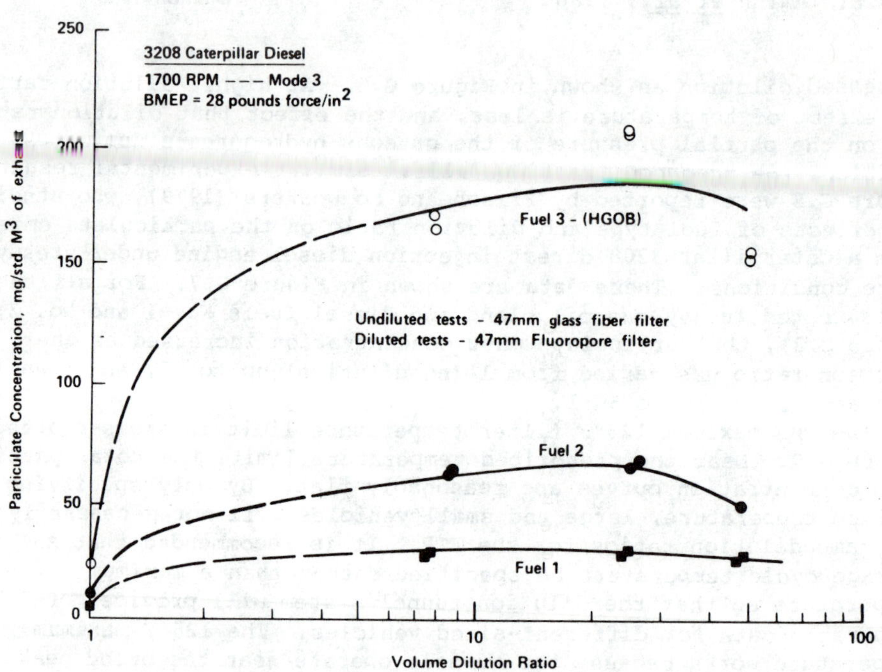

FIGURE G.8 Total particulate emission versus volume dilution ratio for mode 3 and fuels 1, 2, and 3. SOURCE: Frisch *et al.*, 1979.

APPENDIX H:
SITE VISITS AND CONSULTATIONS

Arthur D. Little, Inc.	Cambridge, MA
Amoco Oil Company	Chicago, IL
AVL	Graz, Austria
Battelle Columbus Laboratory	Columbus, OH
Caterpillar Tractor Co.	Mossville, IL
Chevron USA, Inc.	Richmond, CA
Chrysler Corporation	Detroit, MI
Corning Glass Works	Corning, NY
Cummins Engine Co.	Columbus, IN
Daimler-Benz	Stuttgart, Germany
Department of Energy	Bartlesville, OK
Bartlesville Energy Technology Center	
Department of Transportation	Cambridge, MA
Transportation Systems Center	
Drexel Institute of Technology	Philadelphia, PA
E. I. DuPont de Nemours & Co.	Wilmington, DE
Eaton Engineering & Research Center	Southfield, MI
Engelhard Minerals & Chemicals Co.	Menlo Park, NJ
Environmental Protection Agency	Research Triangle Park, NC
Environmental Protection Agency	Ann Arbor, MI
Fiat	Torino, Italy
Ford Motor Co.	Dearborn, MI
Garrett Air Research	Los Angeles, CA
General Motors Corp.	Warren, MI
Hino Motor Company, Ltd.	Tokyo, Japan
Hokkaido University	Saporro, Japan
Institute Francais du Petrole	Paris, France
Isuzu Motors, Ltd.	Tokyo, Japan
Japan Automobile Research Institute	Ibaraki-ken, Japan
Japanese Automobile Manufacturers Association	Tokyo, Japan and Washington, DC
Johnson-Matthey	Reading-Berkshire, England
Lucas, CAV	London, England

Lucas, SGRD   London, England
Massachusetts Institute of Technology   Cambridge, MA
Matthey-Bishop   Wayne, PA
Michigan Technological University   Houghton, MI
Mobil Research and Development Corp.   Paulsboro, NJ
Nissan Motor Company, Ltd.   Tokyo, Japan
Pennsylvania State University   University Park, PA

Peugeot   LaGarenne, France
Princeton University   Princeton, NJ
Ricardo Consulting Laboratories   West Sussex, England
Robert Bosch   Stuttgart, Germany
Sandia National Laboratory   Livermore, CA
Southwest Research Institute   San Antonio, TX
Standard Oil of California   San Francisco, CA
TARADCOM (Tank Automotive Research and   Warren, MI
   Development Command)
Tokyo Institute of Technology   Tokyo, Japan
Toyota Motor Co.   Aichi-ken, Japan
University of California   Berkeley, CA
University of Michigan   Ann Arbor, MI
University of Minnesota   Minneapolis, MN
University of Tokyo   Tokyo, Japan
University of Wisconsin   Madison, WI
Wayne State University   Detroit, MI
Volkswagen   Wolfsburg, Germany
Volkswagen of America   Warren, MI